T0191549

The Cauchy Method of Residues

Mathematics and Its Applications (*East European Series*)

Dragoslav S. Mitrinović and Jovan D. Kečkić

University of Belgrade, Yugoslavia

The Cauchy Method of Residues

Theory and Applications

D. Reidel Publishing Company

A MEMBER OF THE KLUWER ACADEMIC PUBLISHERS GROUP

Dordrecht / Boston / Lancaster

Library of Congress Cataloging in Publication Data

Mitrinović, Dragoslav S.
    The Cauchy method of residues.

    (Mathematics and its applications. East European series)
    Translation of: Cauchyjev račun ostataka sa primenama.
    1.   Analytic functions.   2.   Calculus of residues.   I.   Kečkić, Jovan D.
II.   Title.   III.   Series: Mathematics and its application (D. Reidel Publishing
Company). East European series.
QA331.M65713   1984   515.9   83−24697

Published by D. Reidel Publishing Company,
P.O. Box 17, 3300 AA Dordrecht, Holland.

Sold and distributed in the U.S.A. and Canada
by Kluwer Academic Publishers,
190 Old Derby Street, Hingham, MA 02043, U.S.A.

In all other countries, sold and distributed
by Kluwer Academic Publishers Group,
P.O. Box 322, 3300 AH Dordrecht, Holland.

This monograph is a revised, extended and updated translation
of the book: *Cauchyjev račun ostataka sa primenama.*
Translated by Prof. J. D. Kečkić from the Serbian edition,
published by Naučna Knjiga, Belgrade 1978.

Printed in The Netherlands.

Table of Contents

Editor's Preface

Growing specialization and diversification have brought a host of monographs and textbooks on increasingly specialized topics. However, the "tree" of knowledge of mathematics and related fields does not grow only by putting forth new branches. It also happens, quite often in fact, that branches which were thought to be completely disparate are suddenly seen to be related.

Further, the kind and level of sophistication of mathematics applied in various sciences has changed drastically in recent years: measure theory is used (non-trivially) in regional and theoretical economics; algebraic geometry interacts with physics; the Minkowsky lemma, coding theory and the structure of water meet one another in packing and covering theory; quantum fields, crystal defects and mathematical programming profit from homotopy theory; Lie algebras are relevant to filtering; and prediction and electrical engineering can use Stein spaces. And in addition to this there are such new emerging subdisciplines as "completely integrable systems", "chaos, synergetics and large-scale order", which are almost impossible to fit into the existing classification schemes. They draw upon widely different sections of mathematics.

This program, Mathematics and Its Applications, is devoted to such (new) interrelations as exampla gratia:
- a central concept which plays an important role in several different mathematical and/or scientific specialized areas;
- new applications of the results and ideas from one area of scientific endeavor into another;
- influences which the results, problems and concepts of one field of enquiry have and have had on the development of another.

The Mathematics and Its Applications programme tries to make available a careful selection of books which fit the philosophy outlined above. With such books, which are stimulating rather than definitive, intriguing rather than encyclopaedic, we hope to contribute something towards better communication among the practitioners in diversified fields.

Because of the wealth of scholarly research being undertaken in the Soviet Union, Eastern Europe, and Japan, it was decided to devote special attention to the work emanating from these particular regions.

Thus it was decided to start three regional series under the umbrella of the main MIA programme.

The calculus of residues has been a powerful tool, both conceptually and in actual applications, since its inception in 1814. Almost a hundred years later Lindelöf devoted an entire monograph to the topic, but that was by no means the end of developments, which remained unsurveyed in systematic book form. The topic has, and always has had, a remarkable range of applications spanning most of analysis besides number theory and theoretical physics, and as such I am particularly happy to be able to include a book on it in this series.

The unreasonable effectiveness of mathematics in science. . . .

Eugene Wigner

Well, if you knows of a better 'ole, go to it.

Bruce Bairnsfather

What is now proved was once only imagined.

William Blake

As long as algebra and geometry proceeded along separate paths, their advance was slow and their applications limited.

But when these sciences joined company they drew from each other fresh vitality and thenceforward marched on at a rapid pace towards perfection.

Joseph Louis Lagrange

Amsterdam, April 1983   **Michiel Hazewinkel**

Preface

The subject of this monograph is the branch of complex analysis, usually referred to as the *Calculus of Residues*. It is a revised, extended and updated translation of the book *Cauchyjev račun ostataka sa primenama*, published in Serbian by Naučna knjiga, Beograd 1978.

The foundations of complex analysis in general, and the calculus of residues in particular, were laid down by the French mathematician Augustin-Louis Cauchy (1789–1857) in 1814, when he wrote his first paper on integrals of complex functions. During the next 40 years Cauchy developed his theory of residues to such an extent that it became an independent mathematical discipline. The actual theory of residues is neither very rich in content, nor particularly interesting, but the applications of that theory to the various branches of mathematics found by Cauchy are amazing. They range from the theory of equations, theory of numbers, matrix analysis, evaluation of real definite integrals, summation of finite and infinite series, expansions of functions into infinite series and products, ordinary and partial differential equations, mathematical and theoretical physics, to the calculus of finite differences and difference equations.

Almost every textbook on complex analysis devotes a chapter, or at least a section, to the calculus of residues, but the applications are usually given only to the evaluation of definite integrals and the summation of series. Other applications are rarely ever mentioned. On the other hand, books devoted entirely to the calculus of residues are very few in number. As far as we know there are only five such books (see [27]−[31] in the references for 10.3.1).

This monograph covers all known applications of the calculus of residues. The first two chapters are introductory, and the last is a short historical sketch of Cauchy's life, work, and the development of the calculus of residues. The material presented here can be easily traced from the Contents.

Although there are short sections on residues of nonanalytic functions and multidimensional residues, this book is devoted entirely to the residues of analytic functions in a single variable. In particular, the study of multidimensional residues and their applications would require a separate monograph.

The original mathematical papers used in the preparation of this monograph (apart from Cauchy's papers) stretch through the period from 1834 till 1982. Cauchy's work on residues is easily found in his collected papers, but we have also managed to procure papers written by lesser-known contemporaries of Cauchy. In addition, we mention that this book contains numerous references to papers and books which are not easy to find or which are written in lesser-known languages. We wish here to acknowledge the invaluable help given by various libraries, librarians and mathematicians from Yugoslavia and abroad.

Since this monograph is the first one which embodies the classical, as well as the new results on residues, it was rather difficult to make a satisfactory classification of the collected material. Hence, we are aware that certain results are not always represented according to their significance — some are given more credit than they deserve, and some less.

Although this is believed to be the most complete text written on the calculus of residues, we are well conscious of the fact that we have not collected all the papers which are related to this subject. We therefore invite all those mathematicians who have contributed to this theory and whose results are not presented here, to send us their papers. We would also be obliged to those who will draw our attention to papers concerned with the calculus of residues which are not mentioned in this monograph.

Certain parts of this monograph can be used for graduate and especially for post-graduate courses. The monograph as a whole could serve as a basis for special courses in the theory of functions. Moreover, the authors hope that it will stimulate further research in the classical calculus of residues, as well as in more modern generalizations.

In the preparation of this monograph the authors have received valuable advice and criticism from many mathematicians. In particular, D. D. Adamović, R. P. Boas, D. Dimitrovski, G. Kalajdžić, V. Kocić, M. Merkle, D. Mitrović, M. S. Stanković, and P. M. Vasić read parts of the manuscript and made many valuable suggestions.

We have received invaluable help from Professor D. Tošić and Madame S. Kečkić who have carefully read all the proofs.

The authors also wish to express their appreciation to D. Reidel Publishing Company, well known for their high quality productions, and especially to the member of their staff Mrs. N. Jones, as well as the series editor Professor M. Hazewinkel, for most efficient handling of the publication of this book.

Belgrade,   **D. S. Mitrinović**
March 1983.   **J. D. Kečkić**

Chapter 1

Introduction

1.1. ORGANIZATION AND REFERENCES

1.1.1. *Organization of the Book*

This book is divided into chapters, each chapter into sections, and most sections into subsections. The numeration of theorems, definitions, remarks, examples, and formulas is continuous throughout a subsection, or a section which does not contain subsections.

There are many cross-references in the book. If a theorem referred to belongs to the same subsection only its number is given, while if it belongs to another, the numbers of the chapter, section, and subsection are given. So, for example, 3.1.5 means Chapter 3, Section 1, Subsection 5.

1.1.2. *References*

Bibliographical references are quoted after each subsection if it exists, or after each section, if it does not.

The titles of papers and books written in English, French, German, or Italian are given in the original language. All the other titles are translated into English and the original language is given in parentheses.

The abbreviations of the quoted journals are given according to *Mathematical Reviews*. We also adopt the following abbreviations for references which are frequently quoted:

Copson   E. T. Copson: *An Introduction to the Theory of Functions of a Complex Variable*. First edition: London 1935; Reprinted: 1944, 1946.

Garnir-Gobert H. G. Garnir et J. Gobert: *Fonctions d'une variable complexe*. Louvain-Paris 1965.

Hardy   G. H. Hardy: 'General Theorems in Contour Integration with Some Applications', *Quarterly J. Math.* 32 (1901), 369–384.

Julia   G. Julia: *Exercices d'analyse*, t. 2, deuxième édition. Paris 1947.

Mitrinović   D. S. Mitrinović: *Calculus of Residues*. Groningen 1966.

Oeuvres   *Oeuvres complètes d'Augustin Cauchy*. Paris 1882–1974.

1

Tisserand   F. Tisserand: *Recueil complémentaire d'exercices sur le calcul in-finitésimal*, deuxième édition, augmentée de nouvaux exercices sur les variables imaginaires, par P. Painlevé. Paris 1896, Nouveau tirage 1933.

Titchmarsh   E. C. Titchmarsh: *The Theory of Functions*. First edition: London 1932. Second edition: London 1939. Reprinted: 1964.

Watson   G. N. Watson: *Complex Integration and Cauchy's Theorem*. Cambridge Tracts in Mathematics and Mathematical Physics, No. 15. First edition: London 1914. Reprinted: New York (without the year of publication).

1.2. Notations and Theorems

1.2.1. *Notations*

Since standard notations are used throughout the book, it is believed unnecessary to define them all. Hence, we list only a few of them.

N the set of all positive integers
N_0 the set of all nonnegative integers
Z the set of all integers
R the set of all real numbers
R^+ the set of all positive numbers
R_0^+ the set of all nonnegative real numbers
C the set of all complex numbers
$\mathrm{dg}\, P$ degree of the polynomial P
$\mathrm{int}\, \Gamma$ for a closed contour Γ, int Γ denotes the finite (internal) region bounded by Γ
$\mathrm{ext}\, \Gamma$ for a closed contour Γ, ext Γ denotes the region bounded by Γ which contains the point ∞
∂G the boundary of the region G
v.p. $\int$ the principal value of the integral

1.2.2. *Theorems*

The book begins with the definition of the residue. The reader is therefore assumed to be familiar with complex functions, their derivatives and integrals, with Laurent's expansions and singularities. For the sake of completeness, we give here those theorems which are explicitly used in further text.

THEOREM 1 (Cauchy). *If f is a regular function in a simply connected region G, then*

$$\oint_\Gamma f(z)\ dz = 0,$$

where Γ *($\subset G$) is a closed contour.*

THEOREM 2 (Cauchy). *If f is a regular function on a closed contour Γ and in the region* int Γ, *then*

$$f(a) = \frac{1}{2\pi i} \oint_{\Gamma} \frac{f(z)}{z-a} \, dz,$$

where $a \in$ int Γ.

THEOREM 3. *If f is a regular function on a closed contour Γ and in the region* int Γ, *then*

$$f^{(n)}(a) = \frac{n!}{2\pi i} \oint_{\Gamma} \frac{f(z)}{(z-a)^{n+1}} \, dz,$$

where $a \in$ int Γ.

THEOREM 4. (i) *The principal part of Laurent's expansion of the function f at z=a has no terms if and only if z=a is a regular point or a removable singularity of f.*

(ii) *The principal part of Laurent's expansion of the function f at z=a has a finite number of terms if and only if z=a is a pole of f.*

(iii) *The principal part of Laurent's expansion of the function f at z=a has an infinite number of terms if and only if z=a is an essential singularity of f.*

THEOREM 5. *Suppose that Γ is a closed contour and that a, b $\in$ R. Let $(z, t) \mapsto F(z, t)$ be a continuous function in both variables z and t, when z $\in$ int Γ and $a \le t \le b$. Furthermore, let F be regular in* int Γ *for every t $\in$ $[a, b]$. Then the function f defined by*

$$f(z) = \int_{a}^{b} F(z, t) \, dt$$

is also regular in int Γ *and its derivatives are obtained by differentiating under the integral, i.e.*

$$f^{(n)}(z) = \int_{a}^{b} \frac{\partial^n F(z, t)}{\partial z^n} \, dt \qquad (n=1, 2, \ldots).$$

THEOREM 6 (Cauchy–Poincaré). *Let ω be a holomorphic form of degree n on an analytic variety M of complex dimension n. Then for every (n+1)-dimensional chain $\sigma \subset M$ with a smooth boundary we have*

$$\int_{\partial\sigma} \omega = 0.$$

The following special case is particularly important:

If f is holomorphic in a region $D \subset \mathbf{C}^n$, then for every $(n+1)$-dimensional surface $G \subset D$ with a smooth boundary ∂G, where $\overline{G}$ is a compact belonging to G, we have

$$\int_{\partial G} f \, dz_1 \wedge \ldots \wedge dz_n = 0.$$

Chapter 2

Definition and Evaluation of Residues

2.1. THE RESIDUE OF AN ANALYTIC FUNCTION

2.1.1. *Definition of the Residue*

DEFINITION 1. Suppose that the point $z=a$ is the only singularity of the analytic function f in the region $G = \{z \mid |z-a| < r\}$ and that f is regular on $\Gamma = \partial G$. The residue of f at $z=a$ is defined by

$$(1) \qquad \operatorname*{Res}_{z=a} f(z) = \frac{1}{2\pi i} \oint_{\Gamma} f(z)\, dz.$$

REMARK 1. In view of Cauchy's Theorem 1 from 1.2.2, the region G and its boundary Γ can be replaced by any simply connected region G_1 which is bounded by a closed contour Γ_1.

Suppose that f is represented by the Laurent series

$$f(z) = \sum_{n=-\infty}^{+\infty} A_n (z-a)^n.$$

in the region $\{z \mid 0 < |z-a| < r\}$.

Since Laurent's series can be integrated term by term from (1) we conclude that

$$(2) \qquad \operatorname*{Res}_{z=a} f(z) = A_{-1}.$$

REMARK 2. Formula (2) is often taken as a definition of the residue of f at a.

DEFINITION 2. If f is regular in a neighbourhood of the point ∞ which can be an isolated singularity of f, and if Γ is a circle so large that all other singularities are contained in int Γ, then we define

$$(3) \qquad \operatorname*{Res}_{z=\infty} f(z) = \frac{1}{2\pi i} \int_{\Gamma-} f(z)\, dz.$$

5

REMARK 3. In Definition 2 the integral is taken along negatively oriented contour Γ since in that case the neighbourhood of ∞, i.e. the region ext Γ, remains left from the contour.

The residue at ∞ of f can be connected with the residue at 0 of an other function.

THEOREM 1. *We have*

$$(4) \qquad \operatorname*{Res}_{z=\infty} f(z) = - \operatorname*{Res}_{z=0} \frac{1}{z^2} f\left(\frac{1}{z}\right).$$

Proof. By definition

$$\operatorname*{Res}_{z=\infty} f(z) = \frac{1}{2\pi i} \int_{\Gamma-} f(z)\, dz,$$

where $\Gamma = \{z \mid |z| = R\}$. If we put $z = Re^{-i\theta}$, we get

$$\frac{1}{2\pi i} \int_{\Gamma-} f(z)\, dz = \frac{1}{2\pi i} \int_{0}^{2\pi} f(Re^{-i\theta})\,(-i\,Re^{-i\theta})\, d\theta$$

$$= -\frac{1}{2\pi i} \int_{0}^{2\pi} f\left(\frac{R}{e^{i\theta}}\right) \frac{1}{\left(\frac{1}{R} e^{i\theta}\right)^2}\, d\left(\frac{1}{R} e^{i\theta}\right)$$

$$= -\frac{1}{2\pi i} \oint_{\gamma} \frac{1}{z^2} f\left(\frac{1}{z}\right) dz,$$

where $\gamma = \{z \mid |z| = 1/R\}$ is the circle in which the function $z \mapsto (1/z^2) f(1/z)$ has only one singularity at $z = 0$.

However, in view of Definition 1, we have

$$-\frac{1}{2\pi i} \oint_{\gamma} \frac{1}{z^2} f\left(\frac{1}{z}\right) dz = - \operatorname*{Res}_{z=0} \frac{1}{z^2} f\left(\frac{1}{z}\right),$$

implying (4).

2.1.2. *Evaluation of Residues*

THEOREM 1. *If the point a* ($\neq \infty$) *is a regular point of f, then* $\text{Res}_{z=a}$ $f(z) = 0$.

Proof. The proof follows directly from Definition 1 from 2.1.1 and Theorem 4 from 1.2.2.

THEOREM 2. *If the point a* ($\neq \infty$) *is a removable singularity of f, then* $\text{Res}_{z=a}$ $f(z) = 0$.

Proof. If $z=a$ is a removable singularity, then all the coefficients A_{-n} ($n=1, 2, \ldots$) in the Laurent expansion are zero, and using (2) from 2.1.1, we conclude that $\text{Res}_{z=a} f(z) = 0$.

Theorems 1 and 2 do not hold for the point ∞.

EXAMPLE 1. The function f, defined by $f(z) = 1 + (1/z)$, is regular at ∞. However, $\text{Res}_{z=\infty} f(z) = -1$.

EXAMPLE 2. The point ∞ is a removable singularity of the function f defined by $f(z) = (z^3 - z^2 + 1)/z^3$. However, $\text{Res}_{z=\infty} f(z) = 1$.

THEOREM 3. *If the point a* ($\neq \infty$) *is a pole of order k for the function f, then*

$$(1) \qquad \operatorname*{Res}_{z=a} f(z) = \frac{1}{(k-1)!} \lim_{z \to a} \frac{d^{k-1}}{dz^{k-1}} \left((z-a)^k f(z) \right).$$

Proof. Since a is a pole of order k, f can be written in the form

$$(2) \qquad f(z) = \frac{g(z)}{(z-a)^k},$$

where g is a regular function at a, and also $g(a) \neq 0$.
Hence, by the definition of the residue, from (2) follows

$$(3) \qquad \operatorname*{Res}_{z=a} f(z) = \frac{1}{2\pi i} \oint_{\Gamma} f(z)\, dz = \frac{1}{2\pi i} \oint_{\Gamma} \frac{g(z)}{(z-a)^k}\, dz.$$

Since (Theorem 3 from 1.2.2)

$$g^{(k-1)}(a) = \frac{(k-1)!}{2\pi i} \oint_{\Gamma} \frac{g(z)}{(z-a)^k}\, dz,$$

the equality (3) becomes

$$\operatorname*{Res}_{z=a} f(z) = \frac{1}{(k-1)!}\, g^{(k-1)}(a),$$

and using (2) again, we arrive at (1).

REMARK 1. Formula (1) can also be proved in the following way. In a neighbourhood of its pole a of order k the function f can be written as

$$f(z) = g(z) + \frac{A_{-1}}{z-a} + \cdots + \frac{A_{-k}}{(z-a)^k} \qquad (A_{-k}\neq 0),$$

where g is regular in that neighbourhood. If the identity

$$(z-a)^k f(z) = (z-a)^k g(z) + A_{-1}(z-a)^{k-1} + \cdots + A_{-k}$$

is differentiated $k-1$ times, and if we then put $z=a$, we obtain formula (1).

We now give two useful consequences of Theorem 3.

THEOREM 4. *If* $\lim_{z\to a}(z-a)f(z) = A\ (\neq\infty)$, *then* $\operatorname*{Res}_{z=a} f(z) = A$.
   This is, in fact, a special case of Theorem 3, obtained by putting $k=1$;

THEOREM 5. *If* $\lim_{z\to\infty} zf(z) = A\ (\neq\infty)$, *then* $\operatorname*{Res}_{z=\infty} f(z) = -A$.
   Proof. From Theorem 1 of 2.1.1 follows

(4) $$\operatorname*{Res}_{z=\infty} f(z) = -\operatorname*{Res}_{z=0} \frac{1}{z^2} f\left(\frac{1}{z}\right).$$

On the other hand, the equality $\lim_{z\to\infty} zf(z) = A$ is equivalent to

$$\lim_{z\to 0} \frac{1}{z} f\left(\frac{1}{z}\right) = A,$$

i.e. to

(5) $$\lim_{z\to 0} z\left(\frac{1}{z^2} f\left(\frac{1}{z}\right)\right) = A.$$

Using Theorem 4, we see that (5) implies

$$(6) \qquad \operatorname*{Res}_{z=0} \frac{1}{z^2} f\left(\frac{1}{z}\right) = A,$$

and the statement of the theorem follows from (4) and (6).
The following theorem is also useful.

THEOREM 6. *Let the functions f and g be regular at z=a, and suppose that a is a simple zero of g. Then*

$$\operatorname*{Res}_{z=a} \frac{f(z)}{g(z)} = \frac{f(a)}{g'(a)}.$$

Proof. The point a is a first order pole of the function $z \mapsto f(z)/g(z)$. Hence, by Theorem 3,

$$\operatorname*{Res}_{z=a} \frac{f(z)}{g(z)} = \lim_{z \to a} (z-a) \frac{f(z)}{g(z)}.$$

However, since $g(a)=0$, we have

$$\lim_{z \to a} (z-a) \frac{f(z)}{g(z)} = \lim_{z \to a} \frac{z-a}{g(z)-g(a)} f(z) = \frac{f(a)}{g'(a)}.$$

EXAMPLE 1. We have

$$\operatorname*{Res}_{z=1} \frac{z^2}{(z-1)(z-2)} = \lim_{z \to 1} \frac{z^2}{z-2} = -1,$$

$$\operatorname*{Res}_{z=2} \frac{z^2}{(z-1)(z-2)} = \lim_{z \to 2} \frac{z^2}{z-1} = 4,$$

$$\operatorname*{Res}_{z=\infty} \frac{z^2}{(z-1)(z-2)} = -\operatorname*{Res}_{z=0} \frac{1}{z^2(1-z)(1-2z)} = -\lim_{z \to 0} \left(\frac{1}{(1-z)(1-2z)}\right)' = -3.$$

Notice that the sum of all the residues is equal to 0. See the general Theorem 2 in 3.1.1.

EXAMPLE 2. Using Theorem 6 we find

$$\operatorname*{Res}_{z=k\pi} \cot g\, z = \frac{\cos k\pi}{\cos k\pi} = 1 \qquad (k \in \mathbf{Z}),$$

$$\operatorname*{Res}_{z=k\pi} \frac{1}{\sin z} = \frac{1}{\cos k\pi} = (-1)^k \qquad (k \in \mathbf{Z}).$$

EXAMPLE 3. The function F defined by

$$F(z) = \frac{e^{1/z} z^n}{1+z} \qquad (n \in \mathbb{N}),$$

has a simple pole at $z=-1$, and an essential singularity at $z=0$. Moreover, if we put $z = 1/t$, we get

$$F(z) = F\left(\frac{1}{t}\right) = \frac{e^t}{t^{n-1}(1+t)},$$

and hence $t=0$, i.e. $z=\infty$, is a pole of order $n-1$.

We have

$$\operatorname*{Res}_{z=-1} F(z) = \lim_{z \to -1} e^{1/z} z^n = (-1)^n e^{-1}.$$

In order to evaluate $\operatorname{Res}_{z=0} F(z)$, we look for the coefficient of $1/z$ in the Laurent expansion of F, or equivalently for the coefficient of z^{-n-1} in the expansion of $z \mapsto (e^{1/z})/(1+z)$. Since

$$e^{1/z} = \sum_{k=0}^{+\infty} \frac{1}{k!} \frac{1}{z^k} \quad (0<|z|), \qquad \frac{1}{1+z} = \sum_{k=0}^{+\infty} (-1)^k z^k \quad (|z|<1),$$

multiplying those series, we arrive at the coefficient which is

$$\sum_{k=1}^{+\infty} \frac{(-1)^{k+1}}{(n+k)!}.$$

Hence

$$\operatorname*{Res}_{z=0} F(z) = \sum_{k=1}^{+\infty} \frac{(-1)^{k+1}}{(n+k)!}.$$

In order to evaluate $\operatorname{Res}_{z=\infty} F(z)$, we use Theorem 1 of 2.1.1. Thus

$$\operatorname*{Res}_{z=\infty} F(z) = -\operatorname*{Res}_{z=0} \frac{e^z}{z^{n+1}(1+z)}.$$

and by Theorem 3 and Leibniz' rule

$$\operatorname*{Res}_{z=0} \frac{e^z}{z^{n+1}(1+z)} = \frac{1}{n!} \lim_{z \to 0} \frac{d^n}{dz^n} \left(\frac{e^z}{1+z}\right) = \frac{1}{n!} \lim_{z \to 0} \sum_{k=0}^{n} \binom{n}{k} \frac{(-1)^k k! e^z}{(1+z)^{k+1}}$$

Hence, after simple calculations, we obtain

$$\operatorname*{Res}_{z=\infty} F(z) = \sum_{k=0}^{n} \frac{(-1)^{k+1}}{(n-k)!}.$$

EXAMPLE 4. Suppose that f and g are regular functions in a neighbourhood of a and that $f(a) \neq 0$.

$1°$ If a is a second order zero of g, then

$$\operatorname*{Res}_{z=a} \frac{f(z)}{g(z)} = \frac{6f'(a) g''(a) - 2f(a) g'''(a)}{3g''(a)^2}.$$

$2°$ If a is a simple zero of g, then

$$\operatorname*{Res}_{z=a} \frac{f(z)}{g(z)^2} = \frac{f'(a) g'(a) - f(a) g''(a)}{g'(a)^2}.$$

EXAMPLE 5. If the zeros a_k and the poles b_k of the rational function

$$z \mapsto R(z) = A \frac{\prod\limits_{k=1}^{r} (z-a_k)^{\lambda_k}}{\prod\limits_{k=1}^{s} (z-b_k)^{\mu_k}} \qquad (A \in \mathbf{C}; \lambda_k, \mu_k \in \mathbf{N})$$

are symmetric with respect to a line through $z=0$, at an angle t to the positive x-axis, then

$$\operatorname*{Res}_{z=e^{2ti}\bar{b}_k} R(z) = \frac{A}{\bar{A}} e^{2ti(2m+1)} \overline{\operatorname*{Res}_{z=b_k} R(z)} \qquad \left(2m = \prod_{k=1}^{r} \lambda_k - \prod_{k=1}^{s} \mu_k\right).$$

(We say that poles, or zeros are symmetric with respect to a line if, besides the geometric symmetry, corresponding poles, or zeros, have the same order).

Put

$$R_1(z) = \frac{\prod_{k=1}^{r}(z-\alpha_k)^{\lambda_k}}{\prod_{k=1}^{s}(z-\beta_k)^{\mu_k}}, \quad R_0(z) = R(ze^{it}),$$

where $\alpha_k = a_k e^{-it}$, $\beta_k = b_k e^{-it}$. Then

$$R_0(z) = R(ze^{it}) = A e^{2mit} R_1(z).$$

If

$$R(z) = \sum_p \frac{c_p}{(z-b_k)^p}$$

is the Laurent expansion of R in a neighbourhood of b_k, then

$$\operatorname*{Res}_{z=b_k} R(z) = c_{-1} \quad \text{and} \quad R_0(z) = \sum_p \frac{c_p}{(ze^{it}-b_k)^p} = \sum_p \frac{c_p e^{-ipt}}{(z-\beta_k)^p}.$$

This implies

$$\operatorname*{Res}_{z=\beta_k} R_0(z) = e^{-it} \operatorname*{Res}_{z=b_k} R(z),$$

and hence

(7) $$\operatorname*{Res}_{z=b_k} R(z) = A e^{(2m+1)it} \operatorname*{Res}_{z=\beta_k} R_1(z).$$

In view of the definition of R_1 and the properties of a_k and b_k we see that R_1 is a rational function with real coefficients. Therefore

(8) $$\operatorname*{Res}_{z=\bar\beta_k} R_1(z) = \overline{\operatorname*{Res}_{z=\beta_k} R_1(z)}.$$

The required formula is obtained from (8) when $\operatorname{Res} R_1(z)$ is replaced by $\operatorname{Res} R(z)$, according to (7).

This result is due to D. Ž. Djoković.

EXAMPLE 6. We have

$$\operatorname*{Res}_{z=\infty} e^z \log \frac{z-a}{z-b} = e^a - e^b;$$

$$\operatorname*{Res}_{z=\infty} z^{-2} (2z^2-1) e^{z^2} (\operatorname{arctg} z)^2 = \pi (1-e^{-1}),$$

where in the first case we take the branch of the logarithm which vanishes for $z=1$, and in the second the branch of arctg which takes the value $\pi/2$ for $z=+\infty$.

REFERENCE
Tisserand, pp. 494, 497.

EXAMPLE 7. Consider the meromorphic function f defined by

$$f(z) = \frac{Q(z)}{P(z)^n} \qquad (n \in N),$$

where P is a polynomial of degree m. Let $z = a$ be a simple zero of P, i.e. $z = a$ is a pole of order n for the function f. In order to compute $\operatorname{Res}_{z=a} f(z)$ we write

$$f(z) = \frac{1}{(z-a)^n} \left(\sum_{k=0}^{+\infty} a_k (z-a)^k \right) \left(\sum_{k=0}^{m-1} b_k (z-a)^k \right)^{-n},$$

where

$$a_k = \frac{1}{k!} Q^{(k)}(a), \qquad b_k = \frac{1}{(k+1)!} P^{(k+1)}(a) \qquad (k=0, 1, \ldots, m-1)$$

and $b_k = 0$ for $k > m-1$.

The multinomial formula [1] states that for any real number s we have

$$\left(\sum_{k=0}^{+\infty} b_k (z-a)^k \right)^s = \sum_{k=0}^{+\infty} B_k (z-a)^k,$$

where

$$B_0 = b_0^s, \qquad B_k = \frac{1}{kb_0} \sum_{\nu=1}^{k} (\nu(s+1)-k) b_\nu B_{k-\nu}.$$

Using this formula (for $s=-n$) we obtain

$$f(z) = \frac{1}{(z-a)^n} \left(\sum_{k=0}^{+\infty} a_k (z-a)^k \right) \left(\sum_{k=0}^{+\infty} B_k (z-a)^k \right)$$

$$= \frac{1}{(z-a)^n} \sum_{k=0}^{+\infty} C_k (z-a)^k,$$

where

$$C_k = \sum_{v=0}^{k} A_v B_{k-v}, \qquad B_k = \frac{1}{kb_0} \sum_{v=1}^{k} \left(v(1-n)-k \right) b_v B_{k-v}.$$

Since this is the Laurent expansion of f, we conclude:

$$\operatorname*{Res}_{z=a} f(z) = C_{n-1}.$$

REFERENCES

1. H. W. Gould: 'Coefficient Identities for Powers of Taylor and Dirichlet Series', *Am. Math. Monthly* 81 (1974), 3–14.
2. B. K. Sachdeva and B. Ross: 'Evaluation of Certain Real Integrals by Contour Integration', *Am. Math. Monthly* 89 (1982), 246–249.

EXAMPLE 8. Suppose that h and g are regular functions at $z=a$, and that $z=a$ is a zero of order m of the function g, i.e.

$$h(z) = \sum_{n=0}^{+\infty} a_n(z-a)^n, \qquad g(z) = (z-a)^m \sum_{n=0}^{+\infty} b_n(z-a)^n \qquad (a_0 \neq 0, b_0 \neq 0).$$

The point $z=a$ is a pole of order m for the function f defined by $f(z)=h(z)/g(z)$. Hence, if

$$\frac{\sum_{n=0}^{+\infty} a_n(z-a)^n}{\sum_{n=0}^{+\infty} b_n(z-a)^n} = \sum_{k=0}^{+\infty} c_k(z-a)^k,$$

then $\operatorname{Res}_{z=a} f(z) = c_{m-1}$. From the last equation follows

$$\sum_{n=0}^{+\infty} a_n(z-a)^n = \sum_{j=0}^{+\infty} b_j(z-a)^j \sum_{k=0}^{+\infty} c_k(z-a)^k,$$

which implies

$$a_0 = b_0 c_0$$
$$a_1 = b_1 c_0 + b_0 c_1$$
$$\cdot$$
$$\cdot$$
$$\cdot$$
$$a_{m-1} = b_{m-1} c_0 + \cdots + b_0 c_{m-1}.$$
$$\cdot$$
$$\cdot$$
$$\cdot$$

Since we are only interested in the coefficient c_{m-1}, we only consider the first m equations of this infinite system. The determinant of that system is diagonal and equals b_0^m. Hence,

$$\operatorname*{Res}_{z=a} f(z) = c_{m-1} = \frac{1}{b_0^m} D,$$

where

$$D = \begin{vmatrix} b_0 & 0 & 0 & \cdots & a_0 \\ b_1 & b_0 & 0 & & a_1 \\ b_2 & b_1 & b_0 & & a_2 \\ \vdots & & & & \\ b_{m-1} & b_{m-2} & b_{m-3} & & a_{m-1} \end{vmatrix}$$

The following theorem will be used later.

THEOREM 7. *Let ζ be a point in the extended plane at which the function H is regular and $H'(\zeta) \neq 0$, or H has a simple pole. Let $\omega = H(\zeta)$ be an isolated singularity of the function G. If $w \mapsto h(w)$ is the inverse of $z \mapsto H(z)$ for w in a neighbourhood of ω, then*

$$(9) \qquad \operatorname*{Res}_{z=\zeta} G\big(H(z)\big) = \operatorname*{Res}_{w=H(\zeta)} G(w)\, h'(w).$$

Proof. We give the proof for the case $\zeta \neq \infty \neq \omega$. The remaining cases are easily obtained by simple modifications of the argument.

The hypotheses on H guarantee the existence of a local inverse function h which maps a neighbourhood of $w=\omega$ onto a neighbourhood of $z=\zeta$ in a one-to-one fashion (see Theorem 1 in 4.4.1). By definition

$$(10) \quad \operatorname*{Res}_{z=\zeta} G(H(z)) = \frac{1}{2\pi i} \oint_{\Gamma} G(H(z)) \, dz,$$

where Γ is a positively oriented circle surrounding ζ. If we put $z = h(w)$, or equivalently $w = H(z)$, we obtain

$$(11) \quad \oint_{\Gamma} G(H(z)) \, dz = \oint_{\Gamma'} G(w) h'(w) \, dw,$$

where Γ', the image of Γ under H, is a positively oriented simple closed contour surrounding the point $\omega = H(\zeta)$. However,

$$(12) \quad \operatorname*{Res}_{w=H(\zeta)} G(w) h'(w) = \frac{1}{2\pi i} \oint_{\Gamma'} G(w) h'(w) \, dw.$$

From (10), (11), (12) follows (9).

REFERENCE

R. P. Boas, Jr. and L. Schoenfeld: 'Indefinite Integration by Residues', *SIAM Review* 8 (1966), 173–183.

2.2. RESIDUES OF NONANALYTIC FUNCTIONS

2.2.1. *On Nonanalytic Functions*

We shall give here a few facts concerning nonanalytic functions which will be used later.

DEFINITION 1. A complex function w of a complex variable z, whose real and imaginary parts are differentiable functions in a region G, and such that the Cauchy–Riemann conditions are not satisfied, is called a nonanalytic function.

A nonanalytic function can be considered formally as a function depending on two independent variables z and $\bar{z}$, and therefore for such function we adopt the notation $z \mapsto w(z, \bar{z})$. The following expressions are important in the theory of nonanalytic functions

$$\frac{\partial w}{\partial z} = \frac{1}{2}\left(\frac{\partial u}{\partial x} + \frac{\partial v}{\partial y} + i\left(\frac{\partial v}{\partial x} - \frac{\partial u}{\partial y}\right)\right); \qquad \frac{\partial w}{\partial \bar{z}} = \frac{1}{2}\left(\frac{\partial u}{\partial x} - \frac{\partial v}{\partial y} + i\left(\frac{\partial v}{\partial x} + \frac{\partial u}{\partial y}\right)\right),$$

where $u(x, y) = \operatorname{Re} w(z, \bar{z})$, $v(x, y) = \operatorname{Im} w(z, \bar{z})$ and $z = x + iy$ $(x, y \in \mathbf{R})$.

A number of theorems which hold for analytic functions have their analogies in the theory of nonanalytic functions. We quote two such theorems.

THEOREM 1. *If w is a complex function which has continuous derivatives with respect to x and y in a simply connected region G bounded by a closed contour Γ, then*

$$\oint_{\Gamma} w(z, \bar{z})\, dz = 2i \iint_{G} \frac{\partial w}{\partial \bar{z}}\, dx\, dy,$$

and

$$\oint_{\Gamma} w(z, \bar{z})\, d\bar{z} = \frac{2}{i} \iint_{G} \frac{\partial w}{\partial z}\, dx\, dy.$$

In particular, if w is an analytic function, we have $\partial w/\partial \bar{z} = 0$, and Theorem 1 reduces to Theorem 1 from 1.2.2.

THEOREM 2. *If the function f satisfies the conditions of Theorem 1, and if $a \in G$, then*

$$(1) \qquad f(a, \bar{a}) = \frac{1}{2\pi i} \oint_{\Gamma} \frac{f(z, \bar{z})}{z-a}\, dz - \frac{1}{\pi} \iint_{G} \frac{\partial f}{\partial \bar{z}} \frac{1}{z-a}\, dx\, dy,$$

and

$$f(a, \bar{a}) = -\frac{1}{2\pi i} \oint_{\Gamma} \frac{f(z, \bar{z})}{\bar{z}-\bar{a}}\, d\bar{z} - \frac{1}{\pi} \iint_{G} \frac{\partial f}{\partial z} \frac{1}{\bar{z}-\bar{a}}\, dx\, dy.$$

Notice that in the case when f is analytic, i.e. when $\partial f/\partial \bar{z} = 0$, the formula (1) becomes Cauchy's integral formula (see Theorem 2 from 1.2.2).

2.2.2. Definitions of Residues

If f is an analytic function in a region G where it can have only one isolated singularity z_0, then we define

$$(1) \qquad \operatorname*{Res}_{z=z_0} f(z) = \frac{1}{2\pi i} \oint_\Gamma f(z)\, dz \qquad (\Gamma = \partial G).$$

If f is a nonanalytic function, the integral on the right-hand side of (1) is no longer independent of the path, and the definition loses its sense.

In the paper [1] Poor introduced the following definition.

DEFINITION 1. Let f be a nonanalytic function in a simply connected region G where it can have only one isolated singularity z_0. The residue of f at z_0 is defined by

$$\operatorname*{Res}_{z=z_0} f(z) = \lim_{r\to 0} \frac{1}{2\pi i} \oint_{C_r} f(z, \bar z)\, dz,$$

where $C_r = \{z \mid |z - z_0| = r\}$.

This definition generalizes the definition for analytic functions (see Definition 1 from 2.1.1). Besides, since the residue is defined as the limit when $r \to 0$, it no longer depends on the contour C_r.

It appears that there exists another possibility (which is, in a way, more natural) of defining residues of nonanalytic functions. Namely, the correctness of the definition for analytic functions, expressed by formula (1), rests upon Cauchy's theorem (Theorem 1 from 1.2.2) that for a function f, regular in a region G, we have

$$\oint_\Gamma f(z)\, dz = 0 \qquad (\Gamma = \partial G).$$

In the case of nonanalytic functions, the theorem which corresponds to Cauchy's is Theorem 1 from 2.2.1. Therefore, we define partial residues and the total residue for nonanalytic functions in the following way.

DEFINITION 2. Partial residues of a nonanalytic function f in a simply connected region G are, by definition,

$$R_1(f, G) = \frac{1}{2\pi i} \oint_\Gamma f(z, \bar{z}) \, dz - \frac{1}{\pi} \iint_G \frac{\partial f}{\partial \bar{z}} \, dx \, dy,$$

$$R_2(f, G) = -\frac{1}{2\pi i} \oint_\Gamma f(z, \bar{z}) \, d\bar{z} - \frac{1}{i} \iint_G \frac{\partial f}{\partial z} \, dx \, dy,$$

where $\Gamma = \partial G$, provided that those integrals exist.

The total residue of a nonanalytic function f in the region G is

$$R(f, G) = R_1(f, G) + R_2(f, G).$$

Definition 2 was also introduced by Poor [2].

Notice that partial and total residues of a nonanalytic function are defined for a region, and not for a point, as is the case of analytic functions.

2.2.3. Evaluation of Residues

THEOREM 1. *If f is a continuous function in a simply connected region G, and if there exists finite derivative $\partial f/\partial \bar{z}$ in G, then $R_1(f, G) = 0$.*

THEOREM 2. *If f is a continuous function in a simply connected region G, and if there exists finite derivative $\partial f/\partial z$ in G, then $R_2(f, G) = 0$.*

Theorems 1 and 2 are direct consequences of Theorem 1 from 2.2.1.

We now introduce the concept of a regular nonanalytic function (see [3]).

DEFINITION 1. A nonanalytic function f is said to be regular at the point a if it may be expressed in the form

$$f(z, \bar{z}) = f(a, \bar{a}) + f_z(a, \bar{a})(z - a) + f_{\bar{z}}(\bar{a}, a)(\bar{z} - \bar{a}) + \varepsilon,$$

where $\epsilon = o(z - a)$ as $z \to a$.

We first evaluate the residues of a function which has only one simple pole in the considered region.

THEOREM 3. *Let f be a regular nonanalytic function at every point of G, and suppose that the function g is defined by*

$$g(z, \bar{z}) = \frac{f(z, \bar{z})}{z-a} \qquad (a \in G).$$

Then $R_1(g, G) = f(a, \bar{a})$; $R_2(g, G) = 0$.

Proof. Let $\Gamma = \partial G$. By the definition of R_1 and by Theorem 2 from 2.2.1, we have

$$R_1(g, G) = \frac{1}{2\pi i} \oint_\Gamma g(z, \bar{z}) \, dz - \frac{1}{\pi} \iint_G \frac{\partial g}{\partial \bar{z}} \, dx \, dy$$

$$= \frac{1}{2\pi i} \oint_\Gamma \frac{f(z, \bar{z})}{z-a} \, dz - \frac{1}{\pi} \iint_G \frac{\partial f}{\partial \bar{z}} \frac{1}{z-a} \, dx \, dy$$

$$= f(a, \bar{a}).$$

Let $k = \{z \mid |z-a| = r\} \subset G$ and let G' be the annular region bounded by the contour Γ and the circle k. Then

$$R_2(g, G') = -\frac{1}{2\pi i} \oint_\Gamma \frac{f(z, \bar{z})}{z-a} \, d\bar{z} + \frac{1}{2\pi i} \oint_k \frac{f(z, \bar{z})}{z-a} \, d\bar{z}$$

$$- \frac{1}{\pi} \iint_{G'} \frac{\partial f}{\partial z} \frac{1}{z-a} \, dx \, dy = 0$$

since g has no singularities in G'.

However, since f is a regular nonanalytic function, we have

$$\oint_k \frac{f(z, \bar{z})}{z-a} \, d\bar{z} = f(a, \bar{a}) \oint_k \frac{1}{z-a} \, d\bar{z} + f_z(a, \bar{a}) \oint_k \frac{z-a}{z-a} \, d\bar{z}$$

(1)

$$+ f_{\bar{z}}(\bar{a}, a) \oint_k \frac{\bar{z}-\bar{a}}{z-a} \, d\bar{z} + \oint_k \frac{\varepsilon}{z-a} \, d\bar{z}.$$

The first three integrals on the right of (1) are identically zero while the last tends to zero as $r \to 0$, since $\varepsilon = o(z-a)$, as $z \to a$. Hence, $R_2(g, G') \to 0$ as $G' \to G$.

This theorem can be extended to functions with more than one simple poles.

THEOREM 4. *Let f be a regular nonanalytic function at every point of the region G, and let g be defined by*

$$g(z, \bar{z}) = \frac{f(z, \bar{z})}{(z-a_1)\cdots(z-a_n)},$$

where $a_i \in G$ ($i=1,\ldots,n$) and $a_i \neq a_j$ for $i \neq j$. Then

$$R_1(g, G) = \sum_{\nu=1}^{n} \frac{f(a_\nu, \bar{a}_\nu)}{\prod\limits_{\substack{\mu=1 \\ \mu \neq \nu}}^{n} (a_\nu - a_\mu)}; \qquad R_2(g, G) = 0.$$

EXAMPLE 1. Let $f(z, \bar{z}) = 1/(az + b\bar{z})$, and let G be a region which contains the point $z = 0$. Then

$$1° \quad R_1(f, G) = \frac{1}{a}, \qquad R_2(f, G) = -\frac{1}{b} \qquad (|b| \neq |a|);$$

$$2° \quad R_1(f, G) = \frac{1}{2a}, \qquad R_2(f, G) = \frac{1}{2b} \qquad (|b| = |a|).$$

EXAMPLE 2. If $f(z, \bar{z}) = (z\bar{z}-1)^{-1}$, and if the circle $\{z \mid |z| = 1\}$ is contained in the region G, then $R_1(f, G) = R_2(f, G) = 0$.

REFERENCES

1. V. C. Poor: 'Residues of Polygenic Functions', *Trans. Am. Math. Soc.* 32 (1930), 216–222.
2. V. C. Poor: 'On Residues of Polygenic Functions', *Trans. Am. Math. Soc.* 75 (1953), 244–255.
3. V. C. Poor: 'On the Hamilton Differential', *Bull. Am. Math. Soc.* 51 (1945), 945–948.

2.3. RESIDUES OF FUNCTIONS IN SEVERAL VARIABLES

2.3.1. *Definition of the Residue*

Let f be meromorphic on $D \subset \mathbb{C}^n$ with the polar set P. Consider the form $\omega = f \, dz = f(z_1, \ldots, z_n) \, dz_1 \wedge \cdots \wedge dz_n$, and let $\sigma \subset D \setminus P$ be an n-dimensional cycle. If σ is homologous to zero, then by Cauchy–Poincaré's theorem (Theorem 6 from 1.2.2) we have $\int_\sigma f\,dz = 0$. In other words, if the cycles σ and σ' are homologous in the region $D \setminus P$, then $\int_\sigma f\,dz = \int_{\sigma'} f\,dz$.

Therefore, if we have a basis of n-dimensional homologies $\sigma_1, \ldots, \sigma_p$ of the set $D \setminus P$ and if the cycle σ can be expressed as

$$\sigma \sim \sum_{\nu=1}^{p} k_\nu \sigma_\nu,$$

then the evaluation of the integral of f along σ reduces to the evaluation of that integral along basis homologies, i.e.

$$\int_\sigma f\,dz = \sum_{\nu=1}^{p} k_\nu \int_{\sigma_\nu} f\,dz.$$

We introduce the following definition.

DEFINITION 1. The residue of f with respect to the basis cycle σ_ν is

$$R_\nu = \frac{1}{(2\pi i)^n} \int_{\sigma_\nu} f\,dz.$$

2.3.2. *Form-Residue and Class-Residue*

Suppose that an $(n-1)$-dimensional subvariety P is defined on an n-dimensional complex variety M in the following way. In a neighbourhood $U_{z_0} \subset M$ of an arbitrary point z_0, P is defined as the set of zeros of f_{z_0}, where the function f_{z_0} is holomorphic in U_{z_0} and grad $f_{z_0} \neq 0$ in that neighbourhood. In other words

$$P \cap U_{z_0} = \{z \,|\, z \in U_{z_0},\, f_{z_0}(z) = 0\}.$$

Furthermore, define on $M \setminus P$ a differential form ω of degree p ($0 < p \leq 2n$) and class C^∞, so that in every neighbourhood U_{z_0} of $z_0 \in P$ the product $f_{z_0} \omega$ can be extended to a form of class C^∞ on U_{z_0}. In that case we say that the form ω has a first-order polar singularity on P.

THEOREM 1. *If the closed form $\omega \in C^\infty(M \setminus P)$ has a first-order polar singularity on P, then in any neighbourhood U of an arbitrary point*

$z_0 \in P$ *it can be represented as*

(1) $\qquad \omega = \dfrac{df}{f} \wedge r + \omega_1,$

where the forms r, $\omega_1 \in C^\infty(U)$. The restriction $r\big|_p$ does not depend on the choice of f and is also a closed form.

We are now in a position to define the form-residue and the class-residue.

DEFINITION 1. The form-residue of a closed form $\omega \in C^\infty(M \backslash P)$ which has a first order polar singularity on P is a closed form equal in a neighbourhood of an arbitrary point $z_0 \in P$ to the restriction of the form r from (1) to P, i.e.

$$\text{res } \omega = r\big|_p.$$

REMARK 1. The operator res maps the group $Z^p(M \backslash P)$ of closed forms of degree p and class C^∞ into the group $Z^{p-1}(P)$ of closed forms of degree $p-1$ and class C^∞, i.e. res: $Z^p(M \backslash P) \to Z^{p-1}(P)$.

REMARK 2. If the form ω is holomorphic on $M \backslash P$ then the form res ω is also holomorphic on P.

DEFINITION 2. Let ω be a closed form, the representative of the class $\omega^* \in H^p(M \backslash P)$. The class of cohomologies from $H^{p-1}(P)$ which contains the form res ω is called the class-residue and is denoted by Res ω = Res ω^*.

REMARK 3. The operator Res maps the group of cohomologies $H^p(M \backslash P)$ into the group of cohomologies $H^{p-1}(P)$. Notice that those groups of cohomologies correspond to the groups of homologies mapped by the operator res (see Remark 1).

The form-residue and the class-residue were introduced by Leray (see 10.3.2).

REFERENCES
See the references in 10.3.2.

Chapter 3

Contour Integration

3.1. CAUCHY'S THEOREM AND APPLICATIONS

3.1.1. *Cauchy's Fundamental Theorem on Residues*

The following theorem, proved by Cauchy, is fundamental.

THEOREM 1. *Let f be an analytic function in the region G, bounded by a closed contour Γ, and let f be continuous on Γ. Furthermore, let $a_1, \ldots, a_n$ be isolated singularities of f contained in G. Then*

$$(1) \qquad \oint_\Gamma f(z)\,dz = 2\pi i \sum_{\nu=1}^{n} \operatorname*{Res}_{z=a_\nu} f(z).$$

Proof. Let $K_\nu = \{z \mid |z - a_\nu| < r_\nu\}$ be a disc which does not contain any of the points $a_1, \ldots, a_{\nu-1}, a_{\nu+1}, \ldots, a_n$, and let $k_\nu = \partial K_\nu$ $(\nu = 1, \ldots, n)$. The function f is regular in the region

$$D = G \setminus \bigcup_{\nu=1}^{n} K_\nu,$$

bounded by the contour C, and hence (Theorem 1 from 1.2.2)

$$\oint_C f(z)\,dz = 0,$$

i.e.

$$\oint_\Gamma f(z)\,dz + \sum_{\nu=1}^{n} \int_{k_\nu} f(z)\,dz = 0,$$

or

$$(2) \qquad \oint_\Gamma f(z)\,dz = \sum_{\nu=1}^{n} \oint_{k_\nu} f(z)\,dz.$$

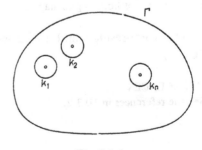

Fig. 3.1.1.

24

·However, by Definition 1 from 2.1.1, we have

$$\oint_{k_\nu} f(z)\,dz = 2\pi i\,\operatorname*{Res}_{z=a_\nu} f(z),$$

and from (2) follows (1).

This theorem is very important both in theory and in applications. In fact, this monograph is concerned with various applications of Theorem 1 in different branches of Mathematics.

Theorem 1 is also interesting because it expresses an integral as a sum of residues, which can often be evaluated by elementary methods, such as differentiation or taking limits.

THEOREM 2. *If an analytic function f has only isolated singularities in the extended plane, then the total sum of residues of f is equal to zero.*

Proof. The number of isolated singularities must· be finite, for in the opposite case there would exist their accumulation point, which would not be an isolated singularity. Denote the singularities in the finite plane by $a_1, \ldots, a_n$. Then there exists a circle $\Gamma = \{z \mid |z| = R\}$, so large that $a_1, \ldots, a_n \in \operatorname{int} \Gamma$. By Definition 2 from 2.1.1 we have

$$(3) \qquad \oint_\Gamma f(z)\,dz = -2\pi i\,\operatorname*{Res}_{z=\infty} f(z),$$

and from Theorem 1 follows

$$(4) \qquad \oint_\Gamma f(z)\,dz = 2\pi i \sum_{\nu=1}^{n} \operatorname*{Res}_{z=a_\nu} f(z).$$

Equalities (3) and (4) imply

$$\operatorname*{Res}_{z=\infty} f(z) + \sum_{\nu=1}^{n} \operatorname*{Res}_{z=a_\nu} f(z) = 0,$$

and the theorem is proved.

3.1.2. *Evaluation of Complex Integrals*

The first, and most obvious, application of Cauchy's residue theorem is to the evaluation of complex integrals. We give a number of examples.

3.1.2.1. Consider the integral

$$\oint_{\Gamma} \frac{z}{(z^2-1)^2\,(z^2+1)}\,dz,$$

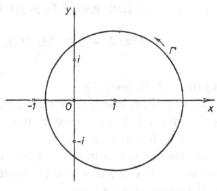

where

$$\Gamma = \{z \mid |z-1| = \sqrt{3}\}.$$

The function f, defined by

$$f(z) = \frac{z}{(z^2-1)^2\,(z^2+1)},$$

Fig. 3.1.2.1.

has four singularities: simple poles at $a_1 = i$, $a_2 = -i$ and second order poles at $a_3 = 1$ and $a_4 = -1$. The first three are inside the contour Γ, while the fourth is outside. Hence

$$\oint_{\Gamma} f(z)\,dz = 2\pi i \sum_{\nu=1}^{3} \operatorname*{Res}_{z=a_\nu} f(z).$$

We have

$$\operatorname*{Res}_{z=i} f(z) = \lim_{z\to i}\,(z-i)\,f(z) = \lim_{z\to i} \frac{z}{(z^2-1)^2\,(z+i)} = \frac{1}{8};$$

$$\operatorname*{Res}_{z=-i} f(z) = \lim_{z\to -i}\,(z+i)\,f(z) = \lim_{z\to -i} \frac{z}{(z^2-1)^2\,(z-i)} = \frac{1}{8};$$

$$\operatorname*{Res}_{z=1} f(z) = \lim_{z\to 1} \frac{d}{dz}\big((z-1)^2\,f(z)\big) = \lim_{z\to 1} \frac{d}{dz}\left(\frac{z}{(z+1)^2\,(z^2+1)}\right)$$

$$= \lim_{z\to 1} \frac{-3\,z^3 - z^2 - z + 1}{(z+1)^3\,(z^2+1)^2} = -\frac{1}{8},$$

and hence

$$2\pi i \sum_{\nu=1}^{3} \operatorname*{Res}_{z=a_\nu} f(z) = 2\pi i \left(\frac{1}{8} + \frac{1}{8} - \frac{1}{8}\right) = \frac{\pi i}{4},$$

implying

$$\oint_{\Gamma} \frac{z}{(z^2-1)^2\,(z^2+1)}\,dz = \frac{\pi i}{4}.$$

REMARK 1. Since

$$\operatorname*{Res}_{z=\infty} f(z) = -\operatorname*{Res}_{z=0} \frac{1}{z^2} f\left(\frac{1}{z}\right) = -\operatorname*{Res}_{z=0} \frac{z^3}{(1-z^2)^2 (1+z^2)} = 0,$$

by Theorem 2 from 3.1.1 we conclude that

$$\oint_C \frac{z}{(z^2-1)^2 (z^2+1)} \, dz = 0,$$

where C is any closed contour which surrounds all the singularities of f. A more general statement is given in 3.1.2.3.

3.1.2.2. Let $\Gamma = \{z \,|\, |z| = r\}$ and let $0 < |a| < r < |b|$. Then

$$(1) \qquad \oint_\Gamma (\bar{z}-a)^{-1} (b-\bar{z})^{-1} (z^2+z^{-2}) \, dz = 2\pi i \cdot \frac{r^8+b^4}{b^4 r^2 (b-a)}.$$

Indeed, on the circle Γ we have $z\bar{z} = r^2$, and the integral in (1) can be written as $\oint_\Gamma f(z) \, dz$, where

$$f(z) = \frac{z^4+1}{(r^2-az)(bz-r^2)}.$$

The function f has two simple poles at r^2/a and r^2/b, but only the last one is inside Γ. Therefore

$$\oint_\Gamma f(z) \, dz = 2\pi i \operatorname*{Res}_{z=r^2/b} f(z) = 2\pi i \lim_{z \to r^2/b} \left(z - \frac{r^2}{b}\right) f(z)$$

$$= 2\pi i \lim_{z \to r^2/b} \frac{z^4+1}{b(r^2-az)} = 2\pi i \frac{r^8+b^4}{b^4 r^2 (b-a)},$$

implying (1).

REFERENCE
Mitrinović, pp. 12–13.

3.1.2.3. Let P and Q be polynomials defined by

$$P(z) = az^m + a_1 z^{m-1} + \cdots + a_m, \qquad Q(z) = bz^n + b_1 z^{n-1} + \cdots + b_n.$$

Suppose that $z_1, \ldots, z_r$ are zeros of order $j_1, \ldots, j_r$ of the polynomial Q, and suppose that P and Q do not have common zeros. If Γ is a simple closed contour which surrounds all the zeros of Q, then

$$(1) \qquad \oint_\Gamma \frac{P(z)}{Q(z)} dz = \begin{cases} \dfrac{2\pi i a}{b} & \text{for } n - m = 1, \\[2mm] 0 & \text{for } n - m \geq 2. \end{cases}$$

In order to prove this, we expand $P(z)/Q(z)$ in partial fractions, to obtain

$$\frac{P(z)}{Q(z)} = \sum_{i=1}^{r} \sum_{k=1}^{j_i} \frac{A_{ik}}{(z - z_i)^k},$$

which implies

$$\lim_{|z| \to +\infty} \frac{zP(z)}{Q(z)} = \sum_{i=1}^{r} A_{i1}.$$

On the other hand

$$A_{i1} = \operatorname*{Res}_{z = z_i} \frac{P(z)}{Q(z)}.$$

Therefore, applying Cauchy's residue theorem we find

$$\oint_\Gamma \frac{P(z)}{Q(z)} dz = 2\pi i \sum_{i=1}^{r} A_{i1} = 2\pi i \lim_{|z| \to +\infty} \frac{zP(z)}{Q(z)},$$

implying (1).

We have therefore evaluated the integral

$$\oint_\Gamma \frac{P(z)}{Q(z)} dz$$

in the case when $n > m$. If $m \geq n$, then

$$\frac{P(z)}{Q(z)} = F(z) + \frac{R(z)}{Q(z)},$$

where F is a polynomial, and R is a polynomial of degree $\leq n-1$. However, F is regular in the finite plane, and so

$$\oint_\Gamma \frac{P(z)}{Q(z)} dz = \oint_\Gamma \frac{R(z)}{Q(z)} dz.$$

REFERENCE
E. Just and N. Schaumberger: 'Contour Integration for Rational Functions', *Am. Math. Monthly* **71** (1964), 546–547.

3.1.2.4. If $n \in N_0$ and if $\Gamma = \{z \mid |z| = (n + 1/2)\, \pi\}$, then

$$\oint_\Gamma \mathrm{ch}\, z \, \mathrm{cotg}\, z \, dz = 2\pi i \, \frac{\mathrm{sh}\left(n + \dfrac{1}{2}\right)\pi}{\mathrm{sh}\,\dfrac{1}{2}\,\pi} \qquad (n \in N_0).$$

The function f, defined by $f(z) = \mathrm{ch}\, z\, \mathrm{cotg}\, z$, has simple poles at $z = k\pi$ ($k \in Z$). Using Theorem 6 from 2.1.2, we conclude that

$$\operatorname*{Res}_{z=k\pi} f(z) = \frac{\mathrm{ch}\, k\pi \cos k\pi}{\cos k\pi} = \mathrm{ch}\, k\pi.$$

The point $k\pi$ lies inside int Γ if $|k| \le n$. Hence

$$\oint_\Gamma \mathrm{ch}\, z\, \mathrm{cotg}\, z\, dz = 2\pi i \sum_{k=-n}^{n} \mathrm{ch}\, k\pi$$

$$= 2\pi i\left(1 + 2\sum_{k=1}^{n} \mathrm{ch}\, k\pi\right) = 2\pi i\, \frac{\mathrm{sh}\left(n + \dfrac{1}{2}\right)\pi}{\mathrm{sh}\,\dfrac{1}{2}\,\pi}.$$

REFERENCE
Mitrinović, pp. 13–14.

3.1.2.5. Let us evaluate the integral

$$\oint_\Gamma \frac{z^2}{e^{2\pi i z^3} - 1}\, dz,$$

where $\Gamma = \{z \mid |z| = r\}$, and $n < r^3 < n+1$ ($n \in N$).

The function f, defined by $f(z) = z^2/(e^{2\pi i z^3} - 1)$, has simple poles at the points z, for which $z^3 = k$ ($k \in Z$). The poles of f inside int Γ are

$$0, \quad \pm k^{1/3}, \quad \pm a k^{1/3}, \quad \pm a^2 k^{1/3} \qquad (k = 1, \ldots, n),$$

where $\alpha^3 = 1$, $\alpha \neq 1$.

Let a be a complex number such that $e^{2\pi i a^3} = 1$. Then

$$\operatorname*{Res}_{z=a} f(z) = \lim_{z \to a} \frac{(z-a)\, z^2}{e^{2\pi i z^3} - 1} = \frac{1}{6\pi i}.$$

We also have

$$\operatorname*{Res}_{z=0} f(z) = \lim_{z \to 0} \frac{z^2}{e^{2\pi i z^3} - 1} = \frac{1}{2\pi i}.$$

Therefore the region int Γ contains the pole at $z=0$, with the residue $1/(2\pi i)$, and $6n$ poles with residues $1/(6\pi i)$ each. Hence

$$\oint_\Gamma \frac{z^2}{e^{2\pi i z^3} - 1} \, dz = 2\pi i \left(\frac{1}{2\pi i} + 6n \frac{1}{6\pi i} \right) = 2n + 1.$$

REFERENCE
Mitrinović, pp. 14–15.

3.1.2.6. Consider the integral

$$J = \oint_\Gamma P(z) \left(\exp \frac{1}{z} + \exp \frac{1}{z-1} + \cdots + \exp \frac{1}{z-k} \right) dz,$$

where P is a polynomial of degree k, and $\Gamma = \{z \mid |z| = r\}$, $r \neq 1, 2, \ldots, k$.
    Let p be a nonnegative integer such that $p < r < p+1$ for $r < k$, and $p=k$ for $r \geq k$. Then

$$J = \sum_{\nu=0}^{p} \oint_\Gamma P(z) \exp \frac{1}{z-\nu} \, dz.$$

The function $z \mapsto P(z) \exp 1/(z-\nu)$ has an essential singularity at $z=\nu$. The residue at that point is equal to the coefficient of $1/(z-\nu)$ in the Laurent's expansion of that function. Since

$$P(z) = P(\nu) + \frac{P'(\nu)}{1!}(z-\nu) + \frac{P''(\nu)}{2!}(z-\nu)^2 + \cdots + \frac{P^{(k)}(\nu)}{k!}(z-\nu)^k,$$

$$\exp \frac{1}{z-\nu} = 1 + \frac{1}{1!(z-\nu)} + \frac{1}{2!(z-\nu)^2} + \cdots$$

we easily obtain

$$\operatorname*{Res}_{z=\nu} P(z) \exp \frac{1}{z-\nu} = \frac{P(\nu)}{0!\,1!} + \frac{P'(\nu)}{1!\,2!} + \cdots + \frac{P^{(k)}(\nu)}{k!\,(k+1)!}.$$

Thus

$$\oint_\Gamma P(z) \exp \frac{1}{z-\nu}\, dz = 2\pi i \sum_{m=0}^{k} \frac{P^{(m)}(\nu)}{m!\,(m+1)!},$$

implying

$$J = \sum_{\nu=0}^{p} \oint_\Gamma P(z) \exp \frac{1}{z-\nu}\, dz = 2\pi i \sum_{\nu=0}^{p} \sum_{m=0}^{k} \frac{P^{(m)}(\nu)}{m!\,(m+1)!}.$$

REFERENCE
Mitrinović, pp. 15–16.

3.1.2.7. Let us evaluate the integral

$$I = \oint_\Gamma \frac{1}{\sqrt{z^2+z+1}}\, dz,$$

where $\Gamma = \{z \mid |z| = 2\}$.

Since the integrand is a multiform function, the value of I will depend on the choice of the branch of that function. We choose the branch which is positive for z real and positive.

The chosen branch of the function $z \mapsto 1/\sqrt{z^2+z+1}$ is uniform and analytic in the region $\{z \mid |z| > 1\}$, and can be expanded into Laurent's series

$$\frac{1}{\sqrt{z^2+z+1}} = \frac{1}{z}\left(1 + \frac{1}{z} + \frac{1}{z^2}\right)^{-1/2} = \frac{1}{z} - \frac{1}{2z^2} + \cdots$$

Therefore,

$$I = \oint_\Gamma \frac{1}{z}\, dz - \oint_\Gamma \frac{1}{2z^2}\, dz + \cdots = 2\pi i.$$

REFERENCE
Julia, pp. 157–162.

3.1.2.8. Let a be a real number and let p be a complex number such that $-(n+1) < \operatorname{Re} p < -n$, where $n \in \mathbf{N}$. If Γ is a positively oriented contour, which, starting from some given point a, makes a circuit round the origin and returns to a, then

$$\oint_\Gamma z^{p-1} f(z)\, dz = (e^{2\pi ip} - 1) \int_0^a x^{p-1}\left(f(x) - \sum_{\nu=0}^n \frac{1}{\nu!} f^{(\nu)}(0) x^\nu\right) dx$$

$$+ (e^{2\pi ip} - 1) \sum_{\nu=0}^n \frac{1}{(p+\nu)\,\nu!} f^{(\nu)}(0)\, a^{p+\nu},$$

where f is an analytic function, regular at $z = 0$.

In particular, for $f(x) = e^{-x}$ we obtain Saalschütz's expression for the gamma function (see [1]).

The above result was proved by Dixon [2].

REFERENCES
1. E. T. Whittaker and G. N. Watson: *A Course of Modern Analysis*, 4th edition, Cambridge 1927, p. 244.
2. A. L. Dixon: 'Note on the Evaluation of Contour Integrals', *Messenger Math*. (2) 33 (1904), 176–178.

3.1.2.9. We now prove the fundamental theorem of algebra by contour integration.

Let P be a nonconstant polynomial; we are to show that $P(z) = 0$ for some z. We may suppose that $P(z) \in \mathbf{R}$ when $z \in \mathbf{R}$. Indeed, otherwise let $\overline{P}$ denote the polynomial whose coefficients are conjugates of those of P, and consider the polynomial $z \mapsto P(z)\overline{P}(z)$.

Therefore, suppose that $P(z)$ is real for real z and is never 0. This means that P does not either vanish or change sign for real z, and so

(1) $$\int_0^{2\pi} \frac{1}{P(2\cos\theta)}\, d\theta \neq 0.$$

However,

(2) $$\int_0^{2\pi} \frac{1}{P(2\cos\theta)}\, d\theta = \frac{1}{i}\oint_C \frac{1}{zP(z+z^{-1})}\, dz = \frac{1}{i}\oint_C \frac{z^{n-1}}{Q(z)}\, dz,$$

where $C = \{z \,|\, |z| = 1\}$ and where $z \mapsto Q(z) = z^n P(z+z^{-1})$ is also a polynomial.

For $z \neq 0$, $Q(z) \neq 0$. Furthermore, if a is the leading coefficient of P, then $Q(0) = a \neq 0$. Hence, $Q(z) \neq 0$ for all z, and so the function $z \mapsto z^{n-1}/Q(z)$ is regular in the disc int C. By Cauchy's residue theorem, we find

$$\oint_C \frac{z^{n-1}}{Q(z)} \, dz = 0,$$

contradicting (1).

REMARK 1. This result was proved by Boas [1]. A similar proof was earlier given by Ankeny [2], who used integration along the real axis, and not around the unit circle.

REMARK 2. See 4.1 and 4.3.2.

REFERENCES

1. R. P. Boas, Jr.: 'Yet Another Proof of the Fundamental Theorem of Algebra', *Am. Math. Monthly* 71 (1964), 180.
2. N. C. Ankeny: 'One More Proof of the Fundamental Theorem of Algebra', *Am. Math. Monthly* 54 (1947), 464.

3.1.2.10. In extensive papers [1] and [2] Dini considered integrals of the form

$$(1) \qquad \oint_\Gamma \frac{f(z)}{u(z)^p} \, dz,$$

where f is an analytic function and u is a meromorphic function in the region int Γ and $p \in \mathbb{N}$. Using the integral (1) he evaluated a number of interesting integrals of complex functions depending on a real variable between the limits 0 and 2π. We give a few examples

$$1° \qquad \int_0^{2\pi} e^{i(p+1)t} \frac{1}{(p_0 + p_1 \cos t + q_1 \sin t)^{p+1}} \, dt = \frac{2^{p+2}\pi}{(p_1 - iq_1)^{p+2}},$$

$$2° \qquad \int_0^{2\pi} e^{i(p+2)t} \frac{1}{(p_0 + p_1 \cos t + q_1 \sin t)^{p+1}} \, dt = -\frac{2^{p+3}(p+1)\pi}{(p_1 - iq_1)^{p+2}},$$

$$3° \qquad \int_0^{2\pi} \log(p_0 + p_1 \cos t + q_1 \sin t) \, dt = 2\pi \log \frac{p+iq}{2} - 2\pi^2 i,$$

$$4° \qquad \int_0^{2\pi} e^{int} \log(p_0 + p_1 \cos t + q_1 \sin t) \, dt = -\frac{2\pi}{n} \quad (n > 1, \ n \in \mathbb{N}),$$

where p_0, p_1, and q_1 are complex numbers.

REFERENCES
1. U. Dini: 'Una applicazione della teoria dei residui delle funzioni di variabile complessa', *Atti R. Acc. Naz. Lincei* (5) 2 (1898), 495–545 ≡ *Opere*, Vol. 2, Roma 1954, pp. 393–464.
2. U. Dini: 'Un'applicazione della teoria dei residui delle funzioni di variabile complessa', *Ann. Mat. Pura Appl.* (3) 1 (1898), 39–76 ≡ *Opere*, Vol. 2, Roma 1954, pp. 465–506.

3.1.3. *Evaluation of Residues by Means of Integrals*

In the previous section we evaluated complex integrals by means of residues. Sometimes, we reverse the procedure and evaluate residues by means of integrals.

EXAMPLE 1. The problem is to find $\operatorname{Res}_{z=1/2} \operatorname{tg}^{2k-1} \pi z$, where $k \in \mathbb{N}$.
    By definition

$$\operatorname*{Res}_{z=1/2} \operatorname{tg}^{2k-1} \pi z = \frac{1}{2\pi i} \oint_{\Gamma} \operatorname{tg}^{2k-1} \pi z \, dz,$$

where Γ is the closed contour which surrounds only the singularity at $z = 1/2$. For Γ we choose the rectangle shown on Figure 3.1.3, where $R > 0$.
    Thus

$$\oint_{\Gamma} \operatorname{tg}^{2k-1} \pi z \, dz = \int_{-iR}^{1-iR} \operatorname{tg}^{2k-1} \pi z \, dz + \int_{1-iR}^{1+iR} \operatorname{tg}^{2k-1} \pi z \, dz$$

$$+ \int_{1+iR}^{iR} \operatorname{tg}^{2k-1} \pi z \, dz + \int_{iR}^{-iR} \operatorname{tg}^{2k-1} \pi z \, dz.$$

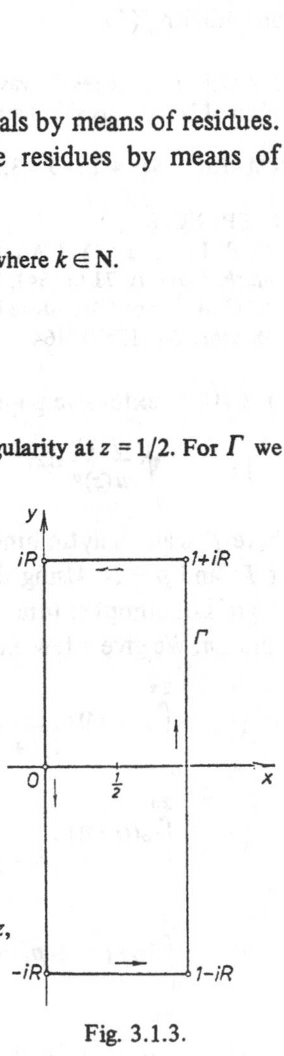

Fig. 3.1.3.

    Since the function $z \mapsto \operatorname{tg} \pi z$ is periodic with period 1, we have

$$\int_{1-iR}^{1+iR} \operatorname{tg}^{2k-1} \pi z \, dz = \int_{-iR}^{iR} \operatorname{tg}^{2k-1} \pi z \, dz = - \int_{iR}^{-iR} \operatorname{tg}^{2k-1} \pi z \, dz,$$

and so

$$\operatorname*{Res}_{z=1/2} \operatorname{tg}^{2k-1} \pi z = \frac{1}{2\pi i} \left(\int_{-iR}^{1-iR} \operatorname{tg}^{2k-1} \pi z \, dz + \int_{1+iR}^{iR} \operatorname{tg}^{2k-1} \pi z \, dz \right).$$

Furthermore,

$$\lim_{y\to+\infty} \operatorname{tg} \pi z = -\frac{1}{i}, \qquad \lim_{y\to-\infty} \operatorname{tg} \pi z = -\frac{1}{i} \qquad (y = \operatorname{Im} z).$$

Hence

$$\lim_{R\to+\infty} \int_{-iR}^{1-iR} \operatorname{tg}^{2k-1} \pi z \, dz = \left(\frac{1}{i}\right)^{2k-1},$$

$$\lim_{R\to+\infty} \int_{1+iR}^{iR} \operatorname{tg}^{2k-1} \pi \cdot z \, dz = -\left(-\frac{1}{i}\right)^{2k-1} = \left(\frac{1}{i}\right)^{2k-1},$$

and we finally obtain

$$\operatorname*{Res}_{z=1/2} \operatorname{tg}^{2k-1} \pi z = \frac{1}{2\pi i}\left(2\left(\frac{1}{i}\right)^{2k-1}\right) = \frac{(-1)^k}{\pi}.$$

REFERENCE
Titchmarsh, p. 108.

3.1.4. *Jordan's Lemmas*

Let C_r be the arc of the circle $K_r = \{z \mid |z - z_0| = r\}$ whose end points P_1 and P_2 are the intersections of half-lines p_1 and p_2 from z_0 with K_r. Furthermore, let p_1 and p_2 subtend the angle α at z_0. We suppose that C_r is negatively oriented and that $\theta_i = \arg P_i$ $(i = 1, 2)$. Denote by D_r the region bounded by the arc C_r and the half-lines p_1 and p_2 (see Figure 3.1.4).

The following three theorems on the integration along the arc C_r have important applications in the evaluation of real definite integrals. Those theorems, and particularly Theorem 3, are often called Jordan's lemmas (see [1]).

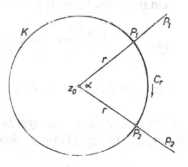

Fig. 3.1.4.

THEOREM 1. *Suppose that f is an analytic function in the region D_ϵ where it can have a finite number of singularities, and suppose that f is regular on $\partial \dot{D}_\epsilon$. If*

(1) $$\lim_{z \to z_0} (z - z_0) f(z) = A,$$

then

(2) $$\lim_{\epsilon \to 0} \int_{C_\epsilon} f(z)\, dz = -iA\, \alpha.$$

Proof. In view of (1) we may write $(z - z_0)f(z) = A + g(z)$, where $\lim_{z \to z_0} g(z) = 0$. We obtain

$$f(z) = \frac{A}{z - z_0} + \frac{g(z)}{z - z_0},$$

and hence

$$\int_{C_\epsilon} f(z)\, dz = A \int_{C_\epsilon} \frac{1}{z - z_0}\, dz + \int_{C_\epsilon} \frac{g(z)}{z - z_0}\, dz.$$

If we put $z = z_0 + \epsilon e^{i\theta}$, we get

(3) $$A \int_{C_\epsilon} \frac{1}{z - z_0}\, dz = iA \int_{\theta_1}^{\theta_2} d\theta = iA\,(\theta_2 - \theta_1) = -iA\,\alpha.$$

On the other hand, if $\epsilon \to 0$, we have

(4) $$\left| \int_{C_\epsilon} \frac{g(z)}{z - z_0}\, dz \right| \le \alpha \sup_{C_\epsilon} |g(z)| = \alpha \sup_{C_\epsilon} |(z - z_0)f(z) - A| \to 0.$$

From (3) and (4) follows (2).

THEOREM 2. *Let f satisfy the conditions of Theorem 1 in the region D_R. If $\lim_{z \to \infty} (z - z_0)f(z) = A$, then*

$$\lim_{R \to +\infty} \int_{C_R} f(z)\, dz = -iA\, \alpha.$$

The proof is analogous to the proof of Theorem 1.

THEOREM 3. *Let f satisfy the conditions of Theorem 1 in the region D_R and let m be a positive number. Furthermore, let $z_0 = 0, \theta_1 = 0, \theta_2 = \pi$. If*

(5) $$\lim_{z \to \infty} f(z) = 0,$$

then

(6) $$\lim_{R \to +\infty} \int_{C_R} e^{miz} f(z) \, dz = 0.$$

Proof. We have

$$\int_{C_R} e^{miz} f(z) \, dz = \int_0^\pi e^{miR(\cos\theta + i\sin\theta)} f(Re^{i\theta}) \, Re^{i\theta} \, i \, d\theta.$$

In view of the hypothesis (5), for a given $\epsilon > 0$ there exists R_0 so that $|f(z)| < \epsilon$ for all z which satisfy $|z| > R_0$ and $0 < \arg z < \pi$. Therefore, since $|e^{miR \cos\theta}| = 1$, for $R > R_0$ we have

$$\left| \int_{C_R} e^{miz} f(z) \, dz \right| \leq \epsilon R \int_0^\pi e^{-mR\sin\theta} \, d\theta$$

$$= \epsilon R \left(\int_0^{\pi/2} e^{-mR\sin\theta} \, d\theta + \int_{\pi/2}^\pi e^{-mR\sin\theta} \, d\theta \right).$$

If we put $\theta = \pi - t$ into the last integral, we find

$$\left| \int_{C_R} e^{miz} f(z) \, dz \right| \leq 2\epsilon R \int_0^{\pi/2} e^{-mR\sin\theta} \, d\theta.$$

Using Jordan's inequality

$$\sin\theta \geq \frac{2\theta}{\pi} \qquad \left(0 \leq \theta \leq \frac{\pi}{2} \right),$$

we obtain

$$2\epsilon R \int_0^{\pi/2} e^{-mR\sin\theta} \, d\theta \leq 2\epsilon R \int_0^{\pi/2} \exp\left(-\frac{2mR\theta}{\pi} \right) d\theta$$

$$= \frac{\epsilon\pi}{m} (1 - e^{-m/R}) < \frac{\epsilon\pi}{m},$$

which implies (5).

REMARK 1. The following generalization of Theorem 3 was given by Dimitrovski [2].

Let $n \in \mathbf{N}$ and let f be an analytic function in the region $D = \{ z \mid 0 \leq \arg z \leq \pi/n \}$ where it can have a finite number of singularities, and let f be regular on ∂D. Suppose further that

$$|f(z)| < AR^{n-2} \qquad \text{as} \qquad R = |z| \to +\infty \qquad (A > 0).$$

If P_n is a polynomial of degree n with positive coefficients, then

$$\lim_{R \to +\infty} \int_{C_R} e^{iP_n(z)} f(z)\, dz = 0,$$

where $C_R = \{ z \mid |z| = R,\ 0 \leq \arg z \leq \pi/n \}$.

For $n = 1$ and $P_n(z) = mz$ ($m > 0$) we obtain Theorem 3.

REFERENCES
1. C. Jordan: *Cours d'analyse de l'École Polytechnique*, t. 2, Paris 1913, p. 331.
2. D. Dimitrovski: 'A Generalization of Jordan's Lemma and Some Applications' (Macedonian), *Bull. Soc. Math. Phys. Macédoine* 12 (1961), 33–43.

3.1.5. *Complex Substitutions in Definite Integrals*

If we know the value of a real definite integral, then in some cases it may be useful to replace the real variable of integration, say x, by an imaginary variable, say it ($t \in \mathbf{R}$), and then after separating the real and imaginary parts in the resulting integral, to obtain values for two new real integrals. We illustrate this by an example.

If $a > 0, p > 0$, then

$$(1) \qquad \int_0^{+\infty} x^{p-1} e^{-ax}\, dx = \frac{1}{a^p} \Gamma(p),$$

where Γ is the gamma function. If we make the formal substitution $x = it$ ($t \in \mathbf{R}$), we get

$$\int_0^{+\infty} (it)^{p-1} e^{-ait} i\, dt = \frac{1}{a^p} \Gamma(p),$$

and, multiplying by $e^{-ip\pi/2}$ and separating real and imaginary parts, we find

$$\int\limits_{0}^{+\infty} t^{p-1}\cos at\, dt = \frac{1}{a^p}\Gamma(p)\cos\frac{1}{2}p\pi,$$

(2)

$$\int\limits_{0}^{+\infty} t^{p-1}\sin at\, dt = \frac{1}{a^p}\Gamma(p)\sin\frac{1}{2}p\pi.$$

However, the theorems on integration by substitution do not cover complex substitutions. The transition from (1) to (2) is justified by an application of Cauchy's residue theorem.

Indeed, start with the integral

$$\oint\limits_{\Gamma} z^{p-1}e^{-az}\, dz,$$

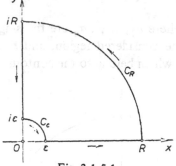

where Γ is the contour consisting of two segments and two circular arcs (Figure 3.1.5.1). If $C_r = \{z \mid |z|=r,\ \operatorname{Re} z \geq 0,\ \operatorname{Im} z \geq 0\}$, then

Fig. 3.1.5.1.

(3)
$$\oint\limits_{\Gamma} z^{p-1}e^{-az}\, dz = \int\limits_{\varepsilon}^{R} x^{p-1}e^{-ax}\, dx + \int\limits_{C_R} z^{p-1}e^{-az}\, dz$$
$$+ \int\limits_{R}^{\varepsilon}(iy)^{p-1}e^{-aiy}\, i\, dy + \int\limits_{C_\varepsilon} z^{p-1}e^{-az}\, dz = 0,$$

because the function $z \mapsto z^{p-1}e^{-az}$ has no singularities in int Γ.

However, if $f(z) = z^{p-1}e^{-az}$, then

$$\lim_{z\to 0} zf(z) = \lim_{z\to 0} z^p e^{-az} = 0 \qquad (p>0),$$

$$\lim_{z\to\infty} zf(z) = \lim_{z\to\infty} z^p e^{-az} = 0 \qquad (a>0),$$

and using Theorems 1 and 2 from 3.1.4, we find

$$\lim_{\varepsilon\to 0} \int\limits_{C_\varepsilon} z^{p-1}e^{-az}\, dz = 0 \quad (p>0), \qquad\qquad \lim_{R\to+\infty} \int\limits_{C_R} z^{p-1}e^{-az}\, dz = 0 \quad (a>0).$$

Hence, if $\epsilon \to 0, R \to + \infty$ in (3), then

$$\int\limits_{0}^{+\infty} x^{p-1} e^{-ax} \, dz = \int\limits_{0}^{+\infty} (iy)^{p-1} e^{-aiy} \, i \, dy,$$

and the transition from (1) to (2) is justified.

In general, if f is an analytic function in the closed region from Figure 3.1.5.2, then

$$\int\limits_{0}^{+\infty} f(x) \, dx = i \int\limits_{0}^{+\infty} f(ix) \, dx - \lim_{R \to +\infty} \int\limits_{C_R} f(z) \, dz + 2\pi i \sum_{\nu=1}^{n} \operatorname*{Res}_{z=a_\nu} f(z) + r,$$

where $a_1, \ldots, a_n$ are the singularities of f which belong to the interior of the considered region, and r is the contribution of eventual singularities of f which belong to the contour, and which has, in that case to be indented.

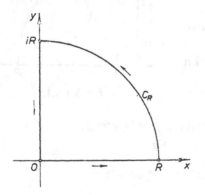
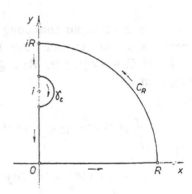

Fig. 3.1.5.2.   Fig. 3.1.5.3.

For example, let $f(z) = e^{iz}/(z^2 + 1)$. The function f has a simple pole at $z = i$ which belongs to the contour of integration. Integrating this function along the contour shown on Figure 3.1.5.3, since f has no singularities in the considered region, we have

$$(4) \int\limits_{0}^{R} \frac{e^{ix}}{x^2+1} \, dx + \int\limits_{C_R} f(z) \, dz + i \int\limits_{R}^{1+\varepsilon} \frac{e^{-y}}{1-y^2} \, dy + \int\limits_{\gamma_\varepsilon} f(z) \, dz + i \int\limits_{1-\varepsilon}^{0} \frac{e^{-y}}{1-y^2} \, dy = 0,$$

where $\gamma_\varepsilon = \{z \,|\, |z - i| = \varepsilon, \ \operatorname{Re} z \geq 0\}$.

Since

$$\lim_{z \to i} (z - i) f(z) = \lim_{z \to i} \frac{e^{iz}}{z + i} = \frac{1}{2ie}, \qquad \lim_{z \to \infty} z f(z) = 0,$$

in view of Theorems 1 and 2 from 3.1.4, we find

$$\lim_{\varepsilon \to 0} \int_{\gamma_\varepsilon} f(z) \, dz = -\frac{\pi}{2e}, \qquad \lim_{R \to +\infty} \int_{C_R} f z) \, dz = 0.$$

Hence, if $\epsilon \to 0$ and $R \to +\infty$, from (4) follows

$$\int_{0}^{+\infty} \frac{e^{ix}}{x^2 + 1} \, dx + i \int_{+\infty}^{0} \frac{e^{-y}}{1 - y^2} \, dy - \frac{\pi}{2e} = 0,$$

i.e.

$$\int_{0}^{+\infty} \frac{e^{ix}}{x^2 + 1} \, dx = i \int_{0}^{+\infty} \frac{e^{-y}}{1 - y^2} \, dy + \frac{\pi}{2e},$$

and, separating real and imaginary parts, we obtain

$$\int_{0}^{+\infty} \frac{\cos x}{x^2 + 1} \, dx = \frac{\pi}{2e}; \qquad \int_{0}^{+\infty} \frac{\sin x}{x^2 + 1} \, dx = \int_{0}^{+\infty} \frac{e^{-y}}{1 - y^2} \, dy.$$

REFERENCES
1. Titchmarsh, pp. 107–108.
2. Garnir–Gobert, pp. 119–135.

3.1.6. *Contour Integration of Goursat's Functions*

DEFINITION 1. Let f and g be analytic functions in a region G. The function w, defined in that region by

(1) $w(z, \bar{z}) = f(z) \bar{z} + g(z),$

is called Goursat's bianalytic function.

REMARK 1. Goursat's functions are not analytic, since $\partial w / \partial \bar{z} = f(z)$.

We shall deduce a theorem on the integration of Goursat's functions of the form (1) with $g(z) \equiv 0$.

If the function f is regular in the region G bounded by a closed contour Γ, then $\oint_\Gamma f(z)\, dz = 0$. It is also easily shown that $\oint_\Gamma d\bar{z} = 0$. Thus

$$\oint_\Gamma \left(\bar{z} f(z)\, dz + \left(\int f(z)\, dz \right) d\bar{z} \right) = 0.$$

Suppose that the function f has in the region G a finite number of isolated poles or essential singularities. Let $z = a_\nu \in G$ be one of those singularities. Then

$$f(z) = \sum_{k=1}^{+\infty} \frac{B_k}{(z-a_\nu)^k} + \sum_{k=0}^{+\infty} A_k\, (z-a_\nu)^k,$$

where in the case of a pole the first sum is finite. Surround now each singularity a_ν by the circle $K_\nu = \{z \,|\, |z-a_\nu| = r\}$, where $r > 0$ is so small that K_ν does not contain other singularities of f. Join the circles K_ν by the lines l_k, and also join the circle K_1 to the contour Γ as shown on Figure 3.1.6. The function f is regular in the region $D = G - \Sigma K_\nu$. Hence, if $\gamma = \partial D$, we have

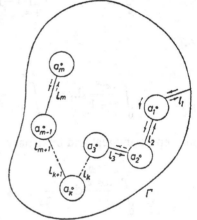

Fig. 3.1.6.

$$\oint_\gamma \left(\bar{z} f(z)\, dz + \left(\int f(z)\, dz \right) d\bar{z} \right) = 0,$$

i.e.

$$\oint_\Gamma \left(\bar{z} f(z)\, dz + \left(\int f(z)\, dz \right) d\bar{z} \right) - \sum_\nu \oint_{K_\nu} \left(\bar{z} f(z)\, dz + \left(\int f(z)\, dz \right) d\bar{z} \right)$$

$$+ \sum_k \int_{l_k^+} \left(\bar{z} f(z)\, dz + \left(\int f(z)\, dz \right) d\bar{z} \right) + \sum_k \int_{l_k^-} \left(\bar{z} f(z)\, dz + \left(\int f(z)\, dz \right) d\bar{z} \right) = 0.$$

However, on the circle K_ν we have $z = a_\nu + re^{i\alpha}$, $\bar z = \bar a_\nu + re^{-i\alpha}$, $dz = ire^{i\alpha}\,d\alpha$, and hence

$$\oint_{K_\nu} \bar z f(z)\,dz = \int_0^{2\pi} (\bar a_\nu + re^{-i\alpha})\left(\sum_{k=1}^{+\infty} \frac{B_k}{(z-a_\nu)^k} + \sum_{k=0}^{+\infty} (A_k (z-a_\nu)^k\right) ire^{i\alpha}\,d\alpha$$

$$= i\int_0^{2\pi} \left(\sum_{k=1}^{+\infty} \frac{B_k}{r^{k-2}} e^{-ik\alpha} + \sum_{k=1}^{+\infty} \frac{\bar a_\nu B_k}{r^{k-1}} e^{-i(k-1)\alpha}\right) d\alpha$$

$$+ i\int_0^{2\pi} \left(\sum_{k=0}^{+\infty} A_k r^{k+2} e^{ik\alpha} + \sum_{k=0}^{+\infty} \bar a_\nu A_k r^{k+1} e^{i(k+1)\alpha}\right) d\alpha$$

$$= i(\bar a_\nu B_1 2\pi + iA_0 r^2 2\pi).$$

Since Laurent's series can be integrated by terms in the region of convergence, we have

$$\oint_{K_\nu} \left(\int f(z)\,dz\right) d\bar z = \oint_{K_\nu} \left(\int \left(\sum_{k=1}^{+\infty} \frac{B_k}{(z-a_\nu)^k} + \sum_{k=0}^{+\infty} A_k (z-a_\nu)^k\right) dz\right) d\bar z$$

$$= \int_0^{2\pi} \left(\sum_{k=2}^{+\infty} \frac{B_k}{(1-k)(z-a_\nu)^{k-1}} + B_1(\log r + i\alpha) + \sum_{k=0}^{+\infty} \frac{A_k}{k+1} r^{k+1} e^{i(k+1)\alpha}\right)(-i)r^{-i\alpha}\,d\alpha$$

$$= i\sum_{k=2}^{+\infty} \frac{B_k}{(k-1)r^{k-2}} \int_0^{2\pi} e^{-ik\alpha}\,d\alpha - B_1 ri \int_0^{2\pi} (\log r + i\alpha) e^{-i\alpha}\,d\alpha$$

$$- i\sum_{k=0}^{+\infty} \frac{A_k}{k+1} r^{k+2} \int_0^{2\pi} e^{ik\alpha}\,d\alpha = -iA_0 r^2 2\pi + B_1 r 2\pi i.$$

and hence

$$\sum_\nu \oint_{K_\nu} \left(\bar z f(z)\,dz + \left(\int f(z)\,dz\right) d\bar z\right) = 2\pi i \sum_\nu (\bar a_\nu B_1 + r^2 A_0 + B_1 r - A_0 r^2).$$

If $r \to 0$, we find

$$\sum_\nu \oint_{K_\nu} \left(\bar z f(z)\,dz + \left(\int f(z)\,dz\right) d\bar z\right) = 2\pi i \sum_\nu \bar a_\nu B_1 = 2\pi i \sum_\nu \bar a_\nu \operatorname{Res}_{z=a_\nu} f(z).$$

Suppose now that $\int f(z)\,dz$ is a uniform function. Then

$$\int\limits_{I_k^+} f(z)\,dz + \int\limits_{I_k^-} f(z)\,dz = 0.$$

and we arrive at the following theorem.

THEOREM 1. *For Goursat's bianalytic functions w defined by*

$$w(z,\ \bar{z}) = \bar{z}f(z),$$

where the function f is analytic in the region G bounded by a closed contour Γ, and where $\int f(z)\,dz$ is a uniform function, we have

$$\oint\limits_{\Gamma} \left(\bar{z}f(z)\,dz + \left(\int f(z)\,dz \right) d\bar{z} \right) = 2\pi i \sum_{\nu} \bar{a}_{\nu} \operatorname*{Res}_{z=a_{\nu}} f(z),$$

where a_{ν} denote the isolated singularities of f in the region G.

REFERENCE
D. S. Dimitrovski: 'A Contribution to the Theory of Generalized Analytic Functions' (Macedonian), *Ann. Fac. Sci. Univ. Skopje, Sec. A* **20** (1970), 25–214.

3.1.7. *An Interpretation of the Theorem of Residues*

The real and imaginary parts of an analytic function satisfy the two-dimensional Laplace equation, and thus may represent the following physical functions: the potential (electric, magnetic, velocity), the flux function, the cartesian components of the velocity in an irrotational flow, the cartesian components of the electric current density (in a steady flow), etc. The poles of functions correspond to physical 'sources' or 'sinks' and the residues are related to the magnitudes of these sources.

For instance, if a liquid is flowing out of a point, uniformly in all directions, and if I is the emission rate, the polar components of the density of flow are

$$\frac{\partial v}{\partial \varrho} = \frac{I}{2\pi\varrho}, \quad \frac{\partial v}{\partial \varphi} = 0.$$

From this we obtain the cartesian components

$$\frac{\partial v}{\partial x} = \frac{\partial v}{\partial \varrho}\cos\varphi = \frac{I\cos\varphi}{2\pi\varrho}, \quad \frac{\partial v}{\partial y} = \frac{I\sin\varphi}{2\pi\varrho},$$

which we now combine into an analytic function

$$\frac{\partial v}{\partial x} - i\frac{\partial v}{\partial y} = \frac{I(\cos\varphi - i\sin\varphi)}{2\pi\varrho} = \frac{I}{2\pi z},$$

where $z = \varrho e^{i\varphi} = x + iy$. Thus a simple source at $z = 0$ corresponds to a simple pole of the 'complex velocity function'. The residue, $I/2\pi$, is very simply related to the emission rate.

Consider now two equal and opposite poles, i.e. a source and a sink, at $z = l/2$ and $z = -l/2$. Then

$$\frac{\partial v}{\partial x} - i\frac{\partial v}{\partial y} = \frac{I}{2\pi\left(z - \frac{1}{2}l\right)} - \frac{I}{2\pi\left(z + \frac{1}{2}l\right)} = \frac{I}{2\pi\left(z^2 - \frac{1}{4}l^2\right)}.$$

Let $l \to 0$ and $I \to \infty$, in such a way that the 'moment', $p = Il$, of the 'dipole' remains constant. Then

$$\frac{\partial v}{\partial x} - i\frac{\partial v}{\partial y} = \frac{p}{2\pi z^2}.$$

and this function has a second order pole at $z = 0$.

A source of the nth order is similarly formed by n simple sources of equal intensity, equispaced on the circumference of an infinitesimal circle, and a simple sink of n times the intensity of each circumferential source. The complex velocity is found to be proportional to z^{-n}, i.e. it is an function with an nth order pole at $z = 0$.

Consider now the integral

$$\oint f(z)\,dz = \oint (v_x - iv_y)(dx + i\,dy) = \oint (v_x\,dx + v_y\,dy) + i\oint (v_x\,dy - v_y\,dx).$$

The real part represents the circulation round the boundary; for the irrotational flow under consideration this should vanish. The imaginary part represents the total rate of flow across the boundary, and must equal the sum of emissions from simple sources inside the boundary, that is, the sum of residues multiplied by 2π. A multiple source in the sense just defined contributes nothing to the integral since the liquid emitted by the simple sources constituting the multiple source is returned to the sink.

Laurent's series is a representation of any given complex velocity by an equivalent combination of sources of various orders situated at $z = a$ and $z = \infty$. The sources at infinity are represented by the nonnegative powers of $z - a$.

A similar situation exists in the case of a solenoidal flow, as in a magnetic field produced by parallel electric current filaments. Around a single filament the magnetic intensity is

$$\frac{\partial H}{\partial \varrho} = 0, \quad \frac{\partial H}{\partial \varphi} = \frac{I}{2\pi\varrho},$$

where I is the current. The circulation of H round the circle of radius ρ, coaxial with the filament, is I; the outflow of H across the circle is zero. In this case

$$\frac{\partial H}{\partial x} - i\frac{\partial H}{\partial y} = \frac{I}{2\pi i z}.$$

If the source distributions in the irrotational and solenoidal fields are the same, the fields are alike except for the interchange of equipotential and stream lines; either type of source illustrates the nature of the poles of functions of a complex variable.

The text of 3.1.7 is taken directly from the book [1].

REFERENCE
1. S. A. Schelkunoff: *Applied Mathematics for Engineers and Scientists*, New York-Toronto-London 1948 and 1951, pp. 313–314.

3.2. THEOREMS ON RESIDUES FOR FUNCTIONS IN SEVERAL VARIABLES

3.2.1. Martinelli's Approach

THEOREM 1. *If the function f is meromorphic in the region $D \subset \mathbf{C}^n$ and if P is its polar set, then for an arbitrary n-dimensional cycle $\sigma \subset D\backslash P$ we have*

$$(1) \qquad \int_\sigma f\,dz = (2\pi i)^n \sum_{\nu=1}^{r} k_\nu R_\nu,$$

where k_ν are the coefficients in the expansion of σ with respect to the basis of n-dimensional homologies, i.e. $\sigma \sim \Sigma^r_{\nu=1} k_\nu \sigma_\nu$ and R_ν are the residues of f with respect to the basis cycles σ_ν.

This theorem generalizes Cauchy's theorem on residues to functions in several variables. However, determination of the basis $\{\sigma_\nu\}$ and the expansion

$\sigma \sim \Sigma \; k_\nu \sigma_\nu$ is a complicated procedure. Martinelli considerably simplified those difficulties. In order to expose his theory we first have to give two definitions and to quote one theorem.

DEFINITION 1. 1° Let the simplexes $S^p = (P, P_1, \ldots, P_p)$ and $T^q = (P, Q_1, \ldots, Q_q)$ be defined on an oriented variety M^r of real dimension r, so that $p + q = r$ and so that P is the only common point of those simplexes. We say that the index of intersection of simplexes S^p and T^q at the point P is $+1$ if the orientation of the r-dimensional simplex $(P, P_1, \ldots, P_p, Q_1, \ldots, Q_q)$ is the same as the orientation of the variety M^r, and -1 in the opposite case.

2° Let

$$(2) \qquad \sigma^p = \Sigma \; a_\nu \, S^p_\nu \quad i \quad \tau^q = \Sigma \; \beta_\nu \, T^q_\nu$$

be two chains of dimensions p and q, respectively, in the r-dimensional variety M^r, so that $p + q = r$. Suppose that pairs of simplexes S^p_ν and T^q_ν have only one common point each. The index of intersection of the chains σ^p and τ^q is the sum of all indices of intersection of the corresponding simplexes, multiplied by the coefficients which stand by those simplexes in (2).

The index of intersection of the chains σ^p and τ^q is denoted by $i(\sigma^p, \tau^q)$.

Suppose that σ^p and τ^{q-1} are disjoint cycles homologous to zero on an oriented variety M^r where $p + q = r$. Then there exists a chain $T^q \subset M^r$ such that $\tau^{q-1} = \partial T^q$. If $T_1^q \subset M^r$ is any chain with boundary τ^{q-1}, then

$$i(\sigma^p, \; T^q) - i(\sigma^p, \; T_1^q) = i(\sigma^p, \; T^q - T_1^q) = 0,$$

since $T^q - T_1^q$ is a cycle (for we have $\partial(T^q - T_1^q) = \tau^{q-1} - \tau^{q-1} = 0$) and σ^p is a cycle homologous to zero.

Hence, the index of intersection of the cycle σ^p with an arbitrary chain $T^q \subset M^r$ with the boundary τ^{q-1} is independent of the choice of the chain, but is defined (for given M^r and σ^p) solely by the cycle τ^{q-1}.

DEFINITION 2. The index of intersection of the cycles σ^p and τ^{q-1} is called the linking coefficient of those cycles and is denoted by $c(\sigma^p, \tau^{q-1})$. Thus

$$c(\sigma^p, \; \tau^{q-1}) = i(\sigma^p, \; T^q),$$

where $\partial T^q = \tau^{q-1}$.

We now quote the following theorem which is called the Alexander–Pontryagin duality principle.

THEOREM 2. *Let S^r be a sphere of real dimension r and let $K \subset S^r$ be a polyhedron. Then for a basis of p-dimensional homologies*

(3) $\sigma_1^p, \ldots, \sigma_k^p$

of the complex $S^r \setminus K$ there exists a basis of (q−1)-dimensional homologies

(4) $\tau_1^{q-1}, \ldots, \tau_k^{q-1}$

of the polyhedron K, such that p+q = r, and such that for every $\mu, \nu = 1, \ldots, k$, we have

$$c\left(\sigma_\mu^p, \tau_\nu^{q-1}\right) = \delta_{\mu\nu},$$

where $\delta_{\mu\nu}$ is the Kronecker delta.
    The basis (4) is dual to the basis (3).

In order to apply Theorem 2 to Theorem 1, suppose that the region D is homeomorphic to a $2n$-dimensional open ball. We map all the points of the boundary ∂D into a point which is then added to D. We thus obtain a $2n$-dimensional closed ball $\overline{D}$. In the same way we map all the points of the intersection of the polar set P of f with ∂D into a point. The set P together with this point is denoted $\overline{P}$. After this $\{\sigma_\nu\}$ is still a basis of n-dimensional homologies of the set $\overline{D} \setminus \overline{P}$, and we still have $\sigma \sim \Sigma\, k_\nu \sigma_\nu$.

In view of Theorem 2, instead of the basis $\{\sigma_\nu\}$ of the set $\overline{D} \setminus \overline{P}$ we can take the basis of $(n-1)$-dimensional homologies $\tau_1, \ldots, \tau_k$ of the polar set P, such that $c(\sigma_\mu, \tau_\nu) = \delta_{\mu\nu}$. The cycles τ_ν are called singular cycles.

It can be proved that in that case $c(\sigma, \tau_\nu) = k_\nu$. In other words, the coefficients k_ν in the formula (1) can be evaluated without knowing the basis $\{\sigma_\nu\}$. The residues R_ν can also be evaluated, though the cycles have not been found. Namely, the residues R_ν can be determined from the system

(5) $\displaystyle \int_{\gamma_\mu} f \, dz = (2\pi i)^n \sum_{\nu=1}^{k} a_{\mu\nu} R_\nu,$

where $\gamma_1, \ldots, \gamma_k$ are any homologically independent n-dimensional cycles which are chosen so that the integrals on the left of (5) can be evaluated, and $a_{\mu\nu} = c(\gamma_\mu, \tau_\nu)$.

Hence, Theorem 1 takes the following form.

THEOREM 3. *Let f be meromorphic in the region $D \subset C^n$ which is homeomorphic to a 2n-dimensional open ball, and let P be the polar set of the function f which is obtained by adding the set $P \cap \partial D$ mapped into a single point, to the set $P \cap D$. Let $\{\tau_1, \ldots, \tau_m\}$ be a basis of (n−1)-dimensional homologies of the set $\overline{P}$ and let R_ν be the residue of f with respect to the singular cycle τ_ν. Then for an arbitrary n-dimensional cycle $\sigma \subset D \backslash P$ we have*

$$\int\limits_\sigma f \, dz = (2\pi i)^n \sum_{\nu=1}^m k_\nu R_\nu,$$

where $k_\nu = c(\sigma, \tau_\nu)$.

3.2.2. *Leray's Approach*

In order to formulate Leray's theorem which generalizes Cauchy's residue theorem, we first have to introduce the so-called Leray's operator δ which to every point $z_0 \in P$ corresponds the homeomorphic image of the circle $\delta z_0 \in M \backslash P$, i.e. raises the real dimension by 1. The operator δ must satisfy the following conditions:

$1°$ In a neighbourhood U_{z_0} there exist local coordinates $z = (z_1, \ldots, z_n)$ with origin z_0, such that the equation of $P \cap U_{z_0}$ with respect to those coordinates is $z_n = 0$, and the equation of $\delta z_0 \in U_{z_0}$ is $|z_n| = 1$;

$2°$ The set $\cup_{z \in P} \delta z$ forms a continuous surface in $M \backslash P$;

$3°$ If $z' \neq z''$, the curves $\delta z'$ and $\delta z''$ do not have a common point.

In fact, the operator δ is a homomorphism of the groups of homologies $\delta: H_{p-1}(P) \to H_p(M \backslash P)$, where only compact homologies are considered (in order to avoid the appearance of improper integrals).

Leray's theorem which generalizes Cauchy's residue theorem reads:

THEOREM 1. *Let σ be a cycle of dimension p−1 which compactly belongs to P and let $\omega \in C^\infty (M \backslash P)$ be a closed form of degree p for which P is the polar set of first order. Then*

$$(1) \qquad \int\limits_{\delta\sigma} \omega = 2\pi i \int\limits_\sigma \text{res } \omega.$$

Notice that in the formula (1) the cycles σ and $\delta\sigma$ can be replaced by arbitrary representatives of classes $h \in H_{p-1}(P)$ and $\delta h \in H_p(M \setminus P)$, so that (1) can be written in the form

$$\int_{\delta h} \omega = 2\pi i \int_h \operatorname{res} \omega.$$

The corresponding formula for the class-residue is

$$\int_{\delta h} \omega^* = 2\pi i \int_h \operatorname{Res} \omega.$$

REFERENCES
See the references in 10.3.2.

Chapter 4

Applications of the Calculus of Residues in the Theory of Functions

4.1. LIOUVILLE'S THEOREM AND APPLICATIONS

The following theorem is called Liouville's theorem.

THEOREM 1. *If f is a regular function in the extended plane, then f is a constant function.*

Proof. Let z_1 and z_2 ($z_1 \neq z_2$) be arbitrary points of z-plane. The function F, defined by

$$F(z) = \frac{f(z)}{(z-z_1)(z-z_2)},$$

has singularities at z_1, z_2, and possibly at ∞.

However, $\lim_{z \to \infty} zF(z) = 0$, and so, by Theorem 5 of 2.1.2, we have $\operatorname{Res}_{z=\infty} F(z) = 0$.

Hence, in view of Theorem 2 from 3.1.1, we conclude that the sum of residues in the finite plane must be zero. Since

$$\operatorname*{Res}_{z=z_1} F(z) = \frac{f(z_1)}{z_1 - z_2}, \quad \operatorname*{Res}_{z=z_2} f(z) = \frac{f(z_2)}{z_2 - z_1},$$

we have

$$\frac{f(z_1)}{z_1 - z_2} + \frac{f(z_2)}{z_2 - z_1} = 0,$$

implying $f(z_1) = f(z_2)$, which means that f is a constant function.

The following theorems are immediate consequences of Liouville's theorem.

THEOREM 2. *Every nonconstant function must have at least one singularity in the extended plane.*

51

THEOREM 3. *If P is a polynomial of order n $(n \geq 1)$ in z, then the equation $P(z) = 0$ has at least one root.*

Proof. If we suppose that the polynomial P does not have a zero, then the function $z \mapsto 1/P(z)$ is regular in the extended z-plane, and by Liouville's theorem, must be constant, which contradicts the hypothesis that P is an nth order polynomial $(n \geq 1)$.

Theorem 3 is often called the fundamental theorem of algebra (see also 3.1.2.9 and 4.2.3).

REFERENCE

T. M. Macrobert: *Functions of a Complex Variable*, 5th edition, London 1962, pp. 68–69.

4.2. THE PRINCIPLE OF THE ARGUMENT AND APPLICATIONS

4.2.1. *The Principle of the Argument*

THEOREM 1. *Let Γ be a closed contour and let g be a regular function in the region $\Gamma \cup$ int Γ, and f a meromorphic function in int Γ. Furthermore, suppose that $n_1, \ldots, n_k$ are zeros of f of order $r_1, \ldots, r_k$, respectively, and that $p_1, \ldots, p_m$ are poles of f of order $s_1, \ldots, s_m$, respectively, such that $n_i \notin \Gamma$ $(i = 1, \ldots, k)$ and $p_i \notin \Gamma$ $(i = 1, \ldots, m)$. Then*

$$(1) \qquad \frac{1}{2\pi i} \oint_{\Gamma} g(z) \frac{f'(z)}{f(z)} \, dz = \sum_{i=1}^{k} r_i g(n_i) - \sum_{i=1}^{m} s_i g(p_i).$$

Proof. In the neighbourhood of n_i the function f can be represented as

$$f(z) = (z - n_i)^{r_i} f_i(z),$$

where f_i is a regular function in that neighbourhood and $f_i(n_i) \neq 0$. The logarithmic derivative of this function is

$$\frac{f'(z)}{f(z)} = \frac{r_i}{z - n_i} + \frac{f_i'(z)}{f_i(z)},$$

and we conclude that n_i is a simple pole of the function

$$z \mapsto g(z) \frac{f'(z)}{f(z)}.$$

Since

$$\lim_{z \to n_i} (z - n_i) g(z) \frac{f'(z)}{f(z)} = r_i g(n_i),$$

we get

$$\operatorname*{Res}_{z = n_i} g(z) \frac{f'(z)}{f(z)} = r_i g(n_i).$$

In a neighbourhood of p_i the function f can be represented as

$$f(z) = \frac{g_i(z)}{(z - p_i)^{s_i}},$$

where g_i is a regular function in that neighbourhood and $g_i(p_i) \neq 0$. The logarithmic derivative is then

$$\frac{f'(z)}{f(z)} = \frac{-s_i}{z - p_i} + \frac{g_i'(z)}{g_i(z)}$$

and we conclude that

$$\operatorname*{Res}_{z = p_i} g(z) \frac{f'(z)}{f(z)} = -s_i g(p_i).$$

Using Cauchy's residue theorem, we obtain (1).

An important consequence of Theorem 1 is contained in the following theorem.

THEOREM 2. *Suppose that f satisfies all the conditions of Theorem 1, and let N be the number of zeros inside the contour (a zero of order m is counted m times) and P the number of poles (a pole of order m is counted m times). Then*

(2) $$\frac{1}{2\pi i} \oint_\Gamma \frac{f'(z)}{f(z)} dz = N - P.$$

Proof. Theorem 2 follows directly from Theorem 1 on setting $g(z) \equiv 1$.

The integral on the left of (2) is called the logarithmic residue of f with

respect to the contour Γ. We now show that this residue has a simple interpretation. We first write it in the form

$$(3) \qquad \frac{1}{2\pi i} \oint_{\Gamma} \frac{f'(z)}{f(z)}\, dz = \frac{1}{2\pi i} \oint_{\Gamma} \frac{d}{dz}(\mathrm{Log}\, f(z))\, dz.$$

Choose a point $z_0 \in \Gamma$ which will be considered as the starting and the
end point. Suppose that the point z moves along Γ in the positive direction.
The value of the function $z \mapsto \mathrm{Log}\, f(z)$ at the initial $z = z_0$ will differ from the
value at the end point $z = z_0$, but the moduli will be equal. Let Φ_0 be the value
of the argument of the function $z \mapsto \mathrm{Log}\, f(z)$ for initial z_0, and let Φ_1 be the
value of the argument for the final z_0. The formula (3) then becomes

$$\frac{1}{2\pi i} \oint_{\Gamma} \frac{f'(z)}{f(z)}\, dz = \frac{1}{2\pi i}\left((\log|f(z_0)| + i\Phi_1) - (\log|f(z_0)| + i\Phi_0)\right) = \frac{\Phi_1 - \Phi_0}{2\pi}.$$

Using (2) we obtain

$$(4) \qquad N - P = \frac{\Phi_1 - \Phi_0}{2\pi}$$

Denote $\Phi_1 - \Phi_0$ by $\Delta_\Gamma \mathrm{Arg}\, f(z)$, where Δ_Γ denotes the variation along the
contour Γ. Using those new notations, the formula (4) takes the form

$$\frac{1}{2\pi} \Delta_\Gamma \mathrm{Arg}\, f(z) = N - P,$$

and is called the principle of the argument.

EXAMPLE 1. We shall determine in which quadrants lie the roots of the equation

$$(5) \qquad f(z) \equiv z^4 + z^3 + 4z^2 + 2z + 3 = 0.$$

This equation has no real roots. Indeed, it is obvious that it cannot have positive
roots. If we put $z = -x$ ($x > 0$), we get

$$(6) \qquad x^4 - x^3 + 4x^2 - 2x + 3 = 0.$$

For $0 < x < 1$, we have $x^4 - x^3 + 4x^2 > 0$ and $-2x + 3 > 0$, which means that (6) has
no roots in the interval $(0, 1)$. For $x > 1$, we have $x^4 - x^3 > 0$ and $4x^2 - 2x + 3 > 0$,

which means that (6) has no roots in the interval $(1, +\infty)$. Since 0 and 1 are not roots of (6), we see that (5) does not have real roots.

If we put $z = iy$ $(y \in R)$ into (5) we get

$$y^4 - iy^3 - 4y^2 + 2iy + 3 = 0,$$

i.e.

(7) $$y^4 - 4y^2 + 3 = 0, \qquad y^3 - 2y = 0,$$

and since the system (7) has no real solutions, we infer that (5) has no purely imaginary roots.

Let

$$C_1 = \{z \mid 0 \leq \mathrm{Re}\, z \leq R\}, \; C_2 = \{z \mid |z| = R, \; \mathrm{Re}\, z \geq 0, \; \mathrm{Im}\, z \geq 0\}, \; C_3 = \{z \mid 0 \leq \mathrm{Im}\, z \leq R\},$$

where R is a sufficiently large positive number. Furthermore, let $C = C_1 \cup C_2 \cup C_3$. Then

$$\Delta_C \arg f(z) = \Delta_{C_1} \arg f(z) + \Delta_{C_2} \arg f(z) + \Delta_{C_3} \arg f(z).$$

First, we have

$$\Delta_{C_1} \arg f(z) = 0.$$

Secondly,

$$\Delta_{C_2} \arg f(z) = \Delta_{C_2} \arg (R^4 e^{4i\theta}) + \Delta_{C_2} \arg (1 + O(R^{-1})) = 2\pi + O(R^{-1}).$$

Finally,

$$\Delta_{C_3} \arg f(z) = \operatorname{arctg} \left(\frac{-y^3 + 2y}{y^4 - 4y^2 + 3} \right).$$

Since the function

$$y \mapsto \frac{-y^3 + 2y}{y^4 - 4y^2 + 3}$$

in the interval $(0, 1)$ increases from 0 to $+\infty$, in the interval $(1, \sqrt{2})$ decreases from $+\infty$ to 0, in the interval $(\sqrt{2}, \sqrt{3})$ increases from 0 to $+\infty$, and in the interval $(\sqrt{3}, +\infty)$ decreases from $+\infty$ to 0, we find

$$\Delta_{C_3} \arg f(z) = -2\pi,$$

and so

$$\Delta_C \arg f(z) = O(R^{-1}) \to 0 \qquad (R \to +\infty).$$

Hence, there are no zeros in the first quadrant. Since zeros occur in conjugate pairs, it follows that there are no zeros in the fourth quadrant. Thus there are two zeros in each of the second and third quadrants.

Any algebraic equation may be treated in the same way.
The above example is taken from Titchmarsh, pp. 117–118.

4.2.2. Applications of the Principle of the Argument

DEFINITION 1. The curves $|f(z)| = c$, where c is a positive constant are called level curves of the function f.

THEOREM 1. *If Γ is a simple closed level curve of f, and if f is regular inside and on it, then f has at least one zero inside Γ.*

Proof. Let $f(z) = u(x, y) + iv(x, y)$. Since on Γ we have $|f(z)| = c$, we put $f(z) = ce^{i\theta}$ for $z \in \Gamma$. Then $c = \sqrt{u^2 + v^2}$, $\theta = \operatorname{arctg}(v/u)$. Let s be the length of Γ measured from some fixed point on it. Then

$$(1) \qquad 0 = \frac{dc}{ds} = \left(u \frac{du}{ds} + v \frac{dv}{ds} \right) \frac{1}{c}, \quad \frac{d\theta}{ds} = \left(u \frac{dv}{ds} - v \frac{du}{ds} \right) \frac{1}{c^2}.$$

We now prove that $d\theta/ds$ cannot vanish on Γ. For if it did, from (1) would follow

$$(u^2 + v^2)\left(\left(\frac{du}{ds} \right)^2 + \left(\frac{dv}{ds} \right)^2 \right) = 0, \text{ tj. } \frac{du}{ds} = 0, \quad \frac{dv}{ds} = 0.$$

On the other hand, since f is analytic, we have

$$(2) \qquad \frac{du}{ds} = \frac{\partial u}{\partial x} \frac{dx}{ds} + \frac{\partial u}{\partial y} \frac{dy}{ds}, \quad \frac{dv}{ds} = \frac{\partial v}{\partial x} \frac{dx}{ds} + \frac{\partial v}{\partial y} \frac{dy}{ds} = -\frac{\partial u}{\partial y} \frac{dx}{ds} + \frac{\partial u}{\partial x} \frac{dy}{ds}.$$

From (2) we obtain

$$(3) \qquad \left(\left(\frac{\partial u}{\partial x} \right)^2 + \left(\frac{\partial u}{\partial y} \right)^2 \right) \left(\left(\frac{dx}{ds} \right)^2 + \left(\frac{dy}{ds} \right)^2 \right) = 0.$$

Since $dx/ds = \cos\psi$, $dy/ds = \sin\psi$, where ψ is the angle between the tangent to the curve Γ and the real axis, we see that the second factor in (3) is 1, and hence from (3) follows $\partial u/\partial x = \partial u/\partial y = 0$, i.e. $f'(z) = 0$.

This is impossible, since Γ is a simple curve. In fact, $f'(z) = 0$ on Γ if and only if Γ has a double point. Indeed, the equation of the level curve is $u^2 + v^2 = c^2$, and it will have a double point if and only if

$$(4) \qquad u \frac{\partial u}{\partial x} + v \frac{\partial v}{\partial x} = 0,$$

$$(5) \qquad u \frac{\partial u}{\partial y} + v \frac{\partial v}{\partial y} = 0.$$

The equations (4) and (5) are satisfied if $f'(z) = 0$.

Conversely, the Equation (5), in view of Cauchy–Riemann equations, can be written as

$$(6) \qquad -u\frac{\partial v}{\partial x} + v\frac{\partial u}{\partial x} = 0,$$

and from (4) and (6) we find

$$(u^2 + v^2)\left(\left(\frac{\partial u}{\partial x}\right)^2 + \left(\frac{\partial v}{\partial x}\right)^2\right) = 0, \quad \text{i.e.} \quad \frac{\partial u}{\partial x} = 0 = \frac{\partial v}{\partial x}, \quad \text{i.e.} \quad f'(z) = 0.$$

Therefore, since $f'(z)$ cannot vanish, $d\theta/ds$ cannot vanish on Γ. This means that $d\theta/ds$ has the same sign at all points of Γ, i.e. that θ increases or decreases on Γ. Hence

$$(7) \qquad \Delta_\Gamma \theta \neq 0.$$

However, we have

$$(8) \qquad \Delta_\Gamma \theta = 2\pi n,$$

where n is the number of zeros of f inside Γ. From (7) and (8) follows that $n \neq 0$, i.e. there is at least one zero of f inside Γ.

THEOREM 2. *Let f and Γ satisfy the conditions of Theorem 1. If f has n zeros inside Γ, then f' has $n-1$ zeros inside Γ.*

Proof. We use the same notations as in the proof of Theorem 1. Let m be the number of zeros of f' inside Γ. We shall prove that $m = n-1$. Since $f(z) = ce^{i\theta}$ on Γ, we have $f'(z) = cie^{i\theta}(d\theta/dz)$, and thus

$$\arg f'(z) = A + \theta + \arg\frac{d\theta}{dz},$$

where A is a constant. This implies

$$\Delta_\Gamma \arg f'(z) = \Delta_\Gamma \arg f(z) + \Delta_\Gamma \arg\frac{d\theta}{dz},$$

and so

$$(9) \qquad 2\pi m = 2\pi n + \Delta_\Gamma \arg\frac{d\theta}{dz}.$$

Furthermore, $d\theta/dz = (d\theta/ds)(ds/dz)$, where, according to Theorem 1, $d\theta/ds$ is real and has constant sign on Γ. Therefore

$$\Delta_\Gamma \arg \frac{d\theta}{dz} = \Delta_\Gamma \arg \frac{ds}{dz}.$$

However, we have

$$\frac{dz}{ds} = \frac{dx}{ds} + i\frac{dy}{ds} = \cos \psi + i \sin \psi = e^{i\psi},$$

which means that

(10) $$\Delta_\Gamma \arg \frac{ds}{dz} = -\Delta_\Gamma \psi = -2\pi.$$

From (9) and (10) follows that $m = n-1$.

THEOREM 3. *Let Γ be a simple closed contour such that f is regular in the region $\Gamma \cup$ int Γ. If Re $f(z) = 0$ at $2k$ distinct points of Γ, then f has at most k zeros inside Γ.*

   Proof. If $f(z) = u(x, y) + iv(x, y)$ and if n is the number of zeros of f inside Γ, then

$$n = \frac{1}{2\pi} \Delta_\Gamma \left(\arctg \frac{v}{u} \right).$$

Starting at a point of Γ where $u \neq 0$, we may take the initial value of arctg (v/u) to lie between $-\pi/2$ and $\pi/2$. We can only pass out of this range, say to $(\pi/2, 3\pi/2)$, if u vanishes, and on to $(3\pi/2, 5\pi/2)$ if u vanishes again. Thus if u vanishes twice on Δ_Γ (arctg v/u) $\leq 2\pi$, and hence $n \leq 1$. The general result obviously follows from the same argument.

REFERENCE
Titchmarsh, pp. 121–123.

4.2.3. *Jensen's Formula and Its Generalizations*

Attempting to determine the distribution of nontrivial zeros of the Riemann zeta-function, Jensen [1] proved the following theorem.

THEOREM 1. *Let f be analytic in the region $G = \{z \mid |z| < R\}$ where it can have a finite number of poles $p_1, \ldots, p_n$, and let f be regular on the*

circle $\Gamma = \{z \,|\, |z| = R\}$. *Denote by* $v_1, \ldots, v_m$ *the zeros of* f *in* G. *If* $|v_i| \neq 0$
$(i = 1, \ldots, m)$ *and* $|p_i| \neq 0$ $(i = 1, \ldots, m)$ *and* $f(z) \neq 0$ *on* Γ, *then*

$$(1) \qquad \frac{1}{2\pi i} \oint_{\Gamma} \frac{\log f(z)}{z} \, dz = \log f(0) + \sum_{k=1}^{m} \log \frac{R}{v_k} - \sum_{k=1}^{n} \log \frac{R}{p_k},$$

where a zero (pole) of order p is counted p times.

If we put $z = Re^{it}$ into (1), and equate the real parts, we obtain

$$(2) \qquad \frac{1}{2\pi} \int_{0}^{2\pi} \log|f(R e^{it})| \, dt = \log|f(0)| + \sum_{k=1}^{m} \log \frac{R}{|v_k|} - \sum_{k=1}^{n} \log \frac{R}{|p_k|}.$$

Formula (2) is called Jensen's formula.

Theorem 1 (and Jensen's formula) were generalized a number of times. The first important generalization was given by Denjoy [2], who proved the following result.

THEOREM 2. *Let* Γ *be a closed contour which surrounds the origin and let* $\zeta \in \Gamma$. *Let* f *and* g *be complex functions such that* $z \mapsto [f'(z)/f(z)]$ *and* $z \mapsto [g'(z)/g(z)]$ *are regular on* Γ *and analytic in* $G = \mathrm{int}\ \Gamma$ *where the first function may have isolated singularities* $r_1, \ldots, r_n$, *and the second* $s_1, \ldots, s_m$. *Then*

$$(3) \qquad \frac{1}{2\pi i} \oint_{\Gamma} \frac{g'(z)}{g(z)} \log f(z) \, dz = \sum_{k=1}^{n} B_{1,k} \log \frac{g(\zeta)}{g(r_k)}$$

$$+ \sum_{k=1}^{m} b_{1,k} \log f(s_k) + \sum_{k=1}^{m} \sum_{i=1}^{+\infty} \frac{b_{i+1,k}}{i!} \frac{d^i}{ds_k{}^i} \log f(s_k)$$

$$- \sum_{k=1}^{n} \sum_{i=1}^{+\infty} \frac{B_{i+1,k}}{i!} \frac{d^i}{dr_k{}^i} \log g(r_k) + \sum_{k=1}^{+\infty} \sum_{j=1}^{m} \sum_{i=1}^{n} \frac{(-1)^k}{k} \left(\frac{B_{1,i}\, b_{k+1,j}}{(r_j - r_i)^k} - \frac{b_{1,j}\, B_{k+1,i}}{(r_i - r_j)^k} \right),$$

where $B_{i,k}$ *are the coefficients of the principal part of Laurent's series of* $z \mapsto [f'(z)/f(z)]$ *in a neighbourhood of* s_k, *and* $b_{i,k}$ *have analogous meaning for the function* $z \mapsto [g'(z)/g(z)]$.

A simpler expression for the integral on the left of (3) was obtained by Angheluţă [3]. His formula does not contain the triple sum on the right-hand side of (3). We give Angheluţă's proof.

From the hypotheses of the theorem follows

$$(4) \quad \frac{f'(z)}{f(z)} = F'(z) + \sum_{k=1}^{n} \sum_{i=1}^{+\infty} \frac{B_{i,k}}{(z-r_k)^i}, \quad \frac{g'(z)}{g(z)} = G'(z) + \sum_{k=1}^{m} \sum_{i=1}^{+\infty} \frac{b_{i,k}}{(z-s_k)^i},$$

where the functions F' and G' are regular inside and on Γ.

From (4) we obtain

$$\log f(z) = F(z) + \sum_{k=1}^{n} B_{1,k} \log(z-r_k) - \sum_{k=1}^{n} \sum_{i=1}^{+\infty} \frac{B_{i+1,k}}{i(z-r_k)^i},$$

$$\log g(z) = G(z) + \sum_{k=1}^{m} b_{1,k} \log(z-s_k) - \sum_{k=1}^{m} \sum_{i=1}^{+\infty} \frac{b_{i+1,k}}{i(z-s_k)^i}.$$

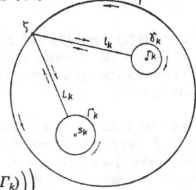

Let $\gamma_k = \{z \,|\, |z-r_k| < \epsilon_k\}$ $(k=1, \dots, n)$ and $\Gamma_k = \{z \,|\, |z-s_k| < \delta_k\}$ $(k=1, \dots, m)$ be sufficiently small circular regions, so that they do not have common points, nor do they intersect with Γ. Let l_k be line segments joining the point ζ with the circle γ_k, and L_k the line segment joining ζ to Γ_k. In the region

$$D = \text{int}\, \Gamma \left\backslash \left(\left(\bigcup_{k=1}^{n} (\gamma_k \cup l_k) \right) \cup \left(\bigcup_{k=1}^{m} (L_k \cup \Gamma_k) \right) \right) \right.$$

Fig. 4.2.3.

the function $z \mapsto [g'(z)/g(z)] \log f(z)$ is uniform and regular, and hence by Cauchy's theorem we have

$$(5) \quad \frac{1}{2\pi i} \oint_{\partial D} \frac{g'(z)}{g(z)} \log f(z)\, dz = 0.$$

However

$$(6) \quad \frac{1}{2\pi i} \oint_{\partial D} \frac{g'(z)}{g(z)} \log f(z)\, dz = \frac{1}{2\pi i} \oint_{\Gamma} \frac{g'(z)}{g(z)} \log f(z)\, dz$$

$$- \sum_{k=1}^{n} \frac{1}{2\pi i} \oint_{\nu_k} \frac{g'(z)}{g(z)} \log f(z)\, dz - \sum_{k=1}^{m} \frac{1}{2\pi i} \oint_{\Gamma_k} \frac{g'(z)}{g(z)} \log f(z)\, dz$$

$$- \sum_{k=1}^{n} \frac{1}{2\pi i} \int_{l_k^+} \frac{g'(z)}{g(z)} \log f(z)\, dz + \sum_{k=1}^{m} \frac{1}{2\pi i} \int_{l_k^-} \frac{g'(z)}{g(z)} \log f(z)\, dz,$$

since the sum of integrals along $L_k{}^+$ and $L_k{}^-$ is annuled, owing to the uniformity of the considered function.

From the second equality from (4) follows

$$\frac{1}{2\pi i}\oint_{\gamma_k}\frac{g'(z)}{g(z)}\log f(z)\,dz=\frac{1}{2\pi i}\oint_{\gamma_k}\left(G'(z)+\sum_{i=1}^{+\infty}\frac{b_{i,k}}{(z-s_k)^i}\right)\log f(z)\,dz$$

$$=\frac{1}{2\pi i}\sum_{i=1}^{+\infty}\oint_{\gamma_k}\frac{b_{i,k}\log f(z)}{(z-s_k)^i}\,dz,$$

since

$$\oint_{\gamma_k} G'(z)\log f(z)\,dz=0,$$

and Laurent's series can be integrated by terms. Using Cauchy's integral formula we find

$$\frac{1}{2\pi i}\oint_{\gamma_k}\frac{b_{i+1,k}\log f(z)}{(z-s_k)^{i+1}}\,dz=\frac{b_{i+1,k}}{i!}\frac{d^i}{ds_k{}^i}\log f(s_k)\qquad (i\geq 1)$$

$$\frac{1}{2\pi i}\oint_{\gamma_k}\frac{b_{1,k}\log f(z)}{z-s_k}\,dz=b_{1,k}\log s_k.$$

Hence

(7) $$\frac{1}{2\pi i}\oint_{\gamma_k}\frac{g'(z)}{g(z)}\log f(z)\,dz=b_{1,k}\log s_k+\sum_{i=1}^{+\infty}\frac{b_{i+1,k}}{i!}\frac{d^i}{ds_k{}^i}\log f(s_k).$$

Similarly, we have

(8) $$\frac{1}{2\pi i}\oint_{\Gamma_k}\frac{g'(z)}{g(z)}\log f(z)\,dz$$

$$=\frac{1}{2\pi i}\oint_{\Gamma_k}\frac{g'(z)}{g(z)}\left(F(z)+B_{1,k}\log(z-r_k)-\sum_{i=1}^{+\infty}\frac{B_{i+1,k}}{i\,(z-r_k)^i}\right)dz+o\,(1)$$

$$=-\frac{1}{2\pi i}\sum_{i=1}^{+\infty}\oint_{\Gamma_k}\frac{B_{i+1,k}}{i\,(z-r_k)^i}\frac{g'(z)}{g(z)}\,dz+o\,(1)$$

$$=-\sum_{i=1}^{+\infty}\frac{B_{i+1,k}}{i!}\frac{d^{i-1}}{dr_k{}^{i-1}}\frac{g'(r_k)}{g(r_k)}+o\,(1)$$

$$=-\sum_{i=1}^{+\infty}\frac{B_{i+1,k}}{i!}\frac{d^i}{dr_k{}^i}\log g\,(r_k)+o\,(1).$$

On the other hand,

$$(9) \qquad \frac{1}{2\pi i} \int\limits_{l_k^+} \frac{g'(z)}{g(z)} \log f(z)\,dz - \frac{1}{2\pi i} \int\limits_{l_k^-} \frac{g'(z)}{g(z)} \log f(z)\,dz$$

$$= -B_{1,k} \int\limits_{\zeta}^{r_k} \frac{g'(z)}{g(z)}\,dz = B_{1,k} \log \frac{g(\zeta)}{g(r_k)}.$$

Finally, from (5)–(9), letting $r_k \to 0$ $(k=1,\ldots,n)$, we find

$$(10) \qquad \frac{1}{2\pi i} \oint\limits_{\Gamma} \frac{g'(z)}{g(z)} \log f(z)\,dz = \sum_{k=1}^{n} B_{1,k} \log \frac{g(\zeta)}{g(r_k)} + \sum_{k=1}^{m} b_{1,k} \log f(s_k)$$

$$+ \sum_{k=1}^{m} \sum_{i=1}^{+\infty} \frac{b_{i+1,k}}{i!} \frac{d^i}{ds_k^i} \log f(s_k) - \sum_{k=1}^{n} \sum_{i=1}^{+\infty} \frac{B_{i+1,k}}{i!} \frac{d^i}{dr_k^i} \log g(r_k).$$

The difference between formulas (10) and (3) is that (10) does not contain the triple sum

$$\sum_{k=1}^{+\infty} \sum_{j=1}^{m} \sum_{i=1}^{n} \frac{(-1)^k}{k} \left(\frac{B_{1,i}\, b_{k+1,j}}{(r_j - r_i)^k} - \frac{b_{1,j}\, B_{k+1,i}}{(r_j - r_i)^k} \right).$$

Therefore, this result of Angheluță is a considerable simplification of Denjoy's result.

In particular, let $\Gamma = \{z \mid |z| = R\}$, $\zeta = R$, $g(z) = z$. Formula (10) becomes

$$(11) \qquad \frac{1}{2\pi i} \oint\limits_{\Gamma} \frac{\log f(z)}{z}\,dz = \sum_{k=1}^{n} B_{1,k} \log \frac{R}{r_k} + \log f(0) + \sum_{k=1}^{n} \sum_{i=1}^{+\infty} \frac{(-1)^i B_{i+1,k}}{i r_k^i},$$

since

$$\frac{d^i}{dr_k^i} \log r_k = (-1)^{k-1} \frac{(k-1)!}{r_k^i}.$$

Putting $z = Re^{it}$ into (11) and equating the real parts, we find

$$(12) \qquad \frac{1}{2\pi} \int\limits_{0}^{2\pi} \log |f(Re^{it})|\,dt$$

$$= \log |f(0)| + \sum_{k=1}^{n} B_{1,k} \log \frac{R}{|r_k|} + \mathrm{Re} \sum_{k=1}^{n} \sum_{i=1}^{+\infty} \frac{(-1)^i B_{i+1,k}}{i r_k^i}.$$

Formulas (11) and (12), though special cases of Angheluṭǎ's formula (10) are generalizations of formulas (1) and (2), since it is permitted that f has essential singularities. Those formulas were rediscovered by Mitrović [4].

It is not difficult to see how (1) can be obtained from (11). Suppose that $v_1, \ldots, v_t$ are the zeros and $p_{t+1}, \ldots, p_n$ the poles of the function f, which has no essential singularities. Then $B_{i+1, k} = 0$ for $i \geq 1$, $B_{1, k}$ $(k = 1, \ldots, t)$ is the order of the zero v_k $(k = 1, \ldots, t)$ and $-B_{1, k}$ $(k = t+1, \ldots, n)$ is the order of the pole p_k $(k = t+1, \ldots, n)$. Therefore

$$\frac{1}{2\pi i} \oint_\Gamma \frac{\log f(z)}{z}\, dz = \log f(0) + \sum_{k=1}^{t} \log \frac{R}{v_k} - \sum_{k=t+1}^{n} \log \frac{R}{p_k},$$

where a zero (pole) of order p is counted p times.

It is worth mentioning that Angheluṭǎ [3] gave the following generalization of his formula (10):

If Γ and ζ are defined as in Theorem 2, and if

$$\frac{f'(z)}{f(z)} = F'(z) + \sum_{k=1}^{n} B_{1, k} \log (z - r_k) + \sum_{k=1}^{n} \sum_{i=1}^{+\infty} \frac{B_{i,k}}{i\,(z - r_k)^i},$$

$$\frac{g'(z)}{g(z)} = G'(z) + \sum_{k=1}^{m} b_{1, k} \log (z - s_k) + \sum_{k=1}^{m} \sum_{i=1}^{+\infty} \frac{b_{i,k}}{i\,(z - s_k)^i},$$

where the functions F' and G' are uniform and regular in int Γ, then

(13) $$\frac{1}{2\pi i} \oint_\Gamma \frac{f'(z)}{f(z)} \frac{g'(z)}{g(z)}\, dz = \sum_{k=1}^{n} B_{1, k} \log \frac{f(\zeta)}{f(r_k)} + \sum_{k=1}^{m} b_{1, k} \log \frac{g(\zeta)}{g(s_k)}$$

$$+ \sum_{k=1}^{n} \sum_{i=1}^{+\infty} \frac{B_{i,k}}{i!} \frac{d^i}{dr_k^i} \log f(r_k) + \sum_{k=1}^{m} \sum_{i=1}^{+\infty} \frac{b_{i,k}}{i!} \frac{d^i}{ds_k^i} \log g(s_k).$$

Formula (13) is not only more general than (10) but also has the advantage of being symmetric with respect to the functions f and g.

Further generalizations of Jensen's formula to the regions which are not circular, but rectangular, or infinite strips, can be found in [5] – [8].

See also [9] where integrals of the form $\int_C f(z) \log g(z)\, dz$ and $\int_C \log f(z) \log g(z)\, dz$ are considered.

64 Chapter 4

REFERENCES

1. J. L. W. V. Jensen: 'Sur un nouvel et important théorème de la théorie des fonctions', *Acta Math.* 22 (1899), 359–364.
2. A. Denjoy: 'Sur une extension de la formule de Jensen', *Mathematica (Cluj)* 7 (1933), 129–135.
3. Th. Angheluţă: 'Sur l'intégrale de M. A. Denjoy', *Mathematica (Cluj)* 12 (1936), 93–98.
4. D. Mitrović: 'Sur les valeurs de certaines intégrales définies', *Glasnik Mat.-Fiz. Astr.* 10 (1955), 259–263.
5. B. Jessen: 'Über die Nullstellen einer analytischen Funktion. Eine Veralgemeinerung der Jensenschen Formel', *Math. Ann.* 108 (1933), 485–516.
6. J. E. Littlewood: 'On the Zeros of the Riemann Zeta Function', *Proc. Camb. Phil. Soc.* 22 (1924), 295–318.
7. A. Pfluger: 'Konforme Abbildung und eine Verallgemeinerung der Jensenschen Formel', *Comment. Math. Helv.* 13 (1944), 284–292.
8. O. Namik: 'Sur une généralisation de la formule de Jensen et quelques applications', *Rev. Fac. Sci. Univ. Istambul (A)* 15 (1950), 289–332.
9. D. S. Dimitrovski: 'Values of Certain Complex Integrals' (Macedonian), *Bull. Soc. Math. Phys. Macédoine* 16 (1965), 27–33.

4.2.4. *Generalizations of Cauchy's and Jensen's Formulas*

In 4.2.1 we proved Cauchy's formula

$$(1) \qquad \frac{1}{2\pi i} \oint_\Gamma \frac{f'(z)}{f(z)} \, dz = N - P,$$

where f is a regular function inside and on the closed contour and N and P are the numbers of zeros and poles of f, respectively; a multiple zero (pole) being repeated according to its order.

In particular, if $\Gamma = \{z \,|\, |z| = r\}$, formula (1) takes the form

$$(2) \qquad \int_0^{2\pi} z \frac{f'(z)}{f(z)} \, dt = 2\pi (N - P) \qquad (z = r\,e^{it}).$$

We shall give here some generalizations of Cauchy's formula (2) and Jensen's formula (2) from 4.2.3.

The following two theorems are due to Petrović [1].

THEOREM 1. *Let g be a continuous function in $(0, 2\pi)$ and orthogonal to the sequence $t \mapsto \cos kt$ $(k = 1, 2, \ldots)$. Furthermore, let f be meromorphic* *ind* $f(z) \in R$ *for* $z \in R$. *Then*

(3) $\displaystyle\int_0^{2\pi} g(t) \, \mathrm{Re} \, F(z) \, dt = (N - P) \int_0^{2\pi} g(t) \, dt,$

where $F(z) = z [f'(z)/f(z)]$ *and* $z = re^{it}$.

THEOREM 2. *Suppose that g is continuous in* $(0, 2\pi)$ *and orthogonal to the sequence* $t \mapsto \cos kt \, (k = 1, 2, \ldots)$. *Let* $v_1, \ldots, v_m$ *be the zeros and* $p_1, \ldots, p_n$ *the poles of f in the region* $\{z \mid |z| < r\}$. *Then*

(4) $\displaystyle\int_0^{2\pi} g(t) \, \mathrm{Re} \, (\log |f(z)|) \, dt$

$$= \left(\log |f(0)| + \sum_{j=1}^{m} \log \frac{r}{|v_j|} - \sum_{j=1}^{n} \log \frac{r}{|p_j|} \right) \int_0^{2\pi} g(t) \, dt.$$

The following theorem was proved by Montel [2].

THEOREM 3. *Let f be regular and nonzero on the circle* $\Gamma = \{z \mid |z| = r\}$, *while in the region* $G = \mathrm{int} \ \Gamma$ *it may have a finite number of poles* $p_1, \ldots, p_n$. *Denote the zeros of f in G by* $v_1, \ldots, v_m$. *If g is a continuous function on* $[0, 2\pi]$, *then*

(5) $\displaystyle\int_0^{2\pi} \left(\left(\frac{g(t) - g(-t)}{2} + a_0 \right) \frac{F(z) + F(\bar{z})}{2} + \left(\frac{g(t) + g(-t)}{2} - a_0 \right) \frac{F(z) - F(\bar{z})}{2} \right) dt$

$$= (N - P) \int_0^{2\pi} g(t) \, dt,$$

where $F(z) = z [f'(z)/f(z)]$, *N and P have the same meaning as before, and*

$$a_0 = \frac{1}{2\pi} \int_0^{2\pi} g(t) \, dt.$$

Proof. The function $z \mapsto [f'(z)/f(z)]$ can be written in the form

$$\frac{f'(z)}{f(z)} = \sum_{k=1}^{m} \frac{1}{z - v_k} - \sum_{k=1}^{n} \frac{1}{z - p_k} + G(z),$$

where $z \mapsto G(z) = \Sigma_{k=0}^{+\infty} A_k z^k$ is regular in the region $\Gamma \cup \text{int } \Gamma$. Then

$$\frac{1}{2\pi} \int_0^{2\pi} e^{ikt} F(z)\, dt = \frac{1}{2\pi i r^k} \oint_\Gamma \frac{z^k f'(z)}{f(z)}\, dz = \frac{1}{r^k}\left(\sum_{j=1}^m v_j^k - \sum_{j=1}^n p_j^k \right).$$

Similarly we obtain the following equalities

$$\frac{1}{2\pi} \int_0^{2\pi} e^{-ikt} F(z)\, dt = \frac{r^k}{2\pi i} \oint_\Gamma \frac{f'(z)}{z^k f(z)}\, dz = A_{k-1} r^k,$$

$$\frac{1}{2\pi} \int_0^{2\pi} e^{ikt} F(\bar z)\, dt = -\frac{r^k}{2\pi i} \oint_\Gamma \frac{f'(\bar z)}{\bar z^k f(\bar z)}\, d\bar z = A_{k-1} r^k,$$

$$\frac{1}{2\pi} \int_0^{2\pi} e^{-ikt} F(\bar z)\, dt = -\frac{1}{2\pi i r^k} \oint_\Gamma \frac{\bar z^k f'(\bar z)}{f(\bar z)}\, d\bar z = \frac{1}{r^k}\left(\sum_{j=1}^m v_j^k - \sum_{j=1}^n p_j^k \right),$$

and we conclude that

$$\frac{1}{2\pi} \int_0^{2\pi} e^{ikt} F(z)\, dt = \frac{1}{2\pi} \int_0^{2\pi} e^{-ikt} F(\bar z)\, dt,$$

$$\frac{1}{2\pi} \int_0^{2\pi} e^{-ikt} F(z)\, dt = \frac{1}{2\pi} \int_0^{2\pi} e^{ikt} F(\bar z)\, dt,$$

i.e.

$$(6)\frac{1}{2\pi} \int_0^{2\pi} \big(F(z) - F(\bar z)\big) \cos kt\, dt = 0; \quad \frac{1}{2\pi} \int_0^{2\pi} \big(F(z) + F(\bar z)\big) \sin kt\, dt = 0.$$

In particular, for $k = 0$ we have (by Cauchy's theorem)

$$\frac{1}{2\pi} \int_0^{2\pi} F(z)\, dt = \frac{1}{2\pi} \int_0^{2\pi} F(\bar z)\, dt = \frac{1}{2\pi} \int_0^{2\pi} \frac{F(z) + F(\bar z)}{2}\, dt = N - P.$$

Since g is, by hypothesis, a continuous function on $[0, 2\pi]$, it can be expanded into a Fourier series which is uniformly convergent and converges to g. Denote by g_n the nth partial sum of that Fourier series, i.e. let

$$g_n(t) = a_0 + g_n^{(c)}(t) + g_n^{(s)}(t),$$

where

$$a_0 = \frac{1}{2\pi} \int_0^{2\pi} g(t)\,dt; \quad g_n^{(c)}(t) = \sum_{k=1}^{n} a_k \cos kt; \quad g_n^{(s)}(t) = \sum_{k=1}^{n} b_k \sin kt.$$

Then, in virtue of (6), we see that

$$(7)\ \frac{1}{2\pi} \int_0^{2\pi} \left((a_0 + g_n^{(c)}(t)) \frac{F(z)+F(\bar{z})}{2} + g_n^{(s)}(t) \frac{F(z)-F(\bar{z})}{2} \right) dt = a_0(N-P).$$

However, we have

$$g_n^{(c)}(t) = \frac{g_n(t)+g_n(-t)}{2} - a_0; \quad g_n^{(s)}(t) = \frac{g_n(t)-g_n(-t)}{2},$$

and formula (7) becomes

$$(8)\ \int_0^{2\pi} \left(\left(\frac{g_n(t)-g_n(-t)}{2} + a_0 \right) \frac{F(z)+F(\bar{z})}{2} + \left(\frac{g_n(t)+g_n(-t)}{2} - a_0 \right) \frac{F(z)-F(\bar{z})}{2} \right) dt$$

$$= a_0(N-P).$$

Finally, if $n \to +\infty$, $g_n(t)$ uniformly converges to $g(t)$, and from (8) follows (5).

Formula (5) is a generalization of Cauchy's formula (2) and Petrović's formula (3). Indeed, if we put $g(t) \equiv 1$ into (5), we obtain Cauchy's formula. On the other hand, if g is orthogonal to the system $t \mapsto \cos kt\,(k = 1, 2, \ldots)$, formula (5) becomes

$$(9)\qquad \int_0^{2\pi} g(t) \frac{F(z)+F(\bar{z})}{2}\,dt = (N-P) \int_0^{2\pi} g(t)\,dt.$$

Introducing the additional hypothesis that $f(z) \in \mathbf{R}$ when $z \in \mathbf{R}$, we get $1/2\,(F(z)+F(\bar{z})) = \operatorname{Re} F(z)$, and (9) is, in fact, Petrović's formula.

In the same paper [2] Montel proved the following theorem, which is given without proof.

THEOREM 4. *Suppose that f satisfies the conditions of Theorem 3, and that g is a continuous function on* $[0, 2\pi]$. *Then*

(10)
$$\int_0^{2\pi} \left(\left(\frac{g(t) - g(-t)}{2} + a_0 \right) \frac{\log|f(z)| + \log|f(\bar{z})|}{2} \right.$$

$$\left. + \left(\frac{g(t) + g(-t)}{2} - a_0 \right) \frac{\log|f(z)| - \log|f(\bar{z})|}{2} \right) dt$$

$$= \left(\log|f(0)| + \sum_{j=1}^m \log \frac{r}{|v_j|} - \sum_{j=1}^n \log \frac{r}{|p_j|} \right) \int_0^{2\pi} g(t) dt.$$

If we put $g(t) \equiv 1$, (10) reduces to Jensen's formula (2) from 4.2.3. If we add the conditions that $f(z)$ is real for real z, and that g is orthogonal to the system $t \mapsto \cos kt$ $(k = 1, 2, \ldots)$, we obtain Petrović's formula (4).

In the considered cases f was a meromorphic function. The case when f has essential singularities was treated by Mitrović [3]. We quote his results.

THEOREM 5. *Let f be a uniform analytic function in the region* $G = \{z \mid |z| \leq r\}$ *in the interior of which it may have a finite number of singularities* b_k $(k = 1, \ldots, n)$. *Let* $v_1, \ldots, v_m$ *be the zeros of f in int G. If the point* $z = 0$ *is neither a zero nor a singularity of f, and if g is continuous and orthogonal to the system* $t \mapsto \cos kt$ $(k = 1, 2, \ldots)$ *on* $[0, 2\pi]$, *then*

$$\int_0^{2\pi} g(t) \frac{F(z) + F(\bar{z})}{2} dt = \left(N + \sum_{k=1}^n B_{1,k} \right) \int_0^{2\pi} g(t) dt,$$

where F is defined as before, $B_{n,k}$ *are the coefficients of the principal part of Laurent's series of the function* $z \mapsto [f'(z)/f(z)]$ *with respect to the point* b_k.

THEOREM 6. *If f and g satisfy the conditions of Theorem 5, then*

(11)
$$\int_0^{2\pi} g(t) \frac{\log|f(z)| + \log|f(\bar{z})|}{2} dt = \left(\log|f(0)| + \sum_{j=1}^m \log \frac{r}{|v_j|} \right.$$

$$\left. + \sum_{j=1}^n B_{1,j} \log \frac{r}{|b_j|} + \text{Re} \sum_{j=1}^n \sum_{i=1}^{+\infty} \frac{(-1)^i B_{i+1,j}}{ib_j^i} \right) \int_0^{2\pi} g(t) dt.$$

In particular, if we suppose that b_k are poles of f, then the coefficients $B_{1, k}$ with the opposite sign, represent the orders of those poles, and the double sum in (11) vanishes. Hence, from Theorems 5 and 6 we immediately obtain Montel's Theorems 3 and 4.

REFERENCES
1. M. Petrovitch: 'Procédé élémentaire d'application des intégrales définies réelles aux équations algébriques et transcendantes', *Nouv. Ann. Math.* (4) 8 (1908), 1–15.
2. P. Montel: 'Sur une formule de M. Michel Petrovitch', *Publ. Math. Univ. Belgrade* 6–7 (1938), 174–182.
3. D. Mitrović: 'Une généralisation de certaines formules de Montel', *Compt. Rend. Acad. Sci. Paris* 256 (1963), 1212–1213.

4.2.5. Nyquist's Criterion for Stability

In network analysis two complex functions are defined: the return difference F and the return ratio T, which are tied by the equation $F = 1 + T$.

Nyquist's criterion gives necessary and sufficient conditions for the stability of a structure. In network analysis the functions F and T are always rational. Suppose further that F is regular on the imaginary axis. Nyquist's curve is constructed by conformal mapping of the contour shown on Figure 4.2.5, with $R \to +\infty$, into the F-plane (or T-plane). Since F and T approach constant values as $R \to +\infty$, this is a mapping of the imaginary axis into F-plane (T-plane).

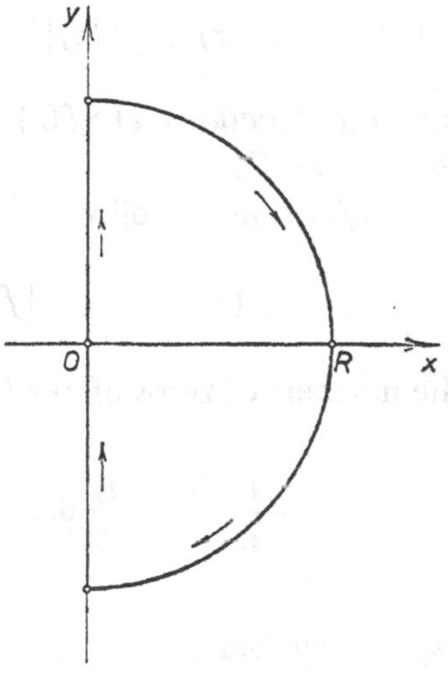

The functions F, or T, will have zeros and poles in the right half-plane if the obtained contours encircle the critical points $(0, 0)$, or $(-1, 0)$, respectively. Applying Cauchy's formula (Theorem 2 of 4.2.1), we find

Fig. 4.2.5.

$$\frac{1}{2\pi i} \oint \frac{F'}{F}\, dz = P - N,$$

where P and N are the numbers of poles and zeros of F, and the integration is performed along the negatively oriented contour.

If $N = 0$, i.e. if F has no zeros with positive real part, the structure will be absolutely stable. This implies the criterion:

A structure is stable if Nyquist's curve in the F-plane (or in the T-plane) encircles the point $(0, 0)$ (or the point $(-1, 0)$) clockwise $-P$ times, where P is the number of poles of F (or T).

REFERENCE
H. W. Bode: *Network Analysis and Feedback Amplifier Designs*, Toronto, New York, London 1952, pp. 151–162.

4.3. ROUCHÉ'S THEOREM AND ITS CONSEQUENCES

4.3.1. *Rouché's Theorem*

THEOREM 1. *If the functions f and g are regular inside a closed contour Γ, and if they are continuous on Γ, and if*

(1) $\qquad |f(z)| > |g(z)| \qquad (z \in \Gamma),$

then the functions $z \mapsto f(z)$ and $z \mapsto g(z) + f(z)$ have the same number of zeros inside Γ.

Proof. From (1) follows

$$|f(z)| > 0, \qquad |f(z) + g(z)| \geq |f(z)| - |g(z)| > 0.$$

The numbers of zeros of $z \mapsto f(z)$ and $z \mapsto f(z) + g(z)$ in int Γ are given by

$$\frac{1}{2\pi i} \oint_\Gamma \frac{f'(z)}{f(z)} \, dz, \qquad \frac{1}{2\pi i} \oint_\Gamma \frac{f'(z) + g'(z)}{f(z) + g(z)} \, dz.$$

respectively. Since

$$f(z) + g(z) = f(z)\left(1 + \frac{g(z)}{f(z)}\right),$$

we get

$$\log\left(f(z) + g(z)\right) = \log f(z) + \log\left(1 + \frac{g(z)}{f(z)}\right),$$

implying

$$(2) \qquad \frac{f'(z)+g'(z)}{f(z)+g(z)} = \frac{f'(z)}{f(z)} + \frac{h'(z)}{h(z)},$$

where

$$h(z) = 1 + \frac{g(z)}{f(z)}.$$

Since

$$|h(z)-1| = \left| \frac{g(z)}{f(z)} \right| < 1,$$

we conclude that if z describes a closed contour Γ, then h also describes a closed contour Γ_1 which belongs to the region $\{h \,|\, |h-1| < 1\}$. Thus, the point $h = 0$ is outside the contour Γ_1 and so

$$\oint_{\Gamma} \frac{h'(z)}{h(z)} \, \mathrm{d}z = \oint_{\Gamma} \frac{1}{h} \, \mathrm{d}h = 0,$$

which together with (2) implies

$$\frac{1}{2\pi i} \oint_{\Gamma} \frac{f'(z)+g'(z)}{f(z)+g(z)} \, \mathrm{d}z = \frac{1}{2\pi i} \oint_{\Gamma} \frac{f'(z)}{f(z)} \, \mathrm{d}z.$$

Theorem 1 was proved by Rouché [1]. It was generalized a number of times. Some generalizations of this important theorem can be found in [2] – [5], and the generalization to functions in several variables in [6].

REFERENCES

1. E. Rouché: 'Mémoire sur la série de Lagrange', *École Polytechn. J.* 22 (1862), 39e cahier, 193–224.
2. S. Sarantopoulos: 'Sur le nombre des racines des fonctions holomorphes dans une courbe donnée', *Compt. Rend. Acad. Sci. Paris* 175 (1922), 1033–1035.
3. D. Pompeiu: 'Sur un théorème, analogue à celui de Rouché relatif aux zéros des fonctions holomorphes', *Compt. Rend. Acad. Sci. Paris* 195 (1932), 855–856.
4. P. Montel: 'Sur un théorème de Rouché', *Compt. Rend. Acad. Sci. Paris* 195 (1932), 1214–1216.
5. W. Tutschke: 'Über eine Verschärfung der Aussage des Satzes von Rouché', *Archiv der Mathematik* 17 (1966), 432–434.
6. A. P. Južakov: *On the Logarithmic Residue in C^n and Some Applications*, (Russian), Voprosy matematiki – Sbornik naučnyh trudov No. 510, Taškent 1976, pp. 79–80.

4.3.2. The Fundamental Theorem of Algebra

Consider the polynomial P, defined by

$$P(z) = a_0 z^n + a_1 z^{n-1} + \cdots + a_n \qquad (n \geq 1)$$

which we write in the form

$$P(z) = f(z) + g(z),$$

where

$$f(z) = a_0 z^n, \quad g(z) = a_1 z^{n-1} + \cdots + a_n.$$

Then

$$\left| \frac{g(z)}{f(z)} \right| = \frac{1}{|a_0 z|} \left| a_1 + \frac{a_2}{z} + \cdots + \frac{a_n}{z^{n-1}} \right|.$$

This quotient can be made smaller than an aribitrary positive number if we take $|z|$ to be large enough. Hence, there exists a circle $K = \{z \mid |z| = R\}$ on which $f(z) \neq 0$, and $|g(z)| < |f(z)|$. Since f has one zero of order n inside K, from Rouché's theorem follows that P has n zeros inside K, where R is large enough.

REMARK 1. Suppose that the zeros of P are denoted by $z_1, \ldots, z_n$. They need not be distinct, i.e. we do not request that $z_i \neq z_j$ for $i \neq j$. Then the function

$$z \mapsto \frac{a_0 z^n + a_1 z^{n-1} + \cdots + a_n}{(z-z_1)(z-z_2) \ldots (z-z_n)}$$

is regular in the extended plane and tends to a_0 as $z \to \infty$. By Liouville's theorem, this function must be constant, and it takes the value a_0 for every z. This gives the factorization of P into linear factors:

$$a_0 z^n + a_1 z^{n-1} + \cdots + a_n = a_0 (z-z_1)(z-z_2) \ldots (z-z_n).$$

REMARK 2. See 3.1.2.9 and 4.1.

4.3.3. Three Theorems on Univalent Functions

THEOREM 1. *If f is a univalent function in a region G, then $f'(z) \neq 0$ in G.*

    Proof. Suppose, on the contrary, that there exists $a \in G$ such that $f'(a) = 0$. Then the function $z \mapsto f(z) - f(a)$ has a zero of order n $(n \geq 2)$ at $z = a$. Since

f is not a constant function, there exists a circle $K = \{z \mid |z-a| = r\}$ on which $f(z) - f(a) \neq 0$, and inside which f' has no zeros except a. Let m be the lower bound of $|f(z) - f(a)|$ on this circle. Then by Rouché's theorem, if $0 < |A| < m$, the function $z \mapsto f(z) - f(a) - A$ has n zeros in the circle (it cannot have a double zero, since f' has no other zeros in the circle). This is contrary to the hypothesis that f is univalent, i.e. that f does not take any value more than once.

THEOREM 2. *Let f be a regular function at $z = 0$ and let $f'(0) \neq 0$. Then f is univalent in the region $\{z \mid |z| \leq r\}$, if r is small enough.*

Proof. We may suppose that $f(0) = 0$. Since $f'(0) \neq 0$, $z = 0$ is a zero of the first order, and hence there exists a circle K, with centre $z = 0$, on which $f(z) \neq 0$, and inside which f has no zero other than $z = 0$. Let m be the lower bound of $|f(z)|$ on K. Since f is continuous and $f(0) = 0$, there exists a circle $k = \{z \mid |z| = r\}$ inside which $|f(z)| < m$. Let us prove that f is univalent in the circle k.

Let w_0 be any number such that $|w_0| < m$. Then, by Rouché's theorem, the number of zeros of the function $z \mapsto f(z) - w_0$ in int K is equal to the number of zeros of f inside K. This means that $z \mapsto f(z) - w_0$ has only one zero in int K. Therefore, there exists just one point $z_0 \in$ int K corresponding to each such value of w_0. The region consisting of these values of z_0 includes the circle k, and hence, f is univalent inside k.

THEOREM 3. *If (f_n) is a sequence of univalent functions in a region G, and if $f_n \to f$ uniformly in G, then the function f is either univalent in G, or is a constant function in G.*

Proof. The function f is regular in G. Suppose first that f is not constant. If f is not univalent, then there exist two points $z_1, z_2 \in G$ such that $f(z_1) = f(z_2) = w_0$. Let k_1 and k_2 be two circles in G, with centres at z_1 and z_2, which do not overlap, and such that $f(z) - w_0 \neq 0$ on either circumference (this is possible if f is not constant). Let m be the lower bound of $|f(z) - w_0|$ on the two circles. We then choose N so large that

$$|f(z) - f_N(z)| < m, \qquad (z \in k_1 \cup k_2).$$

Hence, by Rouché's theorem, it follows that the function

$$z \mapsto f_N(z) - w_0 = (f(z) - w_0) + (f_N(z) - f(z))$$

has the same number of zeros in the circles k_1, k_2 as the function

$z \mapsto f(z) - w_0$, i.e. two. Therefore, the function f_N is not univalent, contrary to the hypothesis.

The sequence (f_n) defined by $f_n(z) = z/n$, shows that f can be a constant function.

REFERENCE
Titchmarsh, pp. 198–201.

4.3.4. A Theorem of Hurwitz

THEOREM 1. *Let (f_n) be a sequence of functions which are regular in the region G, bounded by a closed contour, and let $f_n \to f \not\equiv 0$ uniformly in G. Let $z_0 \in G$. Then z_0 is a zero of f if and only if it is a limit point of the set of zeros of the functions f_n, points which are zeros for an infinity of values of n being counted as limit points.*

Proof. Let z_0 be a zero of f. We choose ϵ so small that the circle $K = \{z \mid |z - z_0| = \epsilon\}$ lies entirely in G, and contains or has on it no zero of f except possibly the point z_0 itself. Then $|f(z)|$ has a positive lower bound on K, say $|f(z)| \geq m > 0$. For fixed ϵ and m, we choose sufficiently large N so that

$$|f_n(z) - f(z)| < m \qquad (n > N)$$

on the circle K. Since

$$f_n(z) = f(z) + (f_n(z) - f(z)),$$

from Rouché's theorem follows that f_n, for $n > N$ has the same number of zeros in K as f, i.e. one. Therefore, it may happen that $f_n(z_0) = 0$ for all $n > N$. If this is not the case, we still have $|z_n - z_0| < \epsilon$, where z_n is a zero of f_n, and then z_0 is the limit point of the sequence (z_n).

If z_0 is not a zero of f, then by Rouché's theorem, the functions f_n $(n > N)$ have no zeros in K.

REFERENCE
Titchmarsh, p. 119.

4.4. INVERSE FUNCTIONS

4.4.1. Existence of the Inverse Function

THEOREM 1. *Let f be a regular function in a neighbourhood of $z = z_0$ and let $f(z_0) = w_0$. The necessary and sufficient condition that the equation*

$f(z) = w$ *should have a unique solution* $z = F(w)$, *regular in a nieghbourhood of* $w = w_0$, *is that* $f'(z_0) \neq 0$.

Proof. The condition is clearly necessary. Indeed, if the function $w \mapsto F(w)$ is regular in a neighbourhood of w_0, then $F'(w_0)$ must be finite. However,

$$F'(w_0) = \frac{1}{f'(z_0)},$$

and $f'(z_0)$ cannot be zero.

Let us now prove that the condition is sufficient. The proof given here is due to Landau [1] (see also [2]).

Suppose that $z_0 = w_0 = 0$, for if they are not, we can make the transformation $z' = z - z_0$, $w' = w - w_0$. By hypothesis, f is regular in a disc $\{z \mid |z| \leq R\}$, and can be expressed there as a convergent Taylor series

$$f(z) = a_1 z + a_2 z^2 + \cdots,$$

where

$$|f'(0)| = |a_1| = a > 0.$$

If z_1 and z_2 are any two points of the disc $\{z \mid |z| \leq \lambda R\}$ where $0 < \lambda < 1$, then

$$\left| \frac{f(z_1) - f(z_2)}{z_1 - z_2} \right| = \left| a_1 + \sum_{n=2}^{+\infty} a_n (z_1^{n-1} + z_1^{n-2} z_2 + \cdots + z_2^{n-1}) \right|$$

$$\geq a - \sum_{n=2}^{+\infty} n |a_n| \lambda^{n-1} R^{n-1}.$$

Hence, if λ is so small that

$$\sum_{n=2}^{+\infty} n |a_n| \lambda^{n-1} R^{n-1} < a,$$

the equation $w = f(z)$ can have at most one root in G. Furthermore, if that root exists, it cannot be a multiple root, for

$$|f'(z)| = \left| a_1 + \sum_{n=2}^{+\infty} n a_n z^{n-1} \right| \geq a - \sum_{n=2}^{+\infty} n |a_n| \lambda^{n-1} R^{n-1} > 0.$$

Since f is, by hypothesis, regular in $\{z\,|\,|z|\le R\}$, we have the following Cauchy's inequalities

$$a\le\frac{M}{R}\ \text{and}\ |a_n|\le\frac{M}{R^n},$$

and therefore

$$\sum_{n=2}^{+\infty} n\,|a_n|\,\lambda^{n-1}\,R^{n-1}\le\sum_{n=2}^{+\infty}\frac{M}{R^n}\,\lambda^{n-1}=\frac{M\lambda(2-\lambda)}{R\,(1-\lambda)^2}<\frac{2M}{R\,(1-\lambda)^2}\,.$$

We now take $\lambda = Ra/4M$, implying $\lambda\le 1/4$, and so

$$\sum_{n=2}^{+\infty} n\,|a_n|\,\lambda^{n-1}\,R^{n-1}<\frac{8a}{9}\,.$$

Thus the equation $w = f(z)$ has at most one simple zero in the region $\{z\,|\,|z|\le R^2 a/4M\}$, and that is the point 0.

If $|z| = R^2 a/4M$, using the inequality $Ra\le M$, we get

$$|f(z)|\ge|a_1\,z|-\sum_{n=2}^{+\infty}|a_n\,z^n|\ge a\,|z|-M\sum_{n=2}^{+\infty}\left|\frac{z}{R}\right|^n$$

$$=\frac{R^2 a^2}{4M}-M\sum_{n=2}^{+\infty}\left(\frac{Ra}{4M}\right)^n=\frac{R^2 a^2}{4M}\left(1-\frac{M}{4M-Ra}\right)\ge\frac{R^2 a^2}{4M}\left(1-\frac{1}{3}\right)=\frac{R^2 a^2}{6M}\,.$$

In view of Rouché's theorem, the function $z\mapsto f(z)-w$ has the same number of zeros as the function f in the region $\{z\,|\,|z|\le R^2 a/4M\}$, provided that $|w|<R^2 a^2/6M$, i.e. precisely one simple zero. Denote this zero by $F(w)$.

Using Theorem 1 from 4.2.1, for $g(z)\equiv z$, we can get the explicit formula for $F(w)$, i.e.

$$(1)\qquad F(w)=\frac{1}{2\pi i}\oint_K\frac{zf'(z)}{f(z)-w}\,dz,$$

where $K=\{z\,|\,|z|=R^2 a/4M\}$. On the circle K we have $z=(R^2 a/4M)\,e^{it}$, where t varies from 0 to 2π, and so the integrand in (1) is continuous. By Theorem 5 from 1.2.2 it follows that $w\mapsto F(w)$ is a regular function in the region $\{w\,|\,|w|<R^2 a^2/6M\}$.

This completes the proof.

REFERENCES
1. E. Landau: 'Über eine Verallgemeinerung des Picardschen Satzes', *Berlin Ber.* 1904, 1118–1133.
2. Copson, pp. 121–123.

4.4.2. *Lagrange's Formula for the Reversion of Series*

In previous section we proved that if the function f, defined by

$$f(z) = w_0 + a_1(z - z_0) + a_2(z - z_0)^2 + \cdots \qquad (a_1 \neq 0),$$

is regular in a neighbourhood of z_0, than there exists a unique function F, defined by

$$F(w) = z_0 + b_1(w - w_0) + b_2(w - w_0)^2 + \cdots,$$

which is regular in a neighbourhood of w_0, such that $z = F(w)$ is the solution of the equation $w = f(z)$.

We shall give here an elegant method for obtaining the coefficients b_1, b_2, ... , which is based upon Cauchy's residue theorem. We use the same notations as in 4.4.1.

We first have

$$F(w) = \frac{1}{2\pi i} \oint_K \frac{z f'(z)}{f(z) - w} \, dz,$$

where $K = \{z \,|\, |z| = R^2 a/4M\}$. Integrating by parts, we obtain

$$F'(w) = \frac{1}{2\pi i} \oint_K \frac{z f'(z)}{(f(z) - w)^2} \, dz = \frac{1}{2\pi i} \oint_K \frac{1}{f(z) - w} \, dz,$$

and so

$$(1) \qquad F'(w) = \frac{1}{2\pi i} \oint_K \left(1 + \sum_{n=1}^{+\infty} \left(\frac{w - w_0}{f(z) - w_0} \right)^n \right) \frac{dz}{f(z) - w_0}.$$

In the previous section we have shown that

$$|f(z) - w_0| \geq \frac{R^2 a^2}{6M},$$

when $z \in K$. Therefore, if

$$|w - w_0| \leq \frac{R^2 a^2}{6 M \lambda} \qquad (0 < \lambda < 1),$$

the series under the integration sign in (1) is uniformly convergent with respect to z, and can be integrated term by term. We get

$$F'(w) = \sum_{n=1}^{+\infty} n b_n (w - w_0)^{n-1},$$

where

$$n b_n = \frac{1}{2 \pi i} \oint_K \frac{1}{(f(z) - w_0)^n} \, dz.$$

Since the function $z \mapsto f(z) - w_0$ has a simple zero at $z = z_0$ and vanishes nowhere else in the region $D = \{z \mid |z - z_0| \leq R^2 a / 4M\}$, for the coefficient $n b_n$ we have

$$n b_n = \operatorname*{Res}_{z = z_0} \frac{1}{(f(z) - w_0)^n}.$$

If we put $f(z) - w_0 = (z - z_0)/g(z)$, the function g is regular in D, and hence

$$n b_n = \frac{1}{2 \pi i} \oint_K \frac{g(z)^n}{(z - z_0)^n} \, dz = \frac{1}{(n-1)!} \lim_{z \to z_0} \frac{d^{n-1}}{dz^{n-1}} (g(z)^n).$$

Substituting this value of b_n into the series for $F'(w)$, we obtain the following result.

THEOREM 1. *If f is a regular function in a neighbourhood of z_0, and if $f(z_0) = w_0$, $f'(z_0) \neq 0$, then the equation $f(z) = w$ has a unique solution, regular in a neighbourhood of w_0, which can be expressed by*

$$(2) \qquad z = z_0 + \sum_{n=1}^{+\infty} \frac{(w - w_0)^n}{n!} \lim_{z \to z_0} \frac{d^{n-1}}{dz^{n-1}} (g(z)^n),$$

where

$$f(z) - w_0 = \frac{z - z_0}{g(z)}.$$

REMARK 1. Formula (2) was obtained by Langrange in 1768, naturally without the use of residues, which were not defined at that time. It is a special case of the more general formula

$$(3) \qquad h(z) = h(z_0) + \sum_{n=1}^{+\infty} \frac{(w-w_0)^n}{n!} \lim_{z \to z_0} \frac{d^{n-1}}{dz^{n-1}} (h'(z) g(z)^n)$$

which is also due to Lagrange. Formula (3) can be proved in a similar manner as formula (2).

Lagrange's formulas (2) and (3) were generalized to functions in several variables a number of times (see, for example, [1], [2], [3], [4], [5], [6], and also the references given there). The most complete generalization so far was given by Južakov ([7], [8]) who generalized Lagrange's expansion to arbitrary implicit functions. The main result of Južakov runs as follows:

Let $(w, z) \mapsto F_j(w, z)$ $(j=1, \ldots, n)$ and $(w, z) \mapsto H(w, z)$ be holomorphic functions of the variables $w = (w_1, \ldots, w_m)$ and $z = (z_1, \ldots, z_n)$ in a neighbourhood of $(0, 0)$, let $F_j(0, 0) = 0$ $(j=1, \ldots, n)$ and let

$$\frac{\partial(F)}{\partial(z)} = \frac{\partial(F_1, \ldots, F_n)}{\partial(z_1, \ldots, z_n)} \bigg|_{(0, 0)} \neq 0.$$

If $w \mapsto z = h(w)$, where $h = (h_1, \ldots, h_n)$ is the implicit vector-valued function (or a single-valued branch of it) defined in a neighbourhood of $(0, 0)$ by the system of equations

$$F_j(w, z) = 0 \qquad (j=1, \ldots, n)$$

then

$$H(w, h(w)) = \sum_{\nu \geq 0} \frac{(-1)^{|\nu|}}{\nu!} \frac{\partial^{|\nu|}}{\partial z^\nu} \left(H(w, z) g(w, z)^\nu \frac{\partial(F)}{\partial(z)} \right) \bigg|_{z=0},$$

where

$$\nu = (\nu_1, \ldots, \nu_n), \ |\nu| = \nu_1 + \cdots + \nu_n, \frac{\partial^{|\nu|}}{\partial z^\nu} = \frac{\partial^{|\nu|}}{\partial z_1^{\nu_1} \cdots \partial z_n^{\nu_n}},$$

and

$$g^\nu = g_1^{\nu_1} \cdots g_n^{\nu_n}, g_j(w, z) = F_j(w, z) - z_j \ (j=1, \ldots, n)$$

and this expansion is uniform and absolutely convergent in a neighbourhood of the origin in w-space.

Južakov proved this result by means of a multidimensional analog of the theorem on logarithmic residues. As a special case a formula is obtained for the inverse of a holomorphic map in C^n.

As examples of application of the above result Južakov gives

$$z = \sum_{\nu=0}^{+\infty} \frac{(3\nu)!}{3^{3\nu+1}\nu!\,(2\nu+1)!}\, w^{3\nu+2}$$

for the branch of the curve $w^3 - 3wz - z^3 = 0$ (the folium of Descartes) tangent to $z = 0$, and he also obtained the expansions for the inverse of the holomorphic mapping

$$(4) \qquad w_1 = z_1 + z_2{}^p, \qquad w_2 = z_2 + z_1{}^q,$$

i.e. the series expansions for z_1 and z_2 defined by (4).

REFERENCES

1. T. J. Stieltjes: 'Sur une généralisation de la série de Lagrange', *Ann. Sci. École Norm. Sup.* (3) 2 (1885), 93–98.
2. H. Poincaré: 'Sur les résidus des intégrales doubles', *Acta Math.* 9 (1887), 321–380.
3. I. J. Good: 'Generalizations to Several Variables of Lagrange's Expansion, with Applications to Stochastic Processes', *Proc. Cambridge Philos. Soc.* 56 (1960), 367–380.
4. J. K. Percus: 'A Note on Extension of the Lagrange Inversion Formula', *Comm. Pure Appl. Math.* 17 (1964), 137–146.
5. R. A. Sack: 'Interpretation of Lagrange's Expansion and its Generalization to Several Variables as Integration Formulas', *SIAM Journal* 13 (1965), 47–59.
6. R. A. Sack: 'Generalization of Lagrange's Expansion for Functions of Several Implicitly Defined Variables', *SIAM J. Math. Anal.* 13 (1965), 913–926.
7. A. P. Južakov: 'A Generalization of the Lagrange Expansion to Arbitrary Implicit Functions', (Russian), *Dokl. Akad. Nauk SSSR* 219 (1974), 822–824. Also appeared in English in: *Soviet Math. Dokl.* 15 (1974), 1694–1697 (1975).
8. A. P. Južakov: 'The Application of the Multiform Logarithmic Residue to the Expansion of Implicit Functions in Power Series', (Russian), *Mat. Sb.* 97 (139) (1975), 177–192. Also appeared in English in: *Math. USSR Sbornik* 26 (1975), 165–179.

4.4.3. *Weierstrass's Representation and Its Application to Implicit Functions*

Suppose that the function $(z, w) \mapsto F(z, w)$ is regular in a neighbourhood of $(z, w) = (0, 0)$ and that $F(0, w)$ is not identically zero. The function F can be written in the form

$$F(z, w) = \sum_{\nu=0}^{+\infty} A_\nu(z) w^\nu,$$

where A_ν are regular functions in a neighbourhood of $z = 0$.

The following theorem is valid (see, for example [1]):

THEOREM. *If* $A_0(0) = \cdots = A_{n-1}(0) = 0, A_n(0) \neq 0$, *then the equation in w*:

$$F(z, w) = 0$$

has exactly n roots in a neighbourhood of w = 0, provided that z is sufficiently near the point z = 0.

Denote those roots by $w_1, \ldots, w_n$ and let $\Gamma = \{w \mid \mid w \mid = r\}$. If z is fixed and such that $|z| < \rho$, consider the integral

$$I = \frac{1}{2\pi i} \int_{\Gamma} \frac{F_w(z, w)}{F(z, w)(w - \omega)} \, dw,$$

where $F_w = \partial F / \partial w$ and $\omega \in \text{int } \Gamma$. By the residue theorem we have

(1) $$I = \sum \text{Res} \frac{F_w(z, w)}{F(z, w)(w - \omega)} = \frac{F_w(z, \omega)}{F(z, \omega)} + \sum_{k=1}^{n} \frac{1}{w_k - \omega} .$$

On the other hand, expanding $1/(w - \omega)$ into a potential series with respect to ω, clearly

(2) $$I = \sum_{k=0}^{+\infty} G_k(z) \omega^k$$

where G_k are regular functions of z.

Therefore, replacing ω by w and setting

$$f(w) = \prod_{k=1}^{n} (w - w_k),$$

from (1) and (2) follows

$$\sum_{k=0}^{+\infty} G_k(z) w^k = \frac{F_w}{F} - \frac{f_w}{f},$$

i.e.

(3) $$H(z, w) = \frac{F_w}{F} - \frac{f_w}{f}$$

where H is a regular function with respect to z and w in a neighbourhood of $(0, 0)$.

However, f is a polynomial in w:

$$w^n + a_1 w^{n-1} + \cdots + a_{n-1} w + a_n,$$

whose coefficients are regular functions of z. In order to see that, consider the integral

$$J = \frac{1}{2\pi i} \int_{\Gamma} w^k \frac{F_w(z, w)}{F(z, w)} \, dw.$$

This integral is a regular function of z; but we also have

$$J = \sum_{\nu=1}^{n} \operatorname*{Res}_{w=w_\nu} w^k \frac{F_w(z, w)}{F(z, w)} = \sum_{\nu=1}^{n} w_\nu^k,$$

which means that sums of powers of $w_1, \ldots, w_n$ are regular functions of z, and hence so are the coefficients $a_1, \ldots, a_n$.

Therefore, integrating (3) we get

$$F(z, w) = C(z)f(z, w)e^{h(z, w)}$$

where C is the constant (with respect to w) of integration and h is again regular. In order to determine C, put $w = w_0$ where $|w_0| < r$. We get

$$F(z, w_0) = C(z)f(z, w_0)e^{h(z, w)},$$

where $F(z, w_0)$ and $f(z, w_0)$ have no zeros in $\{z \mid |z| < \rho\}$. We thus obtain the Weierstrass representation

$$F(z, w) = f(z, w)K(z, w)$$

where K is regular with respect to z and w and has no zeros when $|z| < \rho$ and $|w| < r$, and f is a polynomial in w, whose leading coefficient is 1, and other coefficients are regular functions of z.

REMARK. According to [1], this theorem was, since 1860, read by Weierstrass in his lectures at the University of Berlin. The proof given here is taken from [1], and is due to G. Simart.

This theorem can be nicely applied to the existence of implicit functions. Indeed, consider the equation

(4) $F(z, w) = 0$

where F is regular with respect to z and w in a neighbourhood of $(z, w) = (0, 0)$. Suppose that $F_w(0, 0) \neq 0$. Then, using the Weierstrass representation (in this case we have $n = 1$), we obtain

$$F(z, w) = (w + a_1(z))K(z, w).$$

Since $K(z, w) \neq 0$ in the considered neighbourhood of $(0, 0)$, the equation (4) implies

$$w + a_1(z) = 0,$$

which proves the existence of the implicit function $z \mapsto w(z)$ in a neighbourhood of $z = 0$.

In particular, if $F(z, w) = z - G(w)$, where G is regular in a neighbourhood of $w = 0$, $G(0) = 0$ and $G'(0) \neq 0$, we see that there exists the inverse function $z \mapsto w(z)$ which is regular in a neighbourhood of $z = 0$. This result (with z and w interchanged) was proved in Theorem 1 of 4.4.1.

The Weierstrass representation is easily extended to functions in more than two variables.

REFERENCE
1. É. Picard: *Traité d'analyse*, tome II, 2ème édition, Paris 1905, pp. 261–266.

4.5. EXPANSION OF A FUNCTION AS A SERIES OF RATIONAL FUNCTIONS

4.5.1. *Cauchy's General Theorem*

One of the most important applications of the calculus of residues is concerned with the expansion, under suitable conditions, of an analytic function as a series of rational functions (partial fractions). The result is due to Cauchy [1] (see also [2]).

THEOREM 1. *Suppose that the following conditions are satisfied:*

1° *the function f is analytic in any finite region of the z-plane, where it may have only poles;*

$2°$ *there exists an increasing sequence of positive numbers* $(R_n)_{n=1, 2, \ldots}$, *such that* $\lim_{n \to +\infty} R_n = +\infty$, *and such that the circle* $K_n = \{z \mid |z| = R_n\}$ *passes through no pole of* f, *for any* $n = 1, 2, \ldots$;

$3°$ *the upper bound of* $|f(z)|$ *on* K_n *is itself bounded as* $n \to +\infty$;

$4°$ $|f(R_n e^{i\theta})| \to 0$ $(n \to +\infty)$ *uniformly with respect to* θ, *where* $0 \le \theta \le 2\pi$. *Then, if* ζ *is not a pole of* f, *we have*

$$f(\zeta) = \lim_{n \to +\infty} S_n(\zeta),$$

where $S_n(\zeta)$ *is the sum of residues of the function* $z \mapsto f(z)/(\zeta - z)$ *at the poles of* f *within* K_n.

Proof. Suppose that ζ is any point of the bounded closed region G which lies in $\{z \mid |z| \le R\}$ and which contains no poles of f. Since, by hypothesis $2°$, $R_n \to +\infty$ as $n \to +\infty$, we can choose a positive integer N such that $R_N > R$. The point ζ lies, therefore, within all the circles K_n $(n \ge N)$. Since the singularities of $z \mapsto f(z)/(\zeta - z)$ are ζ and the poles of f, by Cauchy's residue theorem, for $n \ge N$, we have

(1) $\qquad \dfrac{1}{2\pi i} \oint_{K_n} \dfrac{f(z)}{\zeta - z} \, dz = S_n(\zeta) - f(\zeta).$

We now have to prove that the integral on the left of (1) tends to zero as $n \to +\infty$. We have

$$\left| \oint_{K_n} \frac{f(z)}{\zeta - z} \, dz \right| = \left| \int_0^{2\pi} \frac{f(R_n e^{i\theta})}{\zeta - R_n e^{i\theta}} R_n i e^{i\theta} \, d\theta \right|$$

$$\le \int_0^{2\pi} \frac{|f(R_n e^{i\theta})|}{|\zeta - R_n e^{i\theta}|} R_n \, d\theta \le \frac{R_n}{R_n - R} \int_0^{2\pi} |f(R_n e^{i\theta})| \, d\theta \le \frac{R_n}{R_n - R} 2\pi\varepsilon_n,$$

since

$$|\zeta - R_n e^{i\theta}| \ge ||\zeta| - R_n| \ge R_n - |\zeta| \ge R_n - R,$$

and ε_n is the upper bound for $|f(R_n e^{i\theta})|$ when $0 \le \theta \le 2\pi$, and by the hypothesis $4°$, $\varepsilon_n \to 0$ $(n \to +\infty)$. This implies that

$$\lim_{n \to +\infty} \oint_{K_n} \frac{f(z)}{\zeta - z} \, dz = 0,$$

and so

(2) $f(\zeta) = \lim_{n \to +\infty} S_n(\zeta).$

Notice that we have proved that $S_n(\zeta)$ converges uniformly to $f(\zeta)$ when $\zeta \in G$.

Formula (2) can be written in the form of the series

$$f(\zeta) = S_1(\zeta) + \sum_{n=1}^{+\infty} (S_{n+1}(\zeta) - S_n(\zeta)),$$

which converges uniformly in any bounded closed region which contains no poles of f.

However, $S_{n+1}(\zeta) - S_n(\zeta)$ is the sum of the residues of $z \mapsto f(z)/(\zeta - z)$ at the poles of f which lie in the annular region bounded by K_n and K_{n+1}. Since each residue is a rational function of ζ, we conclude that $f(\zeta)$ is expressed as a series of rational functions, which converges uniformly.

REMARK 1. Cauchy generalized Theorem 1 to the case when f satisfies 1°, 2°, and 3°, but not 4°. In that case we consider the function F, defined by $F(z) = f(z)/z$. This function obviously satisfies conditions 1° and 2°. Moreover, on the circle K_n we have $z = R_n e^{i\theta}$, and so

$$|F(z)| = \frac{|f(z)|}{R_n} \leq \frac{M}{R_n},$$

implying that $F(R_n e^{i\theta})$ tends to zero uniformly with respect to θ in $0 \leq \theta \leq 2\pi$, as $n \to +\infty$.

The function F, therefore, satisfies all the conditions of Theorem 1, and so

$$\frac{f(z)}{z} = \lim_{n \to +\infty} S_n(z),$$

where $S_n(z)$ is the sum of the residues of the function $t \mapsto f(t)/t(z-t)$ at the poles of $t \mapsto f(t)/t$ which are inside K_n.

An important case arises when f is regular at the origin and has only simple poles. Let $p_1, p_2, \ldots$ be the poles of f, so that $0 < |p_1| \leq |p_2| \leq \ldots$, and let $\text{Res}_{z=p_i} f(z) = r_i$ $(i = 1, 2, \ldots)$. Then

$$\text{Res}_{t=p_i} \frac{f(t)}{zt - t^2} = \frac{r_i}{p_i(z - p_i)} = \frac{r_i}{z}\left(\frac{1}{z - p_i} + \frac{1}{p_i}\right),$$

and

$$\operatorname*{Res}_{t=0} \frac{f(t)}{zt-t^2} = \frac{f(0)}{z}.$$

We thus obtain the expansion of f in the form

$$\frac{f(z)}{z} = \frac{f(0)}{z} + \sum_{i=1}^{+\infty} \frac{r_i}{p_i(z-p_i)},$$

and finally

$$f(z) = f(0) + \sum_{i=1}^{+\infty} r_i \left(\frac{1}{z-p_i} + \frac{1}{p_i} \right).$$

This is usually referred to as the Mittag-Leffler expansion of a meromorphic function.

REFERENCES
1. *Oeuvres* (2) 7, 324–344.
2. G. Mittag-Leffler: 'En metod att analytisk framstäla en funktion at rational karakter ... Öfrersigt Kongl', *Vetenskaps – Akad. Förhandlinger* **33** (1876), 3–16.
3. Copson, pp. 144–145.

4.5.2. *Representation of an Integral Function as an Infinite Product*

Let F be an integral function such that $F(0) \neq 0$, and which has simple zeros $z_1, z_2, \ldots$ ordered so that $|z_1| \leq |z_2| \leq \cdots$. Since those zeros cannot have a finite limiting point, we conclude that $\lim_{n \to +\infty} |z_n| = +\infty$.

Let $F(z) = (z-z_k)G(z)$, where G is a regular function and $G(z_k) \neq 0$. Then

$$\frac{F'(z)}{F(z)} = \frac{1}{z-z_k} + \frac{G'(z)}{G(z)},$$

which means that the only singularities of $z \mapsto F'(z)/F(z)$ are simple poles at $z = z_k$ and $\operatorname*{Res}_{z=z_k} [F'(z)]/[F(z)] = 1$.

Suppose now that the function $z \mapsto F'(z)/F(z)$ satisfies all the conditions of Theorem 1 from 4.5.1. Then

$$(1) \qquad \frac{F'(z)}{F(z)} = \sum_{k=1}^{+\infty} \frac{1}{z-z_k},$$

and this series is uniformly convergent in any bounded closed region which does not contain zeros of F, provided that the terms are suitably grouped. If we integrate (1) term by term, we find

$$\text{Log } F(z) = \text{Log } F(0) + \sum_{k=1}^{+\infty} \text{Log}\left(1 - \frac{z}{z_k}\right),$$

and hence

$$F(z) = F(0) \prod_{k=1}^{+\infty} \left(1 - \frac{z}{z_k}\right),$$

and this product is uniformly convergent in any bounded closed region which contains no zeros of F.

However, if the function $z \mapsto F'(z)/F(z)$ satisfies only the conditions $1°$, $2°$, and $3°$ of Theorem 1 from 4.5.1, then in view of Remark 1 of that section, we have

$$\frac{F'(z)}{F(z)} = \frac{F'(0)}{F(0)} + \sum_{k=1}^{+\infty} \left(\frac{1}{z - z_k} + \frac{1}{z_k}\right),$$

and after integration

(2) $$\text{Log } F(z) = \text{Log } F(0) + \frac{F'(0)}{F(0)} z + \sum_{k=1}^{+\infty} \left(\text{Log}\left(1 - \frac{z}{z_k}\right) + \frac{z}{z_k}\right).$$

Finally, from (2) follows

$$F(z) = F(0) \exp\left(z \frac{F'(0)}{F(0)}\right) \prod_{k=1}^{+\infty} \left(1 - \frac{z}{z_k}\right) \exp\frac{z}{z_k}.$$

This is usually referred to as the Weierstrass expansion of an integral function.

4.5.3. Expansion of cosec as a Series of Rational Functions

The function $z \mapsto \text{cosec } z$ has simple poles at $z = 0, \pm\pi, \pm 2\pi, \ldots$, and hence the circle $K_n = \{z \mid |z| = (n + \frac{1}{2})\pi\}$ does not pass through a pole of cosec for any value of n.

We must examine the behaviour of cosec on the circle K_n.

Construct circles of radius ϵ, where $\epsilon < \pi/2$, with centres at the poles of $z \mapsto \text{cosec } z$. We shall prove that the function $z \mapsto \text{cosec } z$ is bounded in the region T exterior to all these circles.

First, if $z = x + iy$ ($x, y \in R$), we have

$$|\cosec z| = \frac{2}{|e^{ix-y} - e^{ix+y}|} \leq \frac{2}{|e^{-y} - e^y|} = \cosech |y|,$$

which means that the inequality $|\cosec z| \leq \cosech a$ holds in that part of T which is outside the strip $\{z \mid |\operatorname{Im} z| \leq a\}$. On the other hand, $|\cosec z|$ is clearly bounded in the part of T within the rectangle with vertices $\pm(3\pi/2)$ $\pm ai$, and, by periodicity, is also bounded in the part of T for which $|\operatorname{Im} z| \leq a$. We thus conclude that the inequality

(1) $|\cosec z| \leq M$

holds everywhere in T, where M depends on ϵ. However, since the points of K_n are all at a distance not less than $(\pi/2) - \epsilon$ from the circles defining T, the inequality (1) holds on every circle K_n ($n = 1, 2, \ldots$).

Furthermore, if $z = (n + \frac{1}{2})\pi e^{i\theta}$, we have

$$|\cosec z| \leq \cosech \left|\left(n + \frac{1}{2}\right)\pi \sin\theta\right| \leq \cosech \left(\left(n + \frac{1}{2}\right)\pi \sin\delta\right),$$

for $\delta \leq \theta \leq \pi - \delta$, or $\pi + \delta \leq \theta \leq 2\pi - \delta$, where δ is an arbitrary small positive number. This implies that, when $n \to +\infty$, then $|\cosec z|$ tends uniformly to zero with respect to θ in the two given angles.

We have therefore proved that all the conditions of Theorem 1 from 4.5.1 are satisfied. By an application of that theorem, we obtain

(2) $\displaystyle \cosec \zeta = \lim_{n \to +\infty} \sum_{k=-n}^{+n} \operatorname*{Res}_{z=k\pi} \cosec \frac{z}{\zeta - z}$

$$= \lim_{n \to +\infty} \sum_{k=-n}^{+n} \frac{(-1)^k}{\zeta - k\pi} = \frac{1}{\zeta} + \sum_{k=1}^{+\infty} \frac{2(-1)^k \zeta}{\zeta^2 - k^2 \pi^2},$$

where ζ is not a pole of the function $z \mapsto \cosec z$.

It is easily established that the series in (2) converges uniformly and absolutely when ζ lies in any bounded part of T. Hence, this series can be integrated term by term:

$$\operatorname{Log} \operatorname{tg} \frac{\zeta}{2} = \operatorname{Log} C + \operatorname{Log} \frac{\zeta}{2} + \sum_{k=1}^{+\infty} (-1)^k \operatorname{Log}\left(1 - \frac{\zeta^2}{k^2 \pi^2}\right),$$

i.e.

$$tg\frac{\zeta}{2} = \frac{A\zeta}{2} \prod_{k=1}^{+\infty} \frac{1-\dfrac{\zeta^2}{4k^2 \pi^2}}{1-\dfrac{\zeta^2}{(2k-1)^2 \pi^2}}.$$

Since

$$\lim_{\zeta \to 0} \frac{tg\dfrac{\zeta}{2}}{\dfrac{\zeta}{2}} = 1,$$

we conclude that $A = 1$.

Putting $\zeta = 2z$, we obtain

$$\frac{tg\,z}{z} = \frac{\displaystyle\prod_{k=1}^{+\infty} \left(1 - \frac{z^2}{k^2 \pi^2}\right)}{\displaystyle\prod_{k=1}^{+\infty} \left(1 - \frac{4z^2}{(2k-1)^2 \pi^2}\right)}.$$

4.5.4. Expansion of sin as an Infinite Product

Start with the function F, defined by

$$F(z) = \frac{\sin \pi z}{\pi z} = \sum_{n=0}^{+\infty} (-1)^n \frac{(\pi z)^{2n}}{(2n+1)!} \quad (z \neq 0), \quad F(0) = 1,$$

which has simple zeros at $z = \pm 1, \pm 2, \ldots$. The logarithmic derivative of that function, namely

$$\frac{F'(z)}{F(z)} = \pi \cot g\,\pi z - \frac{1}{z}$$

satisfies all the conditions of Theorem 1 of 4.5.1. Hence

$$\frac{F'(z)}{F(z)} = \lim_{m \to +\infty} \sum_{n=1}^{m} \left(\frac{1}{z-n} + \frac{1}{z+n}\right) = \sum_{n=1}^{+\infty} \frac{2z}{z^2 - n^2}.$$

After the integration, we find

$$F(z) = \prod_{n=1}^{+\infty} \left(1 - \frac{z^2}{n^2}\right),$$

and so

(1) $$\sin \pi z = \pi z \prod_{n=1}^{+\infty} \left(1 - \frac{z^2}{n^2}\right).$$

REMARK 1. Formula (1) cannot be written in the form

$$F(z) = \prod_{\substack{n=-\infty \\ n \neq 0}}^{+\infty} \left(1 - \frac{z}{n}\right),$$

since this last product diverges. However, using Remark 1 of 4.5.1, we obtain

$$\frac{F'(z)}{F(z)} = \lim_{m \to +\infty} \left(\sum_{n=1}^{m} \left(\frac{1}{z-n} + \frac{1}{n}\right) + \sum_{n=1}^{m} \left(\frac{1}{z+n} - \frac{1}{n}\right)\right)$$

$$= \sum_{n=1}^{+\infty} \left(\frac{1}{z-n} + \frac{1}{n}\right) + \sum_{n=1}^{+\infty} \left(\frac{1}{z+n} - \frac{1}{n}\right),$$

where each of those series is uniformly and absolutely convergent in any bounded closed region which contains no zeros of F. From this it follows

$$F(z) = \prod_{n=1}^{+\infty} \left(\left(1 - \frac{z}{n}\right) e^{z/n}\right) \prod_{n=1}^{+\infty} \left(\left(1 + \frac{z}{n}\right) e^{-z/n}\right),$$

and so

$$\sin \pi z = \pi z \prod_{\substack{n=-\infty \\ n \neq 0}}^{+\infty} \left(\left(1 - \frac{z}{n}\right) e^{z/n}\right),$$

where the last product is absolutely convergent.

REMARK 2. If we put $z = \frac{1}{2}$ into (1) we obtain Wallis' formula

$$\frac{\pi}{2} = \frac{2 \cdot 2}{1 \cdot 3} \cdot \frac{4 \cdot 4}{3 \cdot 5} \cdot \frac{6 \cdot 6}{5 \cdot 7} \cdots$$

which can be written as

$$\lim_{n \to +\infty} \frac{2^2 \cdot 4^2 \cdots (2n-2)^2}{3^2 \cdot 5^2 \cdots (2n-1)^2} \cdot 2n = \frac{\pi}{2},$$

and after simple calculation we obtain

$$\lim_{n \to +\infty} \frac{(2n-2)!!}{(2n-3)!! \sqrt{n}} = \sqrt{\pi}.$$

4.6. EXPANSIONS OF FUNCTIONS IN SERIES

4.6.1. *Analysis of the Idea*

Let g be a meromorphic function which has an infinite (but countable) number of simple poles $a_1, a_2, \ldots$ and let $C_R = \{z \mid |z| = R\}$. Furthermore, let $\text{Res}_{z=a_k} g(z) = A_k$ $(k = 1, 2, \ldots)$. From Cauchy's residue theorem we obtain

$$(1) \qquad \frac{1}{2\pi i} \oint_{C_R} g(z)\, dz = \sum_{\nu=1}^{n} A_\nu,$$

where it is supposed that the poles $a_1, \ldots, a_n$ are inside C_R, and that all the other poles are outside C_R.

Suppose now that the function g also depends on t, where t is independent of z. Then the residues of g with respect to the poles $a_1, a_2, \ldots$ will be functions of t, and formula (1) takes the form

$$(2) \qquad \frac{1}{2\pi i} \oint_{C_R} g(z, t)\, dz = \sum_{\nu=1}^{n} A_\nu(t).$$

If $R \to +\infty$, then $n \to +\infty$, and from (2) follows

$$(3) \qquad \lim_{R \to +\infty} \frac{1}{2\pi i} \oint_{C_R} g(z, t)\, dz = \sum_{\nu=1}^{+\infty} A_\nu(t).$$

The expression on the left-hand side of (3) is, under certain conditions, a function of t, $t \mapsto G(t)$ say. Hence,

$$(4) \qquad G(t) = \sum_{\nu=1}^{+\infty} A_\nu(t).$$

Formula (4) gives the expansion of G as a series of functions $A_1, A_2, \ldots$.

4.6.2. *Formula for the Expansion*

Let $g(z, t) = [p(z)/q(z)]\, f(z, t)$, where p and f are regular with respect to z, and q has only simple poles $a_1, a_2, \ldots$. Since

$$\text{Res}_{z=a_\nu} \frac{p(z)}{q(z)} f(z, t) = \lim_{z \to a_\nu} \frac{z - a_\nu}{q(z) - q(a_\nu)} p(z) f(z, t) = \frac{p(a_\nu)}{q'(a_\nu)} f(a_\nu, t),$$

formula (2) of 4.6.1 becomes

(1) $S_R \equiv \dfrac{1}{2\pi i} \oint\limits_{C_R} \dfrac{p(z)}{q(z)} f(z, t)\, dz = \sum\limits_{\nu=1}^{n} \dfrac{p(a_\nu)}{q'(a_\nu)} f(a_\nu, t).$

Let $z = re^{i\theta}$. Then $dz = ire^{i\theta}\, d\theta$, and we get

(2) $S_R = \dfrac{r}{2\pi} \int\limits_{\theta_0}^{\theta_0+2\pi} \dfrac{p(re^{i\theta})}{q(re^{i\theta})} f(re^{i\theta}, t) e^{i\theta}\, d\theta,$

or, putting $\theta_0 = -\pi/2$,

$$S_R = \dfrac{r}{2\pi} \int\limits_{-\pi/2}^{\pi/2} \dfrac{p(re^{i\theta})}{q(re^{i\theta})} f(re^{i\theta}, t) e^{i\theta}\, d\theta + \dfrac{r}{2\pi} \int\limits_{\pi/2}^{3\pi/2} \dfrac{p(re^{i\theta})}{q(re^{i\theta})} f(re^{i\theta}, t) e^{i\theta}\, d\theta$$

$$= \dfrac{r}{2\pi} \int\limits_{-\pi/2}^{\pi/2} \dfrac{p(re^{i\theta})}{q(re^{i\theta})} f(re^{i\theta}, t) e^{i\theta}\, d\theta - \dfrac{r}{2\pi} \int\limits_{-\pi/2}^{\pi/2} \dfrac{p(-re^{i\theta})}{q(-re^{i\theta})} f(-re^{i\theta}, t) e^{i\theta}\, d\theta.$$

The integral S_R will be convergent if the limit

$$\lim_{r\to+\infty} \left(r\dfrac{p(re^{i\theta})}{q(re^{i\theta})} f(re^{i\theta}, t) - r\dfrac{p(-re^{i\theta})}{q(-re^{i\theta})} f(-re^{i\theta}, t) \right)$$

exists independently of θ. If we put $re^{i\theta} = z$, we see that when $\theta \in [-\pi/2, \pi/2]$, then $\operatorname{Re} z \geq 0$, and hence the condition for the convergence of S_R is that, for $\operatorname{Re} z \geq 0$, the limit

$$\lim_{|z|\to+\infty} \left(z\dfrac{p(z)}{q(z)} f(z, t) - z\dfrac{p(-z)}{q(-z)} f(-z, t) \right).$$

exists. We now set

(3) $f(z, t) = \int\limits_{t_0}^{t} e^{z(t-x)} F(x)\, dx,$

where F satisfies Dirichlet's conditions, and we investigate whether for $\operatorname{Re} z \geq 0$ the limit

$$\lim_{|z|\to+\infty} \left(z\dfrac{p(z)}{q(z)} \int\limits_{t_0}^{t} e^{z(t-x)} F(x)\, dx - z\dfrac{p(-z)}{q(-z)} \int\limits_{t_0}^{t} e^{-z(t-x)} F(x)\, dx \right)$$

exists. Notice first that if F satisfies Dirichlet's conditions, then there exist finite limits

$$\lim_{|z|\to+\infty} z \int_{t_0}^{t} e^{-z(x-t_0)} F(x)\,dx \quad \text{and} \quad \lim_{|z|\to+\infty} z \int_{t_0}^{t} e^{-z(t-x)} F(x)\,dx = F(t).$$

Suppose now that

$$(4) \qquad \lim_{|z|\to+\infty} \frac{p(z)}{q(z)} e^{z(t-t_0)} = 0, \qquad \lim_{|z|\to+\infty} \frac{p(-z)}{q(-z)} = c.$$

We then have

$$\lim_{|z|\to+\infty} \left(z \frac{p(z)}{q(z)} \int_{t_0}^{t} e^{z(t-x)} F(x)\,dx - z \frac{p(-z)}{q(-z)} \int_{t_0}^{t} e^{-z(t-x)} F(x)\,dx \right)$$

$$= \lim_{|z|\to+\infty} \left(\frac{p(z)}{q(z)} e^{z(t-t_0)} \right) \left(z \int_{t_0}^{t} e^{-z(x-t_0)} F(x)\,dx \right)$$

$$- \lim_{|z|\to+\infty} \left(\frac{p(-z)}{q(-z)} \right) \left(z \int_{t_0}^{t} e^{-z(t-x)} F(x)\,dx \right) = c F(t).$$

Therefore, the integral S_R is convergent, and we get

$$\lim_{R\to+\infty} S_R = \lim_{R\to+\infty} \frac{1}{2\pi i} \oint_{C_R} \frac{p(z)}{q(z)}\,dz \int_{t_0}^{t} e^{z(t-x)} F(x)\,dx$$

$$= -\frac{1}{2\pi} \int_{-\pi/2}^{\pi/2} F(t) \lim_{|z|\to+\infty} \left(\frac{p(-z)}{q(-z)} \right) d\theta = -\frac{1}{2} c F(t).$$

Finally, if $R \to +\infty$ in (1), which implies that $n \to +\infty$, we find

$$(5) \qquad -\frac{1}{2} c F(t) = \sum_{\nu=1}^{+\infty} \frac{p(a_\nu)}{q'(a_\nu)} \int_{t_0}^{t} e^{a_\nu(t-x)} F(x)\,dx.$$

If F is discontinuous at t, then instead of $F(t)$ we should take $\lim_{\epsilon\to 0} F(t-\epsilon)$.

If instead of the function $z \mapsto [p(z)/q(z)] f(z, t)$ we consider the function $z \mapsto [r(z)/q(z)] f(z, t)$, which instead of (4) satisfies

$$\lim_{|z| \to +\infty} \frac{r(z)}{q(z)} = C, \qquad \lim_{|z| \to +\infty} \frac{r(-z)}{q(-z)} e^{(t_1 - t)} = 0,$$

and if we define the function f not by (3), but by

$$f(z, t) = \int_t^{t_1} e^{z(t-x)} F(x) \, dx,$$

we then find

$$(6) \qquad \frac{1}{2} CF(t) = \sum_{\nu=1}^{+\infty} \frac{r(a_\nu)}{q'(a_\nu)} \int_t^{t_1} e^{a_\nu(t-x)} F(x) \, dx,$$

where $\lim_{\eta \to 0} F(t + \eta)$ should be taken if F is discontinuous at t.

Formulas (5) and (6) are the formulas for the expansion of F into an infinite series. They can be put into a simpler form.

Indeed, suppose that the functions p, q, and r, which appear in (5) and (6) satisfy:

$$p(z) + r(z) = q(z),$$

$$\lim_{|z| \to +\infty} \frac{r(z)}{q(z)} = 1, \quad \lim_{|z| \to +\infty} \frac{r(-z)}{q(-z)} = 0, \quad \lim_{|z| \to +\infty} \frac{p(z)}{q(z)} = 0, \quad \lim_{|z| \to +\infty} \frac{p(-z)}{q(-z)} = 1,$$

$$\lim_{|z| \to +\infty} \frac{p(z)}{q(z)} e^{z(t-t_0)} = 0, \qquad \lim_{|z| \to +\infty} \frac{r(-z)}{q(-z)} e^{z(t-t_1)} = 0.$$

Then the formulas (5) and (6) become

$$(7) \qquad -\frac{1}{2} F(t) = \sum_{\nu=1}^{+\infty} \frac{p(a_\nu)}{q'(a_\nu)} \int_{t_0}^{t} e^{a_\nu(t-x)} F(x) \, dx \qquad (t > t_0)$$

$$(8) \qquad \frac{1}{2} F(t) = \sum_{\nu=1}^{+\infty} \frac{r(a_\nu)}{q'(a_\nu)} \int_t^{t_1} e^{a_\nu(t-x)} F(x) \, dx \qquad (t_1 > t),$$

wherefrom, subtracting (7) from (8) and taking into account that

$$p(a_\nu) + r(a_\nu) = q(a_\nu) = 0 \qquad (\nu \in \mathbb{N}),$$

we find

$$(9) \qquad F(t) = \sum_{\nu=1}^{+\infty} \frac{r(a_\nu)}{q'(a_\nu)} e^{a_\nu t} \int_{t_0}^{t_1} e^{-a_\nu x} F(x)\, dx,$$

where at the points of discontinuity $F(t)$ should be replaced by

$$\frac{1}{2}\left(\lim_{\varepsilon \to 0} F(t+\varepsilon) + \lim_{\eta \to 0} F(t-\eta)\right).$$

4.6.3. *Fourier's Trigonometric Series*

Formula (9) of 4.6.2, obtained by Cauchy, contains as special cases, many important expansions. We shall show how the formula could be used to derive the expansion of a function F into Fourier's trigonometric series.

Put $p(z) = -1$, $r(z) = e^{az}$ $(a > 0)$, so that $q(z) = e^{az} - 1$. The equation $q(z) = 0$ has only simple roots; they are $a_\nu = 2\nu\pi i/a$ $(\nu \in \mathbf{Z})$.

We have

$$\lim_{|z| \to +\infty} \frac{r(z)}{q(z)} = \lim_{|z| \to +\infty} \frac{e^{az}}{e^{az}-1} = 1, \quad \lim_{|z| \to +\infty} \frac{p(z)}{q(z)} = \lim_{|z| \to +\infty} \frac{-1}{e^{az}-1} = 0,$$

$$\lim_{|z| \to +\infty} \frac{r(-z)}{q(-z)} = \lim_{|z| \to +\infty} \frac{e^{-az}}{e^{-az}-1} = 0, \quad \lim_{|z| \to +\infty} \frac{p(-z)}{q(-z)} = \lim_{|z| \to +\infty} \frac{-1}{e^{-az}-1} = 1,$$

$$\lim_{|z| \to +\infty} \frac{r(-z)}{q(-z)} e^{z(t_1-t)} = \lim_{|z| \to +\infty} \frac{e^{-(a-t_1+t)z}}{e^{-az}-1} = 0, \text{ if } t_1 - a < t,$$

$$\lim_{|z| \to +\infty} \frac{p(z)}{q(z)} e^{z(t-t_0)} = \lim_{z| \to +\infty} -\frac{e^{(t-t_0)z}}{e^{az}-1} = 0, \text{ if } t < t_0 + a.$$

Thus, all the conditions which allow the expansion (9) of 4.6.2 are fulfilled if

$$(1) \qquad t_1 - a < t < t_0 + a,$$

and so, if (1) holds, we have

$$F(t) = \sum_{k=-\infty}^{+\infty} \frac{r(a_k)}{q'(a_k)} e^{a_k t} \int_{t_0}^{t_1} e^{-a_k x} F(x)\, dx,$$

i.e.

$$F(t) = \sum_{k=-\infty}^{+\infty} \frac{e^{2k\pi i}}{a e^{2k\pi i}} e^{\frac{2k\pi i}{a}} \int_{t_0}^{t_1} e^{-\frac{2k\pi i}{a}} F(x)\, dx,$$

or

$$F(t) = \frac{1}{a} \sum_{k=-\infty}^{+\infty} e^{\frac{2k\pi i}{a}} \int_{t_0}^{t_1} e^{-\frac{2k\pi i}{a}} F(x)\, dx \qquad (t_1 - a < t < t_0 + a).$$

We take first the term which corresponds to $k = 0$, and then pair off the terms which correspond to $k = -n$ and $k = n$ $(n \in N)$. In this way we get

$$(2) \qquad F(t) = \frac{1}{a} \int_{t_0}^{t_1} F(x)\, dx + \frac{2}{a} \sum_{\nu=1}^{+\infty} \int_{t_0}^{t_1} \cos \frac{2\nu\pi(t-x)}{a} F(x)\, dx.$$

Putting $a = 2l$, $t_1 = l$, $t_0 = -l$ into (2), we find

$$F(t) = \frac{1}{2l} \int_{-l}^{l} F(x)\, dx + \frac{1}{l} \sum_{\nu=1}^{+\infty} \int_{-l}^{l} \cos \frac{\nu\pi(t-x)}{l} F(x)\, dx \qquad (-l < t < l),$$

i.e.

$$(3) \qquad F(t) = \frac{1}{2} a_0 + \sum_{\nu=1}^{+\infty} \left(a_\nu \cos \frac{\nu\pi t}{l} + b_\nu \sin \frac{\nu\pi t}{l} \right) \qquad (-l < t < l),$$

where

$$(4) \qquad \begin{aligned} a_\nu &= \frac{1}{l} \int_{-l}^{l} F(x) \cos \frac{\nu\pi x}{l}\, dx \; (\nu \in N_0), \\[2mm] b_\nu &= \frac{1}{l} \int_{-l}^{l} F(x) \sin \frac{\nu\pi x}{l}\, dx \; (\nu \in N). \end{aligned}$$

Formulas (3) and (4) define the well-known expansion of F into Fourier's trigonometric series.

REFERENCES

1. A.-L. Cauchy: 'Sur les résidus des fonctions exprimées par des intégrales définies', *Oeuvres* (2) 7, 393.
2. É. Picard: *Traité d'analyse*, t. 2, deuxième édition, Paris 1905, pp. 179–195.

4.7. MISCELLANEOUS APPLICATIONS

4.7.1. Let R be a rational function with poles at $a_1, \ldots, a_n$. Then

$$(1) \qquad R(z) = \sum_{k=1}^{n} \operatorname*{Res}_{t=a_k} \frac{R(t)}{z-t}.$$

In particular, put $R(z) = P(z)/Q(z)$, where P and Q are polynomials, and $Q(z) = (z-a_1) \ldots (z-a_n)$. From (1) follows

$$\frac{P(z)}{(z-a_1)\cdots(z-a_n)} = \sum_{k=1}^{n} \operatorname*{Res}_{t=a_k} \frac{P(t)}{(z-t)(t-a_1)\cdots(t-a_n)}$$

$$= \sum_{k=1}^{n} \lim_{t \to a_k} \frac{P(t)}{(z-t)(t-a_1)\cdots(t-a_{k-1})(t-a_{k+1})\cdots(t-a_n)}$$

$$= \sum_{k=1}^{n} \frac{P(a_k)}{(z-a_k)(a_k-a_1)\cdots(a_k-a_{k-1})(a_k-a_{k+1})\cdots(a_k-a_n)},$$

and we obtain Lagrange's interpolation formula

$$P(z) = \sum_{k=1}^{n} \frac{(z-a_1)\cdots(z-a_{k-1})(z-a_{k+1})\cdots(z-a_n)}{(a_k-a_1)\cdots(a_k-a_{k-1})(a_k-a_{k+1})\cdots(a_k-a_n)} P(a_k).$$

REFERENCE
A.-L. Cauchy: 'Sur un nouveau genre de calcul analogue au calcul infinitésimal', *Exercices de mathématiques*, Paris 1826. ≡ *Oeuvres* (2) 6, 23–37.

4.7.2. Let P and Q be polynomials and let dg $P <$ dg Q. If $a_1, \ldots, a_n$ are all the zeros of Q, then

$$(1) \qquad \frac{P(z)}{Q(z)} = \sum_{k=1}^{n} \operatorname*{Res}_{t=a_k} \frac{P(t)}{(z-t)Q(t)}.$$

Formula (1) is the formula for the decomposition of a rational function into partial fractions.

EXAMPLE. We have

$$
\frac{1}{(z+1)(z-1)^2} = \operatorname*{Res}_{t=-1} \frac{1}{(t+1)(t-1)^2(z-t)} + \operatorname*{Res}_{t=1} \frac{1}{(t+1)(t-1)^2(z-t)}
$$

$$
= \lim_{t \to -1} \frac{1}{(t-1)^2(z-t)} + \lim_{t \to 1} \frac{d}{dt} \left(\frac{1}{(t+1)(z-t)} \right)
$$

$$
= \frac{1}{4} \frac{1}{z+1} + \lim_{t \to 1} \left(-\frac{1}{(t+1)^2(z-t)} + \frac{1}{(t+1)(z-t)^2} \right)
$$

$$
= \frac{1}{4} \frac{1}{z+1} - \frac{1}{4} \frac{1}{z-1} + \frac{1}{2} \frac{1}{(z-1)^2}.
$$

REFERENCES

1. A.-L. Cauchy: 'Sur un nouveau genre de calcul analogue au calcul infinitésimal', *Exercices de mathématiques*, Paris 1826. ≡ *Oeuvres* (2) 6, 23–37.
2. B. Tortolini: 'Trattato del calcolo dei residui', *Giornale Arcad.* 63 (1834–35), 86–138. – See, in particular, pp. 102–104.
3. E. Rouché: 'Sur la décomposition des fractions rationelles et la théorie des résidus', *Compt. Rend. Acad. Sci. Paris* 49 (1859), 863–865.

4.7.3. Let P and Q be polynomials of degree m and n, respectively, and let C be a positive constant. Then there exist polynomials U and V of degree $n-1$ and $m-1$, respectively, such that

$$
(1) \qquad P(z) U(z) + Q(z) V(z) = C.
$$

Proof. Let $P(z) = A(z-a_1) \ldots (z-a_m)$ and $Q(z) = B(z-b_1) \ldots (z-b_n)$, and put $R(z) = C/[P(z)Q(z)]$ into the formula (1) of 4.7.1. We find

$$
\frac{C}{P(z)Q(z)} = \sum_{k=1}^{n} \operatorname*{Res}_{t=b_k} \frac{C}{(z-t)P(t)Q(t)} + \sum_{k=1}^{m} \operatorname*{Res}_{t=a_k} \frac{C}{(z-t)P(t)Q(t)},
$$

i.e.

$$
(2) \qquad C = P(z) \sum_{k=1}^{n} \operatorname*{Res}_{t=b_k} \frac{CQ(z)}{(z-t)P(t)Q(t)} + Q(z) \sum_{k=1}^{m} \operatorname*{Res}_{t=a_k} \frac{CP(z)}{(z-t)P(t)Q(t)}.
$$

It is easily verified that the first sum in (2) is a polynomial in z of degree $n-1$, and that the second sum is a polynomial in z of degree $m-1$. Hence, if we put

$$
U(z) = \sum_{k=1}^{n} \operatorname*{Res}_{t=b_k} \frac{CQ(z)}{(z-t)P(t)Q(t)}, \quad V(z) = \sum_{k=1}^{m} \operatorname*{Res}_{t=a_k} \frac{CP(z)}{(z-t)P(t)Q(t)},
$$

we obtain (1).

REFERENCE
A.-L. Cauchy: 'Sur un théorème d'analyse', *Oeuvres* (2) 6, 202–209.

4.7.4. Let f be a meromorphic function in a region G, and let it be regular on the contour $\Gamma = \partial G$. If $|f(z)| < 1$ when $z \in \Gamma$, then the number of solutions of the equation $f(z) = 1$ in G is equal to the number of poles of f in G.

Proof. If the function F is meromorphic in G, and regular and different from zero on the contour $\Gamma = \partial G$, then we have (see formula (5) of 4.2.1)

$$N - P = \frac{1}{2\pi} \Delta_\Gamma \operatorname{Arg} F(z),$$

where N is the number of zeros, and P the number of poles of F in the region G.

We apply this formula to the function $z \mapsto F(z) = f(z) - 1$. As z describes the contour Γ, $f(z) = u + iv$ describes a contour which lies within the region $\{(u, v) \mid u^2 + v^2 < 1\}$, since, by hypothesis, $|f(z)| < 1$ for $z \in \Gamma$. Hence, $f(z) - 1$ describes a contour Γ' which is contained in the region $\{(u, v) \mid (u+1)^2 + v^2 < 1\}$. The origin is outside of this contour, and so $\Delta_{\Gamma'} \operatorname{Arg}(f(z)-1) = 0$. This implies that the number of zeros of the function $z \mapsto f(z) - 1$ is equal to the number of poles of that function in G. However, the functions $z \mapsto f(z)$ and $z \mapsto f(z) - 1$ have the same poles, which proves the above statement.

REFERENCE
Julia, pp. 121–123.

4.7.5. Let g be a meromorphic function in the region G and let it be regular on the contour $\Gamma = \partial G$. If $|g(z)| > 1$ as $z \in \Gamma$, then the equations $g(z) = 0$ and $g(z) = 1$ have equal number of solutions in G.

Proof. Put $f(z) = 1/g(z)$. Then on the contour Γ we have $|f(z)| < 1$, and we can apply the result of 4.7.4. We conclude that the number of solutions of the equation $f(z) = 1$, i.e. the equation $g(z) = 1$ in G is equal to the number of poles of f, i.e. to the number of zeros of g, i.e. to the number of solutions of $g(z) = 0$ in that region.

REFERENCE
Julia, pp. 121–123.

4.7.6. If the function F satisfies Dirichlet's conditions on the segment $[t_0, t_1]$, and if $a > 0, h > 0$, then

$$(1) \quad F(t) = \frac{h}{2(1 + ah)} \int_{t_0}^{t_1} F(x)\,dx + \sum_\nu \frac{2\nu}{2\,a\nu - \sin 2\,a\nu} \int_{t_0}^{t_1} F(x) \cos \nu\,(t - x)\,dx,$$

where the sum is taken over all the positive solutions of the equation in v:

(2)　　　$v \cos(av) + h \sin(av) = 0.$

REMARK 1. The expansion (1) is given in the textbook [1]. It can be obtained by putting

$$p(z) = e^{-az}(z-h), \qquad r(z) = e^{az}(z+h),$$
$$q(z) = z(e^{az} + e^{-az}) + h(e^{az} - e^{-az}) \qquad (a>0, \ h>0).$$

into formula (9) of 4.6.2.

REMARK 2. The expansion (1) is in connection with the following problem of mathematical physics:

Determine the solution of the partial differential equation

(3)　　　$\dfrac{\partial u}{\partial t} = k \dfrac{\partial^2 u}{\partial r^2} + \dfrac{2}{r} \dfrac{\partial u}{\partial r}$

which satisfies the boundary value condition

(4)　　　$\dfrac{\partial u}{\partial r} + hu = 0$

for $r = R$ (=const > 0) and the initial condition

(5)　　　$u(0, r) = F(r),$

where the function F is defined on $[0, R]$.

Solving the problem (3)–(4)–(5), Fourier arrived at the expansion (1), but he did not examine the convergence of the obtained series.

In connection with this problem, see [2], where it is shown, by geometric reasoning, that the Equation (2) has an infinity of solutions.

REFERENCES
1. É. Picard: *Traité d'analyse*, tome II, 2ème édition, Paris 1905, pp. 190–192.
2. D. S. Mitrinović and J. D. Kečkić: *Equations of mathematical physics*, (Serbian), Second edition, Beograd 1978, pp. 176–179.

4.7.7. Denote by $\pi(x)$ the number of prime numbers not greater than x, and let $[x]$ denote the greatest integer not greater than x. If a and b are prime numbers such that $a > b \geq 5$, then

$$\pi(a) - \pi(b-1) = -\frac{1}{2\pi i} \oint_C z\left(\sin \frac{\pi \Gamma(z)}{z}\right)(\sin \pi z)^{-1} dz,$$

where

$$C = \left\{ z \,\middle|\, \left| z - \frac{1}{2}(a+b) \right| = \frac{a-b+1}{2} \right\}.$$

REFERENCE

K. Langmann: 'Eine Formel für die Anzahl der Primzahlen', *Arch. Math. (Basel)* 25 (1974), 40.

4.7.8. Let $F(z, w) = 0$ be an algebraic equation in z and w. If this equation defines an analytic function w depending on z, and also its inverse function, i.e. the analytic function z depending on w, then

$$\Sigma \operatorname{Res} w(z) = \Sigma \operatorname{Res} z(w).$$

EXAMPLE. Suppose that the equation is

$$Azw + Bz + Cw + D = 0 \qquad (A \neq 0, B, C, D \text{ constants}).$$

Then

$$\Sigma \operatorname{Res} w(z) = \Sigma \operatorname{Res} z(w) = \frac{BC - AD}{A^2}$$

REMARK. This result was obtained by Oltramare and was submitted to the Academy in Paris (see [1]). The referees were Cauchy and Sturm, and they recommended the paper for publication in the journal *Recueil des savants étrangers* (see [2]). Oltramare's paper was not, however, published in that journal. It was published 14 years later in Switzerland [3]. In paper [3] Oltramare also evaluated some definite integrals and summed some series by the use of residues (see 5.4.3.23 and 6.7.14).

REFERENCES

1. G. Oltramare: 'Recherches sur le calcul des résidus', *Compt. Rend. Acad. Sci. Paris* 12 (1841), 953–954.
2. C. Sturm, A.-L. Cauchy: 'Rapport sur un Mémoire de M. Oltramare, relatif au calcul des résidus', *Compt. Rend. Acad. Sci. Paris* 13 (1841), 296–298.
3. G. Oltramare: 'Mémoire sur quelques propositions du calcul des résidus', *Mémoires de l'Institut National Genevois* 3, 1855, 15 pp.

4.7.9. Let f and F be regular functions which do not vanish at the origin. Let $a_1, a_2, \ldots$ be the zeros of f, and let $A_1, A_2, \ldots$ be the zeros of F. If

$$\lim_{z \to 0} \frac{f'\left(\frac{t}{z}\right) F'(z)}{f\left(\frac{t}{z}\right) F(z)} = \lim_{z \to \infty} \frac{f'\left(\frac{t}{z}\right) F'(z)}{f\left(\frac{t}{z}\right) F(z)} = 0,$$

then

$$(1) \qquad \prod \frac{f\left(\frac{t}{A_k}\right)}{f(0)} = \prod \frac{F\left(\frac{t}{a_k}\right)}{F(0)},$$

where the product on the left-hand side of (1) is taken over all the zeros of F, and on the right-hand side over all the zeros of f.

Special cases of this result are interesting. For instance, let $f(z) = 1-z$, $F(z) = \sin \pi\sqrt{z}/\pi\sqrt{z}$. We obtain the known formula

$$\sin \pi\sqrt{t} = \pi\sqrt{t} \prod_{k=1}^{+\infty} \left(1 - \frac{t}{k^2}\right).$$

If we take

$$f(z) = \frac{\sin \frac{\pi}{2}\sqrt{z}}{\frac{\pi}{2}\sqrt{z}}, \quad F(z) = \cos \frac{\pi}{2}\sqrt{z},$$

we find

$$\prod_{k=1}^{+\infty} \cos \frac{t}{2k} = \prod_{k=1}^{+\infty} \frac{2k-1}{t} \sin \frac{t}{2k-1}.$$

REMARK. This result, together with the cited examples, is contained in paper [1]. This paper is not signed, and sometimes one finds in literature that the author is anonymous. This paper is, however, published in the second tome of the second series of Cauchy's collected works.

REFERENCE
1. Anonimus: 'Mémoire sur l'application du calcul des résidus à l'évaluation et à la transformation des produits composés d'un nombre infini de facteurs', Bull. Sci. Math. Phys. Chim. Paris 12 (1829), 202–205.

4.7.10. Let $z \in K = \{z \mid |z-a| = r\}$ and let f be an analytic function on that circle. The contour $\Gamma = \{f(z) \mid z \in K\}$ is then closed. A fixed point $z_0 \in K$ and the point $f(z_0) \in \Gamma$ will be called corresponding points. The corresponding elements of the circle K and the contour Γ are the elements of those curves bounded (on both ends) by corresponding points.

With each element of Γ we associate a mass equal to the corresponding element of K, and denote the centroid of that system by z_T. Then

$$z_T = \Sigma \operatorname{Res} \frac{f(z)}{z-a},$$

where the sum is taken over all the singularities of the function $z \mapsto f(z)/(z-a)$ which lie within the circle K. In particular, if f is a regular function, then $z_T = f(a)$.

This result, as well as some more general results, can be found in [1].

REFERENCE
1. E. Amigues: 'Application du calcul des résidus', *Nouv. Ann. Math.* (3) 12 (1893), 142–148.

4.7.11. In paper [1] Schmid represents imaginary points with Cartesian coordinates $(x_1 + ix_2, y_1 + iy_2)$ as pairs of real points (x_1, y_1) and $(x_1 + x_2, y_1 + y_2)$ and claims to have connected this with the calculus of residues in the following way. Let $y(=y_1 + iy_2)$ be an analytic function of the variable x $(=x_1 + ix_2)$ with at most a finite number of singularities in a region G. Furthermore, let $K = (x_1(t), x_2(t))$ be a simple closed oriented curve such that int $K \subset G$ and that y is regular on K. The obvious formula

$$\oint_K y \, dx = 2\pi i \sum_{z \in \operatorname{int} K} \operatorname*{Res}_{x=z} y(x)$$

is proved.

REFERENCE
1. W. Schmid: 'Imaginärgeometrie und die Residuen Cauchys', *Monatsh. Math.* 38 (1931), 167–172.

4.7.12. Cauchy's residue theorem can be used to compute matrix functions. Namely, if f is an analytic function defined on the spectrum of a square matrix A, then (see [1]):

$$(1) \qquad f(A) = \frac{1}{2\pi i} \oint_C f(t)(tI - A)^{-1} \, dt,$$

where I is the unit matrix and C is a closed contour which encircles the spectrum of A.

As an example, we shall determine A^m $(m \in \mathbb{N})$ where

$$A = \begin{Vmatrix} a & b \\ c & d \end{Vmatrix},$$

$a, b, c, d \in \mathbb{C}$. In this case,

$$(tI - A)^{-1} = \frac{1}{P(t)} \begin{Vmatrix} t-d & b \\ c & t-a \end{Vmatrix},$$

where $P(t) = t^2 - (a+d)t + ad - bc = (t-\alpha)(t-\beta)$. Formula (1) with $f(t) = t^m$ becomes

$$A^m = \frac{1}{2\pi i} \oint_C \frac{t^m}{P(t)} \begin{Vmatrix} t-d & b \\ c & t-a \end{Vmatrix} dt$$

$$= \begin{Vmatrix} \dfrac{1}{2\pi i} \oint_C \dfrac{(t-d)t^m}{P(t)} dt & \dfrac{1}{2\pi i} \oint_C \dfrac{bt^m}{P(t)} dt \\[4mm] \dfrac{1}{2\pi i} \oint_C \dfrac{ct^m}{P(t)} dt & \dfrac{1}{2\pi i} \oint_C \dfrac{(t-a)t^m}{P(t)} dt \end{Vmatrix}.$$

Applying Cauchy's theorem on residues we obtain, for $\alpha \neq \beta$:

$$A^m = \frac{1}{\alpha - \beta} \begin{Vmatrix} (\alpha-d)\alpha^m - (\beta-d)\beta^m & b(\alpha^m - \beta^m) \\ c(\alpha^m - \beta^m) & (\alpha-a)\alpha^m - (\beta-a)\beta^m \end{Vmatrix},$$

and for $\alpha = \beta$:

$$A^m = \begin{Vmatrix} (m+1)\alpha^m - md\alpha^{m-1} & mb\alpha^{m-1} \\ mc\alpha^{m-1} & (m+1)\alpha^m - ma\alpha^{m-1} \end{Vmatrix}.$$

Suppose that P is the characteristic polynomial of the $n \times n$ matrix A, and that

$$P(t) = (t-a_1)^{m_1} (t-a_2)^{m_2} \cdots (t-a_r)^{m_r} \qquad (a_i \neq a_j \text{ for } i \neq j)$$

so that $n = m_1 + \cdots + m_r$. In paper [2] Tanaka defined the following

$(n-1)$th degree polynomials $P_{j,\,k}$:

$$P_{j,\,k}(t)=\frac{1}{(m_k-1)!}\binom{m_k-1}{j}\lim_{z\to a_k}\frac{d^{m_k-1-j}}{dz^{m_k-1-j}}\left(\frac{P(z)-P(t)}{z-t}\;\frac{(z-a_k)^{m_k}}{P(z)}\right).$$

Using those polvnomials he was able to obtain explicit formulas for matrix functions. For $f(A)$, instead of (1), he obtained the following expressions:

$$f(A)=\sum_{k=1}^{r}\sum_{j=0}^{m_k-1}f^{(j)}(a_k)P_{j,\,k}(A);$$

$$f(A)=\sum_{k=1}^{r}\left(\sum_{j=0}^{m_k-1}\frac{1}{j!}f^{(j)}(a_k)(A-a_kI)^j\right)P_{0,\,k}(A).$$

We list some special cases, also taken from [2]:

$$e^A=\sum_{k=1}^{r}e^{a_k}\sum_{j=0}^{m_k-1}P_{j,\,k}(A)=\sum_{k=1}^{r}e^{a_k}\left(\sum_{j=0}^{m_k-1}\frac{1}{j!}(A-a_kI)^j\right)P_{0,\,k}(A);$$

$$\sin A=\sum_{k=1}^{r}\sum_{j=0}^{m_k-1}\sin\left(a_k+\frac{\pi j}{2}\right)P_{j,\,k}(A)$$

$$=\sum_{k=1}^{r}\left(\sum_{j=0}^{m_k-1}\frac{1}{j!}\sin\left(a_k+\frac{\pi j}{2}\right)(A-a_kI)^j\right)P_{0,\,k}(A)$$

$$\cos A=\sum_{k=1}^{r}\sum_{j=0}^{m_k-1}\cos\left(a_k+\frac{\pi j}{2}\right)P_{j,\,k}(A)$$

$$=\sum_{k=1}^{r}\left(\sum_{j=0}^{m_k-1}\frac{1}{j!}\cos\left(a_k+\frac{\pi j}{2}\right)(A-a_kI)^j\right)P_{0,\,k}(A).$$

Positive and negative powers of a matrix A, Tanaka expressed as linear combinations of $I, A, \ldots, A^{n-1}$. Namely, if

$$P(t) = (t-a_1)^{m_1} \cdots (t-a_r)^{m_r} = t^n + \alpha_1 t^{n-1} + \cdots + \alpha_n,$$

and if $m \geq n$, then

$$
\begin{aligned}
(2) \quad A^m = x_m A^{n-1} &- x_{m-1}(\alpha_n I + \alpha_{n-1}A + \cdots + \alpha_2 A^{n-2}) \\
&- x_{m-2}(\alpha_n I + \alpha_{n-1}A + \cdots + \alpha_3 A^{n-3})A - \cdots \\
&- x_{m-n+2}(\alpha_n I + \alpha_{n-1}A)A^{n-3} - x_{m-n+1}\alpha_n A^{n-2},
\end{aligned}
$$

where

$$
x_\nu = \frac{1}{2\pi i} \oint_C \frac{z^\nu}{P(z)}\, dz = \sum_{k=1}^{r} \frac{1}{(m_k-1)!} \lim_{z \to a_k} \frac{d^{m_k-1}}{dz^{m_k-1}}\left(\frac{(z-a_k)^{m_k}}{P(z)} z^\nu \right) \quad (\nu \geq 1),
$$

and C is a contour which encircles the spectrum of A. In particular,

$$x_n = -\alpha_1, \; x_{n-1} = 1, \; x_{n-2} = x_{n-3} = \cdots = x_1 = 0.$$

Furthermore, if 0 is not a characteristic value of A (which means that A^{-1} exists), then for $m \geq 1$:

$$
\begin{aligned}
A^{-m} = y_m A^{n-1} &- y_{m+1}(\alpha_n I + \alpha_{n-1}A + \cdots + \alpha_2 A^{n-2}) \\
&- y_{m+2}(\alpha_n I + \alpha_{n-1}A + \cdots + \alpha_3 A^{n-3})A - \cdots \\
&- y_{m+n-2}(\alpha_n I + \alpha_{n-1}A)A^{n-3} - y_{m+n-1}\alpha_n A^{n-2},
\end{aligned}
$$

where

$$
y_\nu = \frac{1}{2\pi i} \oint_C \frac{z^{-\nu}}{P(z)}\, dz = \sum_{k=1}^{r} \frac{1}{(m_k-1)!} \lim_{z \to a_k} \frac{d^{m_k-1}}{dz^{m_k-1}}\left(\frac{(z-a_k)^{m_k}}{P(z)} z^{-\nu} \right)
$$

and C has the same meaning as before.

In particular, if

$$A = \begin{Vmatrix} a & b \\ c & d \end{Vmatrix},$$

then using (2) we have for $m \geq 2$:

1° if $\alpha \neq \beta$, then

$$A^m = \frac{\alpha^m - \beta^m}{\alpha - \beta} A - \alpha\beta \frac{\alpha^{m-1} - \beta^{m-1}}{\alpha - \beta} I;$$

2° if $\alpha = \beta$, then

$$A^m = m\alpha^{m-1} A - (m-1)\alpha^m I.$$

REMARK. All the results in [2] are proved by means of the calculus of residues. They are also valid if P is the minimal polynomial of A.

REFERENCES

1. F. R. Gantmacher: *The Theory of Matrices*, New York 1959.
2. C. Tanaka: 'Some Applications of the Complex Analysis to Matrix Functions', *Mem. School Sci. Engrg. Waseda Univ.* 41 (1977), 103–119.

The first use of the contour integral (1) to represent the matrix function $f(A)$ is found in:

H. Poincaré: 'Sur les groupes continus', *Trans. Cambridge Phil. Soc.* 18 (1899), 200–225.

It should be noted that Gantmacher in the book cited above makes no mention of Poincaré, although his book contains more than 300 references.

Chapter 5

Evaluation of Real Definite Integrals by Means of Residues

5.1. INTRODUCTION

One of the first application of the calculus of residues was the evaluation of real definite integrals. Even to-day almost all text books on Complex Analysis pay special attention to that application of residues. On the other hand, though the method of evaluating real integrals by residues dates back to Cauchy, one can still find contributions to that topic in the contemporary mathematical literature.

This chapter contains classical as well as the latest results on that subject. The classification of material is made according to the limits of the considered integrals. Hence the chapter is divided into sections on integrals of the type

$$\int_{-\infty}^{+\infty} f(x)\,dx, \ \int_{0}^{+\infty} f(x)\,dx, \text{ and } \int_{a}^{b} f(x)\,dx,$$

where $-\infty < a < b < +\infty$. Sections are subdivided into subsections, mainly according to the type of contour used. At the end of each section we give a number of particular integrals which can be evaluated by the exposed methods. Some of them are illustrations of the previous theorems, and some are taken from papers or problems.

At the end of the chapter we give a summary of more important types of integrals considered here.

5.2. INTEGRALS FROM $-\infty$ TO $+\infty$

5.2.1. *Integration Along a Semicircle*

Improper integrals of the form $\int_{-\infty}^{+\infty} F(x)\,dx$, where F satisfies certain conditions, can be successfully evaluated by means of residues. In this section we shall give two rather wide classes of such integrals.

THEOREM 1. *Suppose that the following conditions are satisfied:*

1° *The function f is analytic in the region* $\{z|\ \text{Im}\ z > 0\}$; *its singularities in that region are denoted by* $z_1, \ldots, z_n$;

2° *On the line* $\{z|\ \text{Im}\ z = 0\}$ *the function f may have only simple poles, which are denoted by* $p_1, \ldots, p_m$;

3° $\lim_{z \to \infty} zf(z) = A$ ($A \in \mathbb{C}$).

Then

(1)
$$\text{v.p.} \int_{-\infty}^{+\infty} f(x)\,dx = 2\pi i \sum_{k=1}^{n} \underset{z=z_k}{\text{Res}} f(z) + \pi i \sum_{k=1}^{m} \underset{z=p_k}{\text{Res}} f(z) - \pi i A.$$

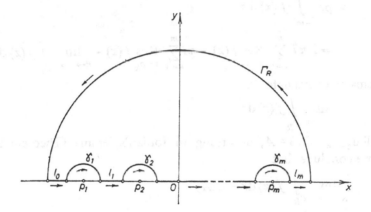

Fig. 5.2.1.

Proof. Suppose that $p_1 < \cdots < p_m$. Let $R > 0$ be sufficiently large so that $R > |p_k|$ ($k = 1, \ldots, m$) and that the singularities $z_1, \ldots, z_n$ belong to the region $\{z|\ \text{Im}\ z > 0, |z| < R\}$. For fixed R, let $r > 0$ be small enough so that the semicircles $\gamma_k = \{z\ |\ |z-p_k| = r, \text{Im}\ z > 0\}$ ($k = 1, \ldots, m$) and $\Gamma_R = \{z\ |\ |z| = R, \text{Im}\ z > 0\}$ are disjoint. We set $l_0 = [-R, p_1 - r]$, $l_k = [p_k + r, p_{k+1} - r]$ ($k = 1, \ldots, m-1$), $l_m = [p_m + r, R]$, and $C = \Gamma_R \cup l_0 \cup \gamma_1 \cup \ldots \cup \gamma_m \cup l_m$.

According to Cauchy's theorem on residues, we have

(2)
$$\oint_C f(z)\,dz = 2\pi i \sum_{k=1}^{n} \underset{z=z_k}{\text{Res}} f(z).$$

However

(3)
$$\oint_C f(z)\,dz = \int_{\Gamma_R} f(z)\,dz + \sum_{k=0}^{m} \int_{l_k} f(z)\,dz + \sum_{k=1}^{m} \int_{\gamma_k} f(z)\,dz.$$

In view of Jordan's lemma (Theorem 1 from 3.1.4) we see that

$$\lim_{r \to 0} \int_{\gamma_k} f(z)\,dz = -\pi i \operatorname*{Res}_{z=p_k} f(z).$$

Hence, if $r \to 0$, from (2) and (3) follows

$$\int_{\Gamma_R} f(z)\,dz + \text{v. p.} \int_{-R}^{R} f(x)\,dx - \pi i \sum_{k=1}^{m} \operatorname*{Res}_{z=p_k} f(z) = 2\pi i \sum_{k=1}^{n} \operatorname*{Res}_{z=z_k} f(z),$$

and letting $R \to +\infty$, we find

(4)
$$\text{v. p.} \int_{-\infty}^{+\infty} f(x)\,dx$$

$$= 2\pi i \sum_{k=1}^{n} \operatorname*{Res}_{z=z_k} f(z) + \pi i \sum_{k=1}^{m} \operatorname*{Res}_{z=p_k} f(z) - \lim_{R \to +\infty} \int_{\Gamma_R} f(z)\,dz.$$

It remains to evaluate

$$\lim_{R \to +\infty} \int_{\Gamma_R} f(z)\,dz.$$

Since $\lim_{z \to \infty} z f(z) = A$, according to Jordan's lemma (Theorem 2 from 3.1.4) we conclude that

$$\lim_{R \to +\infty} \int_{\Gamma_R} f(z)\,dz = \pi i A,$$

and so (4) reduces to (1).

REMARK 1. If P and Q are polynomials such that $\operatorname{dg} Q - \operatorname{dg} P \ge 2$, and if Q has no real zeros then

$$\int_{-\infty}^{+\infty} \frac{P(x)}{Q(x)}\,dx = 2\pi i \sum_{k=1}^{n} \operatorname*{Res}_{z=z_k} \frac{P(z)}{Q(z)},$$

where $z_1, \ldots, z_n$ are the zeros of Q which belong to the upper half-plane. This result can be found, for instance, in book [1], pp. 55–57, or in book [2], pp. 96–97. Notice that the condition $\operatorname{dg} Q - \operatorname{dg} P \ge 2$ ensures the existence of the improper integral $\int_{-\infty}^{+\infty} f(x)\,dx$, and so its principal value need not be taken.

REMARK 2. In [3], pp. 22, the condition 3° of Theorem 1 is replaced by:
There exists a real number $a > 1$, such that $f(z) = 0(|z|^{-a})$ as $z \to \infty$ in the upper half-plane, which again ensures the existence of the improper integral $\int_{-\infty}^{+\infty} f(x)\,dx$.
In the same book, condition 2° is replaced by the more restrictive condition:
The function f is regular on the real axis.

REMARK 3. Instead of the condition 3° of Theorem 1, we may suppose that for sufficiently large $|z|$ there exists $A \in \mathbb{C}$, such that

$$f(z) = \frac{A}{z} + o\left(\frac{1}{z}\right),$$

which implies $\lim_{z \to \infty} zf(z) = A$, i.e. the condition 3°.

EXAMPLE 1. For the function f, defined by

$$f(z) = \frac{z^2 - z + 2}{z^4 + 10z^2 + 9},$$

we have $\lim_{z \to \infty} zf(z) = 0$. Besides, the only singularities of that function in the upper half-plane are $z = i$ and $z = 3i$. Since

$$\operatorname*{Res}_{z=i} f(z) = \frac{-1-i}{16}, \quad \operatorname*{Res}_{z=3i} f(z) = \frac{3-7i}{48},$$

we find

$$\int_{-\infty}^{+\infty} \frac{x^2 - x + 2}{x^4 + 10x^2 + 9} \, dx = \frac{5\pi}{12}.$$

THEOREM 2. *Suppose that the following conditions are satisfied:*
   1° *The function f is analytic in the region $\{z| \operatorname{Im} z > 0\}$; its singularities in that region are denoted by $z_1, \ldots, z_n$;*
   2° *On the line $\{z| \operatorname{Im} z = 0\}$ the function f may have only simple poles $p_1, \ldots, p_m$;*
   3° $\lim_{|z| \to +\infty} f(z) = 0.$
If $a > 0$, then

$$(5) \quad \text{v.p.} \int_{-\infty}^{+\infty} e^{aix} f(x) \, dx = 2\pi i \sum_{k=1}^{n} \operatorname*{Res}_{z=z_k} e^{aiz} f(z) + \pi i \sum_{k=1}^{m} \operatorname*{Res}_{z=p_k} e^{aiz} f(z).$$

   Proof. Define R, r, γ_k, l_k, Γ_R, and C as in the proof of Theorem 1. According to Cauchy's theorem on residues we have

$$(6) \quad \oint_C e^{aiz} f(z) \, dz = 2\pi i \sum_{k=1}^{n} \operatorname*{Res}_{z=z_k} e^{aiz} f(z).$$

However,

$$(7) \quad \oint_C e^{aiz} f(z)\,dz = \int_\Gamma e^{aiz} f(z)\,dz + \sum_{k=0}^{m} \int_{l_k} e^{aix} f(x)\,dx + \sum_{k=1}^{m} \int_{\gamma_k} e^{aiz} f(z)\,dz.$$

Let $r \to 0$. From (6) and (7) follows

$$(8) \quad \int_\Gamma e^{aiz} f(z)\,dz + \text{v. p.} \int_{-R}^{+R} e^{aix} f(x)\,dx + \sum_{k=1}^{m} \lim_{\varepsilon \to 0} \int_{\gamma_k} e^{aiz} f(z)\,dz$$

$$= 2\pi i \sum_{k=1}^{n} \operatorname*{Res}_{z=z_k} e^{aiz} f(z).$$

If we put

$$(9) \qquad r_k = \operatorname*{Res}_{z=p_k} e^{aiz} f(z).$$

then

$$e^{aiz} f(z) = \frac{r_k}{z - p_k} + H_k(z),$$

where H_k is continuous in an ϵ_k-neighbourhood of p_k, and hence it is bounded there, i.e. $|H_k(z)| \le M_k$. Therefore, if $r < \epsilon_k$, we have

$$\left| \int_{\gamma_k} e^{aiz} f(z)\,dz - r_k \int_{\gamma_k} \frac{1}{z - p_k}\,dz \right| = \left| \int_{\gamma_k} H_k(z)\,dz \right| \le \pi \varepsilon M_k.$$

Put $z = p_k + re^{i\theta}$, to obtain

$$\int_{\gamma_k} e^{aiz} f(z)\,dz = r_k \int_\pi^0 \frac{rie^{i\theta}}{re^{i\theta}}\,d\theta + 0(r) = -\pi i r_k + 0(r),$$

and hence

$$(10) \qquad \lim_{r \to 0} \int_{\gamma_k} e^{aiz} f(z)\,dz = -\pi i r_k.$$

In view of (9) and (10), (8) becomes

(11) $\displaystyle\int_{\Gamma} e^{aiz} f(z)\, dz + \text{v.p.} \int_{-R}^{+R} e^{aix} f(x)\, dx$

$$= 2\pi i \sum_{k=1}^{n} \operatorname*{Res}_{z=z_k} e^{aiz} f(z) + \pi i \sum_{k=1}^{m} \operatorname*{Res}_{z=p_k} e^{aiz} f(z).$$

Furthermore, since $a > 0$, according to Jordan's lemma (Theorem 3 from 3.1.4) and the condition $3°$, we have

$$\lim_{R\to+\infty} \int_{\Gamma_R} e^{aiz} f(z)\, dz = 0,$$

and therefore from (11) follows (5).

REMARK 4. A somewhat more special form of Theorem 2 is given in [2] or [3].

REMARK 5. If $a < 0$, then

$$\text{v.p.} \int_{-\infty}^{+\infty} e^{aix} f(x)\, dx = -2\pi i \sum_{k=1}^{s} \operatorname*{Res}_{z=\zeta_k} e^{aiz} f(z) - \pi i \sum_{k=1}^{m} \operatorname*{Res}_{z=p_k} e^{aiz} f(z),$$

where $\zeta_1, \ldots, \zeta_s$ are the singularities of f in the lower half-plane.

REMARK 6. From (5) we immediately obtain

$$\text{v.p.} \int_{-\infty}^{+\infty} f(x) \cos ax\, dx = \operatorname{Re}\left(2\pi i \sum_{k=1}^{n} \operatorname*{Res}_{z=z_k} e^{aiz} f(z) + \pi i \sum_{k=1}^{m} \operatorname*{Res}_{z=p_k} e^{aiz} f(z) \right),$$

$$\text{v.p.} \int_{-\infty}^{+\infty} f(x) \sin ax\, dx = \operatorname{Im}\left(2\pi i \sum_{k=1}^{n} \operatorname*{Res}_{z=z_k} e^{aiz} f(z) + \pi i \sum_{k=1}^{m} \operatorname*{Res}_{z=p_k} e^{aiz} f(z) \right)$$

provided that $f(x) \in \mathbf{R}$.

EXAMPLE 2. If $a > 0, b > 0$, we have

$$\text{v.p.} \int_{-\infty}^{+\infty} \frac{e^{aix}}{x^2 + b^2}\, dx = \frac{\pi}{b} e^{-ab},$$

implying

$$\text{v.p.} \int_{-\infty}^{+\infty} \frac{\cos ax}{x^2 + b^2}\, dx = \frac{\pi}{b}\, e^{-ab}, \qquad \text{v.p.} \int_{-\infty}^{+\infty} \frac{\sin ax}{x^2 + b^2}\, dx = 0.$$

REFERENCES
1. Watson, pp. 55–58.
2. Garnir-Gobert, pp. 99–102.
3. Mitrinović, pp. 36–42.

5.2.2. Integration Along a Rectangle

If the integrand contains a periodic function, or a function which satisfies a condition similar to the condition of periodicity, then it is often useful to choose a rectangle for the contour of integration, so that one of its sides lies on the real axis. We give two classes of such integrals.

THEOREM 1. *Suppose that the following conditions are fulfilled:*
   1° *The functions f and g are analytic in the region* $D = \{z \mid 0 < \text{Im } z < b\}$; *the only singularities of the function* $z \mapsto f(z)g(z)$ *are* $z_1, \dots, z_n$;
   2° *The functions f and g are continuous in* $D \cup \partial D$ *(except for* $z = z_1, \dots, z = z_n$);
   3° $f(z) = f(z + bi)$ $(b \in \mathbf{R}^+)$;
   4° $\lim_{z \to \infty} f(z)g(z) = 0$.
Then

$$(1) \qquad \int_{-\infty}^{+\infty} f(x)\big(g(x) - g(x + bi)\big)\, dx = 2\pi i \sum_{k=1}^{n} \operatorname*{Res}_{z=z_k} f(z)\, g(z).$$

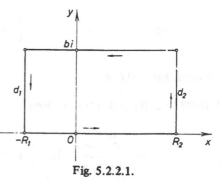

Fig. 5.2.2.1.

Proof. Let $d_1 = \{z \mid \text{Re } z = -R_1,\ 0 \le \text{Im } z \le b\}$, $d_2 = \{z \mid \text{Re } z = R_2,\ 0 \le \text{Im } z \le b\}$, where R_1 and R_2 are sufficiently large positive numbers so that all the singularities of f and g are inside the region $\{z \mid -R_1 < \text{Re } z < R_2\}$. Furthermore, let $\Gamma = d_1 \cup d_2 \cup [-R_1, R_2] \cup \{z \mid -R_1 \le \text{Re } z \le R_2, \text{Im } z = b\}$. According to Cauchy's theorem on residues, we have

(2) $\oint\limits_{\Gamma} f(z)\,g(z)\,dz = 2\pi i \sum\limits_{k=1}^{n} \operatorname*{Res}_{z=z_k} f(z)\,g(z).$

However,

(3) $\oint\limits_{\Gamma} f(z)\,g(z)\,dz = \int\limits_{-R_1}^{R_2} f(x)\,g(x)\,dx + \int\limits_{d_2} f(z)\,g(z)\,dz$

$\qquad\qquad - \int\limits_{-R_1}^{R_2} f(x+bi)\,g(x+bi)\,dx + \int\limits_{d_1} f(z)\,g(z)\,dz$

$\qquad = \int\limits_{-R_1}^{R_2} f(x)\,\big(g(x)-g(x+bi)\big)\,dx + \int\limits_{d_2} f(z)\,g(z)\,dz + \int\limits_{d_1} f(z)\,g(z)\,dz.$

In view of the condition $4°$, for arbitrary $\epsilon > 0$, there exists a positive number N, such that $|f(z)g(z)| < \epsilon$ for $|z| \geq N$. This implies

$\left| \int\limits_{d_2} f(z)\,g(z)\,dz \right| < b\epsilon, \qquad \text{i.e.} \qquad \lim\limits_{R_2 \to +\infty} \int\limits_{d_2} f(z)\,g(z)\,dz = 0.$

Similarly,

$\lim\limits_{R_1 \to +\infty} \int\limits_{d_1} f(z)\,g(z)\,dz = 0.$

Therefore, if $R_1 \to +\infty$, and $R_2 \to +\infty$, from (2) and (3) follows (1).

REMARK 1. A version of this method is given in book [1].

REMARK 2. The integral

(4) $\int\limits_{-\infty}^{+\infty} f(e^x)\,(g\,(x)-g\,(x+2\pi i))\,dx$

was considered in [2]. Since $f(e^z) = f(e^{z+2\pi i})$, (4) is a special case of the integral from Theorem 1.

EXAMPLE 1. Let

$I = \int\limits_{-\infty}^{+\infty} \frac{e^x}{e^{2x}+e^{2a}} \frac{1}{x^2+\pi^2}\,dx.$

Since

$\frac{1}{z^2+\pi^2} = \frac{i}{2\pi\,(z+\pi i)} - \frac{i}{2\pi\,(z-\pi i)},$

we set

$$f(x) = \frac{e^x}{e^{2x} + e^{2a}}, \qquad g(x) = \frac{-i}{2\pi(x - \pi i)}.$$

Applying formula (1), we obtain

$$I = \frac{2\pi e^{-a}}{4a^2 + \pi^2} \cdot \frac{1}{1 + e^{2a}}.$$

THEOREM 2. *Let f satisfy the following conditions:*

1° *f is analytic in the region* $D = \{z \mid 0 \le \operatorname{Im} z \le a\}$ (a *is a positive number*). *The only singularities of f inside D are* $z_1, \ldots, z_n$;

2° *On the real axis* $\{z \mid \operatorname{Im} z = 0\}$ *the function f can have only simple poles* $p_1, \ldots, p_m$;

3° $f(z) = c + o(1)$, *as* $|z| \to +\infty$ *in* D;

4° $f(z + ai) = bf(z)$ \qquad $(b \ne 1)$.

Then

$$(1) \qquad \text{v.p.} \int_{-\infty}^{+\infty} f(x)\,dx = \frac{\pi i}{1 - b}\left(2\sum_{k=1}^{n} \operatorname*{Res}_{z = z_k} f(z) + (1 + b)\sum_{k=1}^{m} \operatorname*{Res}_{z = p_k} f(z)\right).$$

The above theorem can be proved by a direct integration of the function f along the closed contour shown on Figure 5.2.2.2.

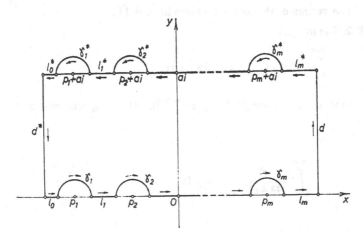

Fig. 5.2.2.2.

EXAMPLE 2. The function $z \mapsto f(z) = e^{tz}/\mathrm{ch}\, z$ where $|\operatorname{Re} t| < 1$, satisfies all the conditions of Theorem 2 with $a = \pi$, $b = -e^{\pi ti}$, $c = 0$. Hence

$$\int_{-\infty}^{+\infty} \frac{e^{tx}}{\mathrm{ch}\, x} \, dx = \frac{2\pi i}{1 + e^{\pi ti}} \, e^{\pi ti/2} = \frac{\pi}{\cos \dfrac{\pi t}{2}},$$

implying, for example,

$$\int_{-\infty}^{+\infty} \frac{\mathrm{ch}\, tx}{\mathrm{ch}\, x} \, dx = \frac{\pi}{\cos \dfrac{\pi t}{2}}.$$

REFERENCES

1. M. A. Evgrafov: *Analytic functions* (Russian), Moscow 1968, p. 249.
2. Mitrinović, pp. 53–61.

5.2.3. *Miscellaneous Integrals*

5.2.3.1 Let f be a regular function in the half-plane $\{z|\operatorname{Im} z \geq 0\}$, where the following inequality also holds:

$$|f(z)| < Ce^{|z|^k} \qquad (C = \mathrm{const};\ k < 1).$$

If the integrals $\int_{-\infty}^{0} f(x)\, dx$ and $\int_{0}^{+\infty} f(x)\, dx$ $(x \in \mathbf{R})$ exist, then $\int_{-\infty}^{+\infty} f(x)\, dx = 0$.

REFERENCE

G. Hoheisel: 'Auswertung bestimmter Integrale', *J. Reine Angew. Math.* **156** (1926), 67–68.

5.2.3.2. If $a_1, \ldots, a_n$ are real, and $c_1, \ldots, c_n, p, q$ positive numbers, then

$$\int_{-\infty}^{+\infty} \frac{\cos}{\sin} \left(px - \frac{q}{x} - \sum_{\nu=1}^{n} \frac{c_\nu}{x - a_\nu} \right) \frac{1}{1 + x^2} \, dx$$

$$= \pi \exp\left(-p - q - \sum_{\nu=1}^{n} \frac{c_\nu}{1 + a_\nu^2} \right) \frac{\cos}{\sin} \sum_{\nu=1}^{n} \frac{a_\nu c_\nu}{1 + a_\nu^2} \, ;$$

$$\mathrm{v.p.} \int_{-\infty}^{+\infty} \frac{\cos}{\sin} \left(px - \frac{q}{x} - \sum_{\nu=1}^{n} \frac{c_\nu}{x - a_\nu} \right) \frac{1}{1 - x^2} \, dx$$

$$= \pi \sin\left(p - q - \sum_{\nu=1}^{n} \frac{c_\nu}{1 - a_\nu^2} \right) \frac{\cos}{\sin} \sum_{\nu=1}^{n} \frac{a_\nu c_\nu}{1 - a_\nu^2}$$

REFERENCE
Hardy.

REMARK. In the quoted paper Hardy first gives a usual proof of some standard theorems implicitly stated by Cauchy and then derives a number of interesting special cases. Hence, Hardy's paper is quoted several times in this book.

5.2.3.3. If $a, s \in \mathbf{R}^+$ and $r \in \mathbf{R}$, then

$$1° \int_{-\infty}^{+\infty} \frac{1}{(1+x^2)(1-iax)^s} \, dx = \frac{\pi}{(1+a)^s} \; ; \quad 2° \int_{-\infty}^{+\infty} \frac{\cos(s \arctan ax)}{(1+x^2)(1+a^2 x^2)^{s/2}} \, dx = \frac{\pi}{(1+a)^s} \; ;$$

$$3° \int_{-\infty}^{+\infty} \frac{e^{r \arctan ax} + e^{-r \arctan ax}}{1+x^2} \cos\left(\frac{r}{2}\log(1+a^2 x^2)\right) dx = 2\pi \cos\left(r \log(1+a)\right);$$

$$4° \int_{-\infty}^{+\infty} \frac{e^{r \arctan ax} + e^{-r \arctan ax}}{1+x^2} \sin\left(\frac{r}{2}\log(1+a^2 x^2)\right) dx = 2\pi \sin\left(r \log(1+a)\right);$$

$$5° \int_{-\infty}^{+\infty} \frac{\cos(s \arctan ax)}{(1+x^2)(1+a^2 x^2)^{s/2}} \log(1+a^2 x^2) \, dx = \frac{2\pi}{(1+a)^s} \log(1+a).$$

REFERENCE
A.-L. Cauchy: 'Sur les intégrales multiples', *Compt. Rend. Acad. Sci. Paris* 11 (1840), 1008–1019. ≡ *Oeuvres* (1) 6, 5–16.

5.2.3.4. Let F be a polynomial, trigonometric or exponential function of x, and let hyp denote the hyperbolic sine or cosine. Glasser [1] gave a method for evaluating integrals of the form

$$(1) \qquad \int_{-\infty}^{+\infty} \frac{F(x)}{\prod_{\nu=1}^{N} \text{hyp}(n_\nu x + a_\nu)} \, dx,$$

where $n_1, \ldots, n_N \in \mathbf{N}$ and $a_1, \ldots, a_N \in \mathbf{R}$.

The method will be illustrated by an example. Consider the integral $I = \int_{-\infty}^{+\infty} f(x) \, dx$, where

$$f(x) = \frac{e^{-px}}{\text{ch } x \, \text{ch}(x+a) \, \text{ch}(x+b)} \qquad (a \neq b; \; a, b \neq 0, \; |p| < 3).$$

The integrand has simple poles at $x = x_1, x_2, x_3$, where

$$x_1 = \frac{1}{2}\pi i, \quad x_2 = \frac{1}{2}\pi i - a, \quad x_3 = \frac{1}{2}\pi i - b.$$

The corresponding residues are:

$$\operatorname*{Res}_{x=x_1} f(x) = \frac{ie^{-i\pi p/2}}{\operatorname{sh}a\,\operatorname{sh}b}, \quad \operatorname*{Res}_{x=x_2} f(x) = \frac{ie^{-i\pi p/2}\,e^{ap}}{\operatorname{sh}a\,\operatorname{sh}(a-b)}, \quad \operatorname*{Res}_{x=x_3} f(x) = \frac{ie^{-i\pi p/2}\,e^{bp}}{\operatorname{sh}b\,\operatorname{sh}(b-a)}.$$

By integrating around the contour extending from $-\infty$ to $+\infty$ along the real axis and $+\infty$ to $-\infty$ along the line $\{x\mid \operatorname{Im} x = \pi i\}$, since $\operatorname{hyp}(u + \pi i) = -\operatorname{hyp} u$, we see that

$$(1 + e^{-\pi pi})\, I = 2\pi i \sum_{\nu=1}^{3} \operatorname*{Res}_{x=x_\nu} f(x),$$

implying

$$\int_{-\infty}^{+\infty} \frac{e^{-px}}{\operatorname{ch}x\,\operatorname{ch}(x+a)\,\operatorname{ch}(x+b)}\, dx = \frac{\pi}{\cos\dfrac{\pi p}{2}}\left(\frac{e^{ap}}{\operatorname{sh}a\,\operatorname{sh}(b-a)} + \frac{e^{bp}}{\operatorname{sh}b\,\operatorname{sh}(a-b)} - \frac{1}{\operatorname{sh}a\,\operatorname{sh}b}\right).$$

REMARK. Integrals of the form (1) with $N = 2$ arise in transport theory, and isolated examples have been evaluated by a variety of techniques in [2].

REFERENCES
1. M. L. Glasser: 'Evaluation of a Class of Definite Integrals', *Univ. Beograd. Publ. Elektrotehn. Fak. Ser. Mat. Fiz.* 498—541 (1975), 49—50.
2. M. L. Glasser and V. E. Wood: 'Evaluation of Some Transport Integrals. II', *J. Appl. Phys.* 37 (1966), 4364—4366.

5.2.3.5. Let p and q be real numbers such that $2q > 2p + 1$. Then

$$\text{v.p.} \int_{-\infty}^{+\infty} \frac{x^{2p}}{1-x^{2q}}\, dx = \frac{\pi}{2q}\cot g\frac{\pi}{2q}(2p+1).$$

REFERENCE
E. Schmid: *Die Cauchy'sche Methode der Auswertung bestimmter Integrale zwischen reellen Grenzen*, Stuttgart 1903.

5.2.3.6. For the functions f and g we define their convolution as the integral

$$f(x) * g(x) = \int_{-\infty}^{+\infty} f(x-t)\, g(t)\, dt,$$

where the functions f and g must satisfy certain conditions which ensure the convergence of the considered integral.

The following results are proved in [1] :

1° If $a \notin \mathbf{R}$ and $b \notin \mathbf{R}$, then

$$\frac{m!}{(x-a)^{m+1}} * \frac{n!}{(x-b)^{n+1}} = 2\pi i \varepsilon \frac{(m+n)!}{(x-a-b)^{m+n+1}},$$

where $m, n \in \mathbf{N_0}$ and $\varepsilon = 1, -1$ or 0, depending on the sign of Im a and Im b.

2° If $a \in \mathbf{R}^+$ and $b \in \mathbf{R}^+$, then

(1) $\dfrac{a}{\pi} \dfrac{1}{x^2+a^2} * \dfrac{b}{\pi} \dfrac{1}{x^2+b^2} = \dfrac{a+b}{\pi} \dfrac{1}{x^2+(a+b)^2}.$

The convolution (1) arises in probability theory.

REFERENCE
1. L. Schwartz: *Analyse mathématique, cours proféssé à l'École Polytechnique*, Paris 1967.

5.2.3.7. In note [1] Boas and Friedman noticed that the semi-circular contour normally used for the evaluation of integrals of the form $\int_{-\infty}^{+\infty} e^{ix} f(x) \, dx$ can be replaced by a simpler triangular contour. We give an example from [1].

Suppose that f is a rational function, regular on the real axis, and such that $\lim_{z \to \infty} f(z) = 0$. Integrate the function $z \mapsto e^{iz} f(z)$ around the triangle T with vertices $-R_1, R_2, (R_1 + R_2)i$, where $R_1, R_2 \in \mathbf{R}^+$ are sufficiently large so that T contains all the singularities of f in the upper half-plane. Let L_1 be the segment joining $-R_1$ and $(R_1 + R_2)i$, and let L_2 be the segment joining R_2 to $(R_1 + R_2)i$. By hypothesis, $|f(z)| < K|z|^{-1} < \sqrt{2} R_2^{-1}$, when $z \in L_2$, where K is a constant. Besides, when $z \in L_2$ we have $|dz| = (1 + R_2^2/(R_1 + R_2)^2)^{1/2} \, dy \leq \sqrt{2} \, dy$, and hence

(1) $\left| \int\limits_{L_2} f(z) e^{iz} \, dz \right| < M_2 R_2^{-1} \int\limits_{0}^{R_1+R_2} e^{-y} \, dy \leq M_2 R_2^{-1}$ $(M_2 = \text{const}).$

Similarly, we prove

(2) $\left| \int\limits_{L_1} f(z) e^{iz} \, dz \right| \leq M_1 R_1^{-1}$ $(M_1 = \text{const}).$

If $R_1, R_2 \to + \infty$, we get

$$(3) \qquad \int\limits_{-\infty}^{+\infty} f(x) e^{ix} \, dx = 2 \pi i \, \Sigma \, \text{Res} \, f(z) \, e^{iz},$$

where the summation is taken over all the poles of f in the upper half-plane.

By using this method, formula (3) and inequalities (1) and (2) are derived in a simpler way than by using the standard integration along a semicircle.

REFERENCE

1. H. P. Boas and E. Friedman: 'A Simplification in Certain Contour Integrals', *Am. Math. Monthly* **84** (1977), 467–468.

5.2.3.8. Consider the integral

$$(1) \qquad J_n = \frac{1}{2\pi} \int\limits_{-\infty}^{+\infty} \frac{g_n(i\omega)}{h_n(i\omega) h_n(-i\omega)} \, d\omega,$$

where $\omega \in \mathbf{R}$ and g_n and h_n are real polynomials defined by

$$(2) \qquad g_n(t) = b_0 t^{2n-2} + b_1 t^{2n-4} + \cdots + b_{n-1};$$

$$h_n(t) = a_0 t^n + a_1 t^{n-1} + \cdots + a_n,$$

so that the polynomial h_n has only simple zeros which all lie in the left half-plane.

If we put $i\omega = y$ into (1), we get

$$J_n = \frac{1}{2\pi i} \int\limits_{-i\infty}^{+i\infty} \frac{g_n(y)}{h_n(y) h_n(-y)} \, dy.$$

Let Γ be the closed contour consisting of the semicircle $C_R = \{z \mid |z| = R, \text{Re } z \le 0\}$ and the line segment joining the points iR and $-iR$ on the imaginary axis, where $R > 0$ is large enough. Then using the residue theorem, we find

$$(3) \quad \frac{1}{2\pi i} \int_{\Gamma} \frac{g_n(z)}{h_n(z)h_n(-z)} dz = \frac{1}{2\pi i} \int_{-iR}^{iR} \frac{g_n(y)}{h_n(y)h_n(-y)} dy + \frac{1}{2\pi i} \int_{C_R} \frac{g_n(z) \, dz}{h_n(z)h_n(-z)}$$

$$= \sum_{k=1}^{n} \operatorname*{Res}_{z=y_k} \frac{g_n(z)}{h_n(z)h_n(-z)},$$

where $y_1, \ldots, y_n$ are the zeros of the polynomial h_n. If $R \to +\infty$, according to Jordan's lemma (Theorem 2 from 3.1.4) the integral along C_R tends to zero, and from (3) follows

$$\frac{1}{2\pi i} \int_{-i\infty}^{+i\infty} \frac{g_n(y)}{h_n(y)h_n(-y)} dy = \sum_{k=1}^{n} \operatorname*{Res}_{z=y_k} \frac{g_n(z)}{h_n(z)h_n(-z)}.$$

Since $y_1, \ldots, y_n$ are the zeros of h_n, we have

$$h_n(y) = a_0 \prod_{k=1}^{n} (y-y_k), \qquad h_n(-y) = a_0 \prod_{k=1}^{n} (-y-y_k),$$

and the rational function

$$y \mapsto \frac{g_n(y)}{h_n(y)h_n(-y)}$$

has the partial fraction representation

$$(4) \quad \frac{g_n(y)}{h_n(y)h_n(-y)} = \sum_{k=1}^{n} A_k \left(\frac{1}{y-y_k} - \frac{1}{y+y_k} \right),$$

wherefrom it follows that

$$A_k = \operatorname*{Res}_{z=y_k} \frac{g_n(z)}{h_n(z)h_n(-z)}.$$

and so

$$J_n = \sum_{k=1}^{n} A_k.$$

In order to calculate the coefficients A_k, we write (4) in the form

$$(5) \qquad g_n(y) = \sum_{k=1}^{n} A_k \left(\frac{h_n(y)}{y-y_k} h_n(-y) + \frac{h_n(-y)}{-y-y_k} h_n(y) \right) .$$

If we let

$$\frac{h_n(y)}{y-y_k} = \sum_{r=1}^{n} B_{rk} y^{n-r}, \qquad \frac{h_n(-y)}{-y-y_k} = \sum_{r=1}^{n} (-1)^{n-r} B_{rk} y^{n-r},$$

then

$$(6) \qquad \frac{h_n(y)}{y-y_k} h_n(-y) + \frac{h_n(-y)}{-y-y_k} h_n(y) =$$

$$= \sum_{s=0}^{n} \sum_{r=1}^{n} (-1)^{n-s} a_s B_{rk} y^{2n-s-r} + \sum_{s=0}^{n} \sum_{r=1}^{n} a_s B_{rk} (-1)^{n-r} y^{2n-s-r}$$

$$= 2 \sum_{m=1}^{n} \sum_{r=1}^{2m} a_{2m-r} B_{rk} (-1)^{n+r} y^{2n-2m},$$

where $a_{2m-r} = 0$ if $2m-r > n$ or $2m-r < 0$.

On the basis of (2) and (6), we see that (5) can be written as

$$g_n(y) = \sum_{m=1}^{n} b_{m-1} y^{2n-2m}$$

$$= \sum_{k=1}^{n} A_k \sum_{m=1}^{n} \sum_{r=1}^{2m} 2(-1)^{n+r} a_{2m-r} B_{rk} y^{2n-2m}$$

$$= 2(-1)^n \sum_{m=1}^{n} y^{2n-2m} \sum_{k=1}^{n} A_k \sum_{r=1}^{2m} (-1)^r a_{2m-r} B_{rk},$$

and equating the corresponding coefficients we find

$$\sum_{k=1}^{n} A_k \sum_{r=1}^{2m} a_{2m-r}(-1)^r B_{rk} = (-1)^n \frac{b_{m-1}}{2} \qquad (m=1,\ldots,n),$$

or

(7) $$\sum_{r=1}^{2m} a_{2m-r}(-1)^r \sum_{k=1}^{n} A_k B_{rk} = (-1)^n \frac{b_{m-1}}{2} \qquad (m=1,\ldots,n).$$

Put

(8) $$c_r = \frac{(-1)^{r+1}}{a_0} \sum_{k=1}^{n} A_k B_{rk} \qquad (r=1,\ldots,n).$$

For $r = 1$, we have $B_{1k} = a_0$ for all k, and so

$$c_1 = \sum_{k=1}^{n} A_k = J_n.$$

Using the notation (8), the equations (7) take the form

$$\sum_{r=1}^{2m} a_{2m-r} c_r = (-1)^{n+1} \frac{b_{m-1}}{2a_0} \qquad (m=1,\ldots,n)$$

and this is a linear system in $c_1, \ldots, c_n$. Hence, by Cramer's rule we finally obtain

(9) $$J_n = (-1)^{n+1} \frac{N_n}{2a_0 D_n},$$

where

$$D_n = \begin{vmatrix} a_1 & a_0 & \cdots & 0 \\ a_3 & a_2 & & 0 \\ \cdot & & & \\ \cdot & & & \\ \cdot & & & \\ 0 & 0 & & a_n \end{vmatrix}$$

is the determinant of the system, and N_n is the determinant obtained from D_n by replacing the first column by $b_0, b_1, \ldots, b_{n-1}$.

REMARK 1. The determinant D_n is the leading Hurwitz determinant of the polynomial h_n. Since all the zeros of h_n lie in the left half-plane, we have $D_n \neq 0$, and the integral J_n always exists.

REMARK 2. Formula (9) remains valid also in the case when h_n has multiple zeros.

REMARK 3. Integrals of the form (1) are encountered in evaluating the least-mean-squares error of control systems subjected to random disturbances.

REFERENCE

V. A. Ivanov, V. S. Medvedev, B. K. Čemodanov, and A. C. Juščenko: *Mathematical Foundations of the Theory of Authomatic Regulation*, (Russian), tome 1, Moscow 1977.

5.3. INTEGRALS FROM 0 TO $+\infty$

5.3.1. *Integration Around a Circle*

In this section we first give two important classes of improper integrals. In both cases we integrate an analytic branch of a multiform function, and hence it is necessary to make the appropriate cuts in the complex plane. Finally, we give some generalizations of the mentioned results.

THEOREM 1. *Suppose that the following conditions are fulfilled:*

1° *The function f is analytic in the complex plane; its singularities are denoted by* $z_1, \ldots, z_n$;

2° *On the half-line* $\{z \mid \operatorname{Im} z = 0, \operatorname{Re} z \geq 0\}$ *the function f is regular;*

3° $\lim_{z \to 0} z^p f(z) = \lim_{z \to \infty} z^p f(z) = 0 \qquad (p > 1)$.

Then

$$(1) \qquad \int_0^{+\infty} f(x)\, dx = -\sum_{k=1}^{n} \operatorname*{Res}_{z=z_k} f(z) \log(-z),$$

where $\log z$ *denotes the principal value of the logarithm, i.e.* $-\pi < \operatorname{Im}(\log z) < \pi$.

Proof. Let

$$\Gamma = \{z \mid |z| = R\}, \quad \gamma = \{z \mid |z| = r\}, \quad d = \{z \mid \operatorname{Im} z = 0, \ r \leq \operatorname{Re} z \leq R\},$$

where $R > 0$ is sufficiently large so that the circle Γ contains all the singularities of f. Let $C = \gamma \cup d \cup (-d) \cup \Gamma$ (see Figure 5.3.1.1). According to

Cauchy's residue theorem we have

$$(2) \qquad \oint_C f(z) \log(-z)\,dz = 2\pi i \sum_{k=1}^{n} \operatorname*{Res}_{z=z_k} f(z) \log(-z).$$

However,

$$\oint_C f(z) \log(-z)\,dz = \int_\Gamma f(z) \log(-z)\,dz + \int_R^r (\log x + i\pi) f(x)\,dx$$

$$+ \int_\gamma f(z) \log(-z)\,dz + \int_r^R (\log x - i\pi) f(x)\,dx,$$

since the integrand on the upper side of the cut is $(\log x - i\pi)f(x)$, and on the lower $(\log x + i\pi)f(x)$.

Therefore.

$$(3) \qquad \oint_C f(z) \log(-z)\,dz = \int_\Gamma f(z) \log(-z)\,dz$$

$$+ \int_\gamma f(z) \log(-z)\,dz - 2\pi i \int_r^R f(x)\,dx.$$

Having in mind the condition $3°$ of the theorem, we see that

$$\lim_{R\to+\infty} \int_\Gamma f(z) \log(-z)\,dz = \lim_{r\to 0} \int_\gamma f(z) \log(-z)\,dz = 0.$$

and hence, if $r\to 0$ and $R\to+\infty$, from (2) and (3) follows

$$-2\pi i \int_0^{+\infty} f(x)\,dx = 2\pi i \sum_{k=1}^{n} \operatorname*{Res}_{z=z_k} f(z) \log(-z),$$

implying (1).

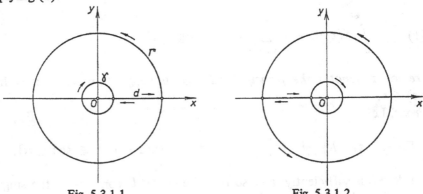

Fig. 5.3.1.1.   Fig. 5.3.1.2.

REMARK 1. The integral $\int_0^{+\infty} f(x)\,dx$ can also be evaluated by integrating the function $z \mapsto f(-z)\log z$ around the contour shown on Figure 5.3.1.2. In that case the condition $2°$ of Theorem 1 must be replaced by the condition:

$2^{i°}$ The function f is regular on the half-line $\{z \mid \operatorname{Im} z = 0, \operatorname{Re} z \le 0\}$,

and we obtain the formula

$$(4) \qquad \int_0^{+\infty} f(x)\,dx = \sum_{k=1}^{n} \operatorname*{Res}_{z=z_k} f(-z)\log z.$$

REMARK 2. Formulas (1) and (4) are standard and can be found in the given, or in a somewhat more special form in most text books. For example, in [1], p. 58, the following result is given:

If P and Q are polynomials in x such that $Q(x) \ne 0$ for $x \ge 0$, and the degree of P is lower than that of Q by at least 2, then

$$\int_0^{+\infty} \frac{P(x)}{Q(x)}\,dx = \sum_{k=1}^{n} \operatorname*{Res}_{z=z_k} \log(-z)\frac{P(z)}{Q(z)},$$

where $z_1, \ldots, z_n$ are the zeros of Q and $-\pi < \operatorname{Im}\log(-z) < \pi$.

The following result can be found in [2], p. 74:

If P and Q are polynomials of degree m and n respectively, with $m < n-2$, and if Q does not vanish for positive or zero values of x, then

$$(5) \qquad \int_0^{+\infty} \frac{P(x)}{Q(x)}\,dx = - \sum_{k=1}^{s} \operatorname*{Res}_{z=z_k} \frac{P(z)}{Q(z)}\log z,$$

where $z_1, \ldots, z_n$ are the zeros of Q and $0 < \arg z < 2\pi$.

The same result is given in [3], p. 208, and, as a special case of a more general result, in [4], pp. 102–104.

REMARK 3. The integral $\int_0^{+\infty} f(x)\,dx$ can be evaluated under weaker hypotheses for f (see Theorem 4).

EXAMPLE 1. Consider the integral $\int_0^{+\infty} (x/(1+x^6))\,dx$. The function f, defined by $f(z) = z/(1+z^6)$, has singularities at $z = z_k = \exp[(2k-1)\pi i/6]$ for $k = 1, \ldots, 6$. Since

$$\operatorname*{Res}_{z=z_k} \frac{z\log z}{1+z^6} = \frac{(2k-1)\pi i}{36}\exp\left(-2(2k-1)\frac{\pi i}{3}\right),$$

using formula (5), we obtain

$$\int_0^{+\infty} \frac{x}{1+x^6}\,dx = -\frac{\pi i}{36}\sum_{k=1}^{6}(2k-1)\exp\frac{-(4k-2)\pi i}{3} = \frac{\pi\sqrt{3}}{9}.$$

THEOREM 2. *Suppose that the following conditions are satisfied:*
   1° *The function f is analytic in the complex plane, where it can have a finite number of singularities* $z_1, \ldots, z_n$;
   2° *On the nonnegative real axis the function f is regular;*
   3° $\lim_{z \to 0} z^{a+1} f(z) = \lim_{z \to \infty} z^{a+1} f(z) = 0$, *where* $a \neq 0$ *and* $|\operatorname{Re} a| < 1$.
Then

$$(6) \qquad \int\limits_0^{+\infty} x^a f(x)\,dx = \frac{2\pi i}{1 - e^{2\pi i a}} \sum_{k=1}^{n} \operatorname*{Res}_{z=z_k} z^a f(z).$$

Proof. Let Γ, γ, d, and C be defined as in the proof of Theorem 1. According to Cauchy's residue theorem we have

$$(7) \qquad \oint\limits_C z^a f(z)\,dz = 2\pi i \sum_{k=1}^{n} \operatorname*{Res}_{z=z_k} z^a f(z).$$

However,

$$(8) \qquad \oint\limits_C z^a f(z)\,dz = \int\limits_\Gamma z^a f(z)\,dz + \int\limits_d x^a f(x)\,dx$$
$$- e^{2\pi i a} \int\limits_d x^a f(x)\,dx + \int\limits_\gamma z^a f(z)\,dz.$$

Since

$$\lim_{z \to 0} z^{a+1} f(z) = \lim_{z \to \infty} z^{a+1} f(z) = 0,$$

in view of Jordan's lemmas (Theorems 1 and 2 from 3.1.4) we conclude that

$$\lim_{r \to 0} \int\limits_\gamma z^a f(z)\,dz = 0, \qquad \lim_{R \to +\infty} \int\limits_\Gamma z^a f(z)\,dz = 0.$$

Hence, if $r \to 0$ and $R \to +\infty$, from (7) and (8) follows

$$(1 - e^{2\pi i a}) \int\limits_0^{+\infty} x^a f(x)\,dx = 2\pi i \sum_{k=1}^{n} \operatorname*{Res}_{z=z_k} z^a f(z),$$

implying (6).

REMARK 1. Integrals of the form

$$(9) \qquad \int\limits_{-\infty}^{+\infty} e^{ax} f(e^{bx})\,dx,$$

where $0 < \mathrm{Re}\, a < b$ and $\lim_{x \to \pm \infty} x^2 e^{ax} f(e^{bx}) = 0$, and where f is regular on the nonnegative real axis, can be reduced, by means of the transformation $e^{bx} = t$ to the form

$$\frac{1}{b} \int\limits_0^{+\infty} t^{\frac{a}{b}-1} f(t)\, dt.$$

See [4], p. 109.

REMARK 2. Integrals of the form

$$\int\limits_0^{+\infty} x^a f(x) \log^m x \, dx,$$

where

$$m \in \mathbb{N},\ a \in \mathbb{R},\ \lim_{x \to 0} x^{a+1} f(x) = \lim_{x \to +\infty} x^{a+1} f(x) = 0,$$

can be reduced to the form $\int x^a f(x)\, dx$, since

$$\int\limits_0^{+\infty} x^a f(x) \log^m x \, dx = \frac{\partial^m}{\partial a^m} \int\limits_0^{+\infty} x^a f(x)\, dx.$$

Differentiation under the integral sign can easily be justified. See [4], p. 110.

REMARK 3. Integrals of the form

$$\int\limits_{-\infty}^{+\infty} x^n e^{ax} f(e^{bx})\, dx,$$

where $0 < \mathrm{Re}\, a < b, n \in \mathbb{N}, \lim_{x \to \pm \infty} f(x) = 0$, can be reduced to the form (9), since

$$\int\limits_{-\infty}^{+\infty} e^{ax} f(e^{bx})\, dx = \frac{\partial^n}{\partial a^n} \int\limits_{-\infty}^{+\infty} x^n e^{ax} f(e^{bx})\, dx.$$

Again, it is not difficult to justify differentiation under the integral sign. See [4], p. 110.

REMARK 4. The integral $\int_0^{+\infty} x^a f(x)\, dx$ can be evaluated under weaker hypotheses for the function f (see Theorem 5).

EXAMPLE 2. Start with the function f, defined by $f(z) = z^a/[(z+c)^2]$. Applying the formula (6) we get

$$I = \int\limits_0^{+\infty} \frac{x^a}{(x+c)^2}\, dx = \frac{2\pi i}{1-e^{2\pi ai}}\, \frac{d}{dz}(z^a)\Big|_{z=-c} = c^{a-1}\frac{\pi a}{\sin \pi a}.$$

Some special cases are listed below:

$$a = 0: \qquad \int\limits_0^{+\infty} \frac{1}{(x+c)^2}\, dx = \frac{1}{c};$$

$$a = -\frac{1}{2}: \qquad \int\limits_0^{+\infty} \frac{1}{\sqrt{x}\,(x+c)^2}\, dx = \frac{\pi}{2c^{3/2}};$$

$$a = \frac{1}{2}: \qquad \int\limits_0^{+\infty} \frac{\sqrt{x}}{(x+c)^2}\, dx = \frac{\pi}{2\sqrt{c}}.$$

If I is differentiated n times with respect to a, we obtain the integral

$$\int\limits_0^{+\infty} \frac{x^a \log^n x}{(x+c)^2}\, dx.$$

So, for example, we have

$$\int\limits_0^{+\infty} \frac{x^a \log x}{(x+c)^2}\, dx = \frac{\pi}{c}\, \frac{\partial}{\partial a}\left(\frac{ac^a}{\sin \pi a}\right) = \frac{\pi}{c^{1-a}}\left(\frac{1+a\log c}{\sin \pi a} - \pi\frac{a\cos \pi a}{\sin^2 \pi a}\right).$$

The following theorem, which generalizes Theorems 1 and 2, is proved in [5]:

THEOREM 3. *Suppose that:*

    $1°$ *The function g is regular in the complex plane, and there exist constants $a \neq 0$ and b, such that for every $z \in \mathbf{C}$ we have $g(z + \pi i) = ag(z) + b$;*

    $2°$ *The function f is analytic in the complex plane; its singularities which do not belong to the positive real axis are denoted by $z_1, \ldots, z_n$;*

$3°$ *On the positive real axis the function f may have only simple poles*
$a_1, \ldots, a_m$;
    $4°$ $\lim_{z \to 0} z f(-z) g (\log z) = \lim_{z \to \infty} z f(-z) g (\log z) = 0$.
Then

(10)
$$\text{v.p.} \int_0^{+\infty} \left(\left(a - \frac{1}{a} \right) f(x) g (\log x) + b \left(1 + \frac{1}{a} \right) f(x) \right) dx$$

$$= 2 \pi i \sum_{k=1}^{n} \operatorname*{Res}_{z=z_k} f(-z) g (\log z) - \pi i \left(a + \frac{1}{a} \right) \sum_{k=1}^{m} \operatorname*{Res}_{z=a_k} f(z) g (\log z)$$

$$- \pi i b \left(1 - \frac{1}{a} \right) \sum_{k=1}^{m} \operatorname*{Res}_{z=a_k} f(z).$$

The proof of this theorem is rather long, and so we shall only briefly sketch it.

Let $a_1 < \ldots < a_m$. Then $-a_1, \ldots, -a_m$ are the singularities of $z \mapsto f(-z)$ on the negative real axis. Let $\Gamma_0 = \{z \mid |z| = R\}$ and $\gamma_0 = \{z \mid |z| = r\}$ be concentric circles where $R > 0$ is so large and $r > 0$ is so small that all the singularities of f (except $z = 0$) lie in the annulus $\{z \mid r < |z| < R\}$. Let γ_k $(k = 1, \ldots, m)$ be negatively oriented semicircle $\{z \mid \operatorname{Im} z \geq 0, |z - a_k| = r\}$, and let γ_k^* $(k = 1, \ldots, m)$ be positively oriented semicircle $\{z \mid \operatorname{Im} z \leq 0,$ $|z - a_k| = r\}$, where r is chosen so that the circle $\gamma_k \cup \gamma_k^*$ does not contain any singularities of f except a_k, and so that those circles do not intersect one another or the circles Γ_0, γ_0. Let $l_k = [-a_{k+1} + r, -a_k - r]$ $(k = 0, 1, \ldots, m)$, where $a_0 = 0$, $a_{m+1} = R + r$, and let l be the directed line segment from $re^{i\theta}$ to $Re^{i\theta}$, where θ is chosen so that no singularities of f are on l, and so that $0 \leq \theta < \pi/2$. Let Γ be the positively oriented

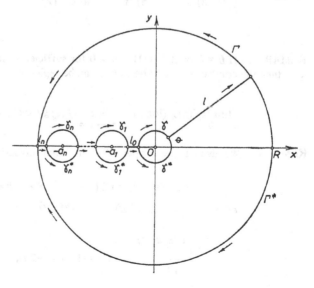

Fig. 5.3.1.3.

portion of the circle Γ_0 defined by $\Gamma = \{z \mid z = Re^{it}, \theta \leq t \leq \pi\}$ and let Γ^* be the negatively oriented portion of this circle defined by $\Gamma^* = \{z \mid z = Re^{it}, -\pi \leq t \leq \theta\}$. Similarly, let γ be the negatively oriented portion of the circle γ_0: $\gamma = \{z \mid z = re^{it}, \theta \leq t \leq \pi\}$, and let γ^* be the positively oriented portion of that circle: $\gamma^* = \{z \mid z = re^{it}, -\pi \leq t \leq \theta\}$. Let

$$C = \Gamma \cup l_n \cup \gamma_n \cup \ldots \cup \gamma_1 \cup l_0 \cup \gamma \cup l \quad \text{and} \quad C^* = \Gamma^* \cup l_n \cup \gamma_n^* \cup \ldots \cup \gamma_1^* \cup l_0 \cup \gamma^* \cup l.$$

Then C is positively oriented and C^* is negatively oriented.

Suppose that the complex plane is cut along the nonpositive imaginary axis. We thus obtain the cut plane ρ. Since $0 \notin \rho$, there is a regular branch of the logarithm, $z \mapsto L(z)$, in ρ such that $L(1) = 0$ and $L(z) = \log z$ for $(-\pi/2) < \arg z < \pi$. Let ρ^* be the plane cut along the nonnegative imaginary axis. There exists a regular branch of the logarithm $z \mapsto L^*(z)$ in ρ^* such that $L^*(1) = 0$ and $L^*(z) = \log z$ for $-\pi < \arg z < \pi/2$.

Define the functions G and G^* by $G(z) = f(-z)g(L(z))$ and $G^*(z) = f(-z)g(L^*(z))$ in ρ and ρ^* respectively, where they are analytic. Note that G is regular on C, and G^* is regular on C^*.

The formula (10) is deduced by an application of Cauchy's residue theorem to the integrals

$$\oint_C G(z)\,dz \quad \text{and} \quad \oint_{C^*} G^*(z)\,dz.$$

REMARK 5. If $b = 0$ or if $|g(z)| \geq \eta > 0$ for sufficiently large $|z|$ satisfying $-\pi \leq \operatorname{Im} z \leq \pi$, then the condition 4° of Theorem 3 can be replaced by

$$\lim_{|z| \to 0} zf(z)g(\log z) = \lim_{|z| \to +\infty} zf(z)g(\log z) = 0.$$

REMARK 6. We give a few examples of entire functions satisfying the condition 1°:

$$e^{2(z+c)}, \; \operatorname{sh} 2(z+c), \; \operatorname{ch} 2(z+c) \quad (a = 1, \; b = 0);$$

$$e^{z+c} \quad (a = -1, \; b = 0); \qquad e^{d(z+c)} \quad (a = e^{d\pi i}, \; b = 0);$$

$$\frac{\dfrac{d}{dz}\theta(z, e^{-\pi})}{\theta(z, e^{-\pi})} \quad (a = -1, \; b = -2\,i),$$

where $z \mapsto \theta(z, q)$ is one of the four theta functions (see [6], pp. 463–464).

The following two theorems are direct consequences of Theorem 3.

THEOREM 4. *Suppose that the function f satisfies the conditions:*
1° *is analytic in the complex plane; the singularities of f which do not belong to the positive real axis are $z_1, \ldots, z_n$;*
2° *the singularities of f on the positive real axis are simple poles $a_1, \ldots, a_m$;*
3° $\lim_{z \to 0} zf(z) \log z = \lim_{z \to \infty} zf(z) \log z = 0$.
Then

$$(11) \quad \text{v.p.} \int_0^{+\infty} f(x)\,dx = \sum_{k=1}^{n} \underset{z=z_k}{\text{Res}} f(-z) \log z - \sum_{k=1}^{m} \underset{z=a_k}{\text{Res}} f(z) \log z.$$

Proof. If $g(z) = z$ in Theorem 3, then $a = 1$, $b = \pi i$, and (10) becomes (11). Theorem 4 is a generalization of Theorem 1. It can be found in [7].

THEOREM 5. *Let f be such that:*
1° *it is analytic in the whole plane; it singularities which do not belong to the positive real axis are denoted by $z_1, \ldots, z_n$;*
2° *the singularities of f on the positive real axis are simple poles $a_1, \ldots, a_m$;*
3° $\lim_{z \to 0} z^{c+1} f(z) = \lim_{z \to \infty} z^{c+1} f(z) = 0$, *where $c \neq 0$.*
Then

$$(12) \quad \text{v.p.} \int_0^{+\infty} x^c f(x)\,dx = \frac{\pi}{\sin \pi c} \sum_{k=1}^{n} \underset{z=z_k}{\text{Res}}\, z^c f(z)$$

$$- \pi \cot g\, \pi c \sum_{k=1}^{m} \underset{z=a_k}{\text{Res}}\, z^c f(z).$$

Proof. If $g(z) = e^{cz}$ in Theorem 3, then $a = e^{c\pi i}$, $b = 0$, and (10) becomes (12). Theorem 5 is a generalization of Theorem 2.

Theorem 3 was further generalized by Schoenfeld [8]. His result, which we give without a proof, reads:

THEOREM 6. *Suppose that the following conditions are satisfied:*
1° *g is an entire function;*
2° *f is an analytic function with at most a finite number of singularities;*
3° $\lim_{z \to 0} zf(-z)g(\log z) = A$, $\lim_{z \to \infty} zf(-z)g(\log z) = B$.

Then

$$\text{v.p.} \int\limits_{0}^{+\infty} f(x)\,\big(g\,(\log x + \pi i) - g\,(\log x - \pi i)\big)\,\mathrm{d}x$$

$$= 2\,\pi i \sum_{k=1}^{n} \operatorname*{Res}_{z=z_k} f(-z)g\,(\log z)$$

$$- 2\,\pi i \sum_{k=1}^{m} \operatorname*{Res}_{z=a_k} f(z)g\,(\log z + \pi i) + 2\,(A-B)\,\pi i,$$

where $z_1, \ldots, z_n$ are the singularities of the function $z \mapsto f(-z)g(\log z)$ not on the non-positive real axis, while $a_1, \ldots, a_m$ are the singularities of the function $z \mapsto f(z)g(\log z + \pi i)$ on the positive real axis.

From Theorem 6 follows Theorem 4 if we suppose that:

$1°$ $g(z + \pi i) = ag(z) + b$ $(a \neq 0)$;

$2°$ the function f has only simple poles on the positive real axis;

$3°$ $A = B = 0$.

REFERENCES

1. Watson, p. 58.
2. T. M. Macrobert: *Functions of a Complex Variable*, London 1st edition 1917, 5th edition 1966, p. 74.
3. Th. Angheluţă: *A Course of the Theory of Functions of a Complex Variable*, (Romanian), Bucureşti 1957, p. 208.
4. Garnir-Gobert, pp. 102–104, 109–110.
5. J. T. Baskin: 'Some General Theorems for Evaluating Definite Integrals by Cauchy's Residue Theorem', M. A. Thesis, The Pennsylvania State University, U.S.A., 1962.[1]
6. E. T. Whittaker and G. N. Watson: *A Course of Modern Analysis*, Fourth Edition, Cambridge 1952, pp. 463–464.
7. R. P. Boas and L. Schoenfeld: 'Indefinite Integration by Residues', *SIAM Review* 8 (1966), 173–183.
8. L. Schoenfeld: 'The Evaluation of Certain Definite Integrals', *SIAM Review* 5 (1963), 358–369.

5.3.2. *Integration Along a Semicircle*

As we saw in 5.2.1, a semicircle together with its diameter (which is a part of the real axis) is a convenient contour for evaluating integrals with limits $-\infty$ and $+\infty$. The exist, however, wide classes of integrals from 0 to $+\infty$ which can also be evaluated by integration along a semicircle, but the integrand should usually be an even or an odd function.

---

[1] According to a private communication by Mr. Baskin, the results from this thesis are not published, and so we have alloted them somewhat more space.

We start with some theorems which are direct consequences of theorems from 5.2.1.

THEOREM 1. *Let f satisfy all the conditions of Theorem 1 from 5.2.1, and, in addition, let f be an even function. Then*

$$(1) \qquad \text{v.p.} \int_0^{+\infty} f(x)\,dx = \pi i \sum_{k=1}^n \operatorname{Res}_{z=z_k} f(z) + \frac{1}{2}\pi i \sum_{k=1}^m \operatorname{Res}_{z=p_k} f(z) - \frac{1}{2}\pi i\,A.$$

Proof. If f is an even function, then

$$\int_{-\infty}^{+\infty} f(x)\,dx = 2\int_0^{+\infty} f(x)\,dx,$$

and (1) follows directly from (1) of 5.2.1.

THEOREM 2. *Let f satisfy all the conditions of Theorem 2 from 5.2.1, and, in addition, let f be an even function. Then*

$$(2) \quad \text{v.p.} \int_0^{+\infty} f(x)\cos ax\,dx = \pi i \sum_{k=1}^n \operatorname{Res}_{z=z_k} e^{aiz} f(z) + \frac{1}{2}\pi i \sum_{k=1}^m \operatorname{Res}_{z=p_k} e^{aiz} f(z).$$

Proof. We have

$$\int_{-\infty}^{+\infty} e^{aix} f(x)\,dx = \int_0^{+\infty} \left(e^{aix} f(x) + e^{-aix} f(-x) \right) dx.$$

Hence, if f is an even function, then

$$(3) \qquad \int_{-\infty}^{+\infty} e^{aix} f(x)\,dx = 2\int_0^{+\infty} f(x)\cos ax\,dx.$$

Formula (5) from 5.2.1, and the equality (3) imply (2).

THEOREM 3. *Let f satisfy all the conditions of Theorem 2 from 5.2.1, and let f be an odd function. Then*

$$\text{v.p.} \int_0^{+\infty} f(x)\sin ax\,dx = \pi i \sum_{k=1}^n \operatorname{Res}_{z=z_k} e^{aiz} f(z) + \frac{1}{2}\pi i \sum_{k=1}^m \operatorname{Res}_{z=p_k} e^{aiz} f(z).$$

The proof of this theorem is similar to the proof of Theorem 2.

The next theorem enables us, under certain conditions, to evaluate integrals of the form $\int_0^{+\infty} f(x)\log x\,dx$.

THEOREM 4. *Suppose that:*

1° *The function f is analytic in the upper half-plane* $\{z|\operatorname{Im} z>0\}$; *its singularities are denoted by* $z_1,\ldots,z_n$;

2° *The function f is regular on the real axis;*

3° $\lim_{z\to 0} zf(-z)\log z = \lim_{z\to\infty} zf(-z)\log z = 0.$

Then

$$(4)\ \int_0^{+\infty} \left(f(x)+f(-x)\right)\log x\,dx + \pi i \int_0^{+\infty} f(x)\,dx = 2\pi i \sum_{k=1}^{n} \operatorname*{Res}_{z=z_k} f(-z)\log z.$$

Proof. Let Γ and γ be the semicircles

$$\Gamma = \{z\,|\,|z|=R,\ \operatorname{Im} z>0\},$$

$$\gamma = \{z\,|\,|z|=r,\ \operatorname{Im} z>0\},$$

where R is so large and r so small, that all the singularities of f from the upper half-plane lie in the region G, where

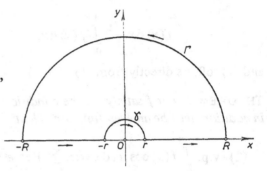

Fig. 5.3.2.

$$\partial G = C = [r,\ R]\cup\Gamma\cup[-R,\ -r]\cup\gamma,$$

From Cauchy's theorem on residues follows

$$(5)\qquad \oint_C f(-z)\log z\,dz = 2\pi i \sum_{k=1}^{n} \operatorname*{Res}_{z=z_k} \left(f(-z)\log z\right).$$

However, we have

$$\oint_C f(-z)\log z\,dz = \int_\gamma f(-z)\log z\,dz + \int_r^R f(-x)\log x\,dx$$

$$+ \int_\Gamma f(-z)\log z\,dz + \int_{-R}^{-r} f(-x)\log x\,dx.$$

In view of the condition 3°, using Jordan's lemmas (Theorems 1 and 2 from 3.1.4) we conclude that

$$\lim_{r\to 0} \int_\gamma f(-z)\log z\,dz = 0,\qquad \lim_{R\to +\infty} \int_\Gamma f(-z)\log z\,dz = 0.$$

Therefore, if $r \to 0$ and $R \to +\infty$, we get

$$\oint_C f(-z) \log z \, dz = \int_0^{+\infty} (f(x) + f(-x)) \log x \, dx + \pi i \int_0^{\infty} f(x) \, dx,$$

which, together with (5), implies (4).

REMARK 1. The above result, under the condition that f is a rational function without poles on the real axis, is given in [1], p. 154.

REMARK 2. The condition 2° of Theorem 4 can be replaced by a weaker condition
2_1° The function f has a finite number of simple poles $p_1, \ldots, p_m$ on the real axis. In that case we obtain the formula

$$\text{v.p.} \int_0^{+\infty} \left((f(x) + f(-x)) \log x + \pi i f(x) \right) dx$$

(6)

$$= 2\pi i \sum_{k=1}^{n} \operatorname*{Res}_{z=z_k} f(-z) \log z + \pi i \sum_{k=1}^{m} \operatorname*{Res}_{z=p_k} f(-z) \log z.$$

Baskin [2] proved a result equivalent to (6).

REMARK 3. If the function f which satisfies the conditions of Theorem 4 is also odd, then from (4) follows

$$\int_0^{+\infty} f(x) \, dx = 2 \sum_{k=1}^{n} \operatorname*{Res}_{z=z_k} f(-z) \log z,$$

or, if 2° is replaced by 2_1°, from (6) follows

$$\text{v.p.} \int_0^{+\infty} f(x) \, dx = 2 \sum_{k=1}^{n} \operatorname*{Res}_{z=z_k} f(-z) \log z + \sum_{k=1}^{m} \operatorname*{Res}_{z=p_k} f(-z) \log z.$$

REMARK 4. If f satisfies the conditions of Theorem 4, and also $f(z) \in \mathbf{R}$ when $z \in \mathbf{R}$, then separating the real and imaginary parts in (4), we obtain

(7) $$\int_0^{+\infty} (f(x) + f(-x)) \log x \, dx = 2 \operatorname{Im} \left(\sum_{k=1}^{n} \operatorname*{Res}_{z=z_k} f(-z) \log z \right).$$

as well as

$$\int\limits_0^{+\infty} f(x)\, dx = 2\operatorname{Re} \sum_{k=1}^{n} \operatorname*{Res}_{z=z_k} f(-z) \log z.$$

If, in addition, the function f is even, then from (7) we get

$$\int\limits_0^{+\infty} f(x) \log x\, dx = \operatorname{Im}\left(\sum_{k=1}^{n} \operatorname*{Res}_{z=z_k} f(-z) \log z\right).$$

Analogous formulas are easily deduced for the case when the condition 2° is replaced by 2_1°.

EXAMPLE 1. The function f, defined by $f(z) = (1 + z^2)^{-2}$ satisfies the conditions of Theorem 1. Since f has only one singularity in the upper half-plane, a pole of order 2 at $z = i$, we get

$$\operatorname*{Res}_{z=i} f(-z) \log z = \frac{i}{8}\,(2 - \pi i),$$

and from (1) follows

$$\int\limits_0^{+\infty} \frac{\log x}{(1+x^2)^2}\, dx = -\frac{\pi}{4}\,, \qquad \int\limits_0^{+\infty} \frac{1}{(1+x^2)^2}\, dx = \frac{\pi}{4}\,.$$

Theorem 4 was generalized by Glasser [3]. He proved that under the conditions of that theorem, we have

$$(8) \quad \int\limits_0^{+\infty} \big(f(x) + f(-x)\big) \log^m x\, dx + \sum_{k=1}^{m} \binom{m}{k} (\pi i)^k \int\limits_0^{+\infty} f(x) \log^{m-k} x\, dx$$

$$= 2\pi i \sum_{k=1}^{n} \operatorname*{Res}_{z=z_k} \big(f(-z) \log^m z\big).$$

Introduce the following notations:

$$I_j = \int\limits_0^{+\infty} f(x) \log^{j-1} x\, dx \qquad\qquad (j \in \mathbf{N}),$$

$$R_t = \sum_{k=1}^{m} \operatorname*{Res}_{z=z_k} \big(f(-z) \log^t z\big) \qquad (t \in \mathbf{N}).$$

As a consequence of (8) we obtain the following recurrent relations:
1° If f is an odd function, then

$$I_m = \frac{1}{m}\left(2\,R_m - \sum_{k=2}^{m}\binom{m}{k}(\pi i)^{k-1}I_{m+1-k}\right).$$

2° If f is an even function, then

$$I_{m+1} = \frac{\pi i}{2}\left(2\,R_m - \sum_{k=1}^{m}\binom{m}{k}(\pi i)^{k-1}I_{m+1-k}\right).$$

EXAMPLE 2. The function f, defined by $f(z) = 1/(1+z^2)$ has only one simple pole at $z = i$. Let

$$I_m = \int\limits_{0}^{+\infty}\frac{\log^{m-1}x}{1+x^2}\,dx.$$

Since $R_0 = 1/(2i)$, $R_1 = \pi/4$, $R_2 = \pi^2 i/8$, we have $I_1 = \pi/2$, $I_2 = 0$, $I_3 = \pi^3/8$, etc.

THEOREM 5. *Suppose that:*
 1° *The function f is analytic in the region $\{z \mid \operatorname{Im} z > 0\}$, where it can have a finite number of singularities $z_1, \ldots, z_n$;*
 2° *On the real axis f has only simple poles $a_1, \ldots, a_m$;*
 3° *For real x we have $f(x) = -f(-x)$;*

 4° $\quad f(x) = o\left(\dfrac{1}{z\log|z|}\right) \qquad (|z| \to +\infty).$

Then

(9) $\qquad \text{v.p.} \int\limits_{0}^{+\infty} f(x)\operatorname{arctg} x\,dx = -\pi\sum_{k=1}^{n}\operatorname*{Res}_{z=z_k} f(z)\log(i+z)$

$$-\frac{\pi}{2}\sum_{k=1}^{m}\operatorname*{Res}_{z=a_k} f(z)\log(i+z).$$

Proof. For real x we have

(10) $\qquad \operatorname{arctg} x = \frac{1}{2i}\left(\log(i-x) - \log(i+x)\right).$

Indeed, for $|\operatorname{Im} z| < 1$, the functions $z \mapsto \log(i-z)$ and $z \mapsto \log(i+z)$ are regular, and hence the function H defined by

$$H(z) = \frac{1}{2i}\left(\log(i-z) - \log(i+z)\right),$$

is a regular function. Hence

$$z \mapsto H'(z) = \frac{1}{2i}\left(-\frac{1}{i-z} - \frac{1}{i+z}\right) = \frac{1}{1+z^2}$$

is also a regular function. Since $H(0) = 0$, we conclude that $H(x) = \text{arctg } x$ for real x.

Using the condition $3°$ we see that

$$\int_0^R f(x)\,\text{arctg}\,x\,dx = \frac{1}{2}\int_{-R}^R f(x)\,\text{arctg}\,x\,dx,$$

which together with (10) implies

$$\text{v.p.}\int_{-R}^R f(x)\,\text{arctg}\,x\,dx = \frac{1}{2i}\text{v.p.}\int_{-R}^R f(x)\log(i-x)\,dx$$

$$-\frac{1}{2i}\,\text{v.p.}\int_{-R}^R f(x)\log(i+x)\,dx.$$

Replacing x by $-x$, and again using $3°$ we find

$$(11) \qquad \text{v.p.}\int_{-R}^R f(x)\,\text{arctg}\,x\,dx = i\,\text{v.p.}\int_{-R}^R f(x)\log(i+x)\,dx.$$

In order to evaluate the last integral use the same contour and the same notations as in Theorem 1 from 5.2.1. Since the function $z \mapsto \log(i+z)$ is regular in the closed region $\{z|\ \text{Im } z \geq 0\}$, by Cauchy's residue theorem we obtain

$$\oint_C f(z)\log(i+z)\,dz = 2\pi i \sum_{k=1}^n \operatorname*{Res}_{z=z_k} f(z)\log(i+z).$$

However,

$$\oint_C f(z)\log(i+z)\,dz$$

$$= \int_\Gamma f(z)\log(i+z)\,dz + \sum_{k=0}^m \int_{l_k} f(z)\log(i+z)\,dz + \sum_{k=1}^m \int_{\gamma_k} f(x)\log(i+x)\,dx.$$

Let $r \to 0$. Then

$$\int_\Gamma f(z)\log(i+z)\,dz + \sum_{k=1}^m \lim_{r\to 0}\int_{\gamma_k} f(z)\log(i+z)\,dz$$

$$+ \text{v.p.}\int_{-R}^R f(x)\log(i+x)\,dx = 2\pi i\sum_{k=1}^n \operatorname*{Res}_{z=z_k} f(z)\log(i+z).$$

The obvious equality

$$\lim_{\substack{r\to 0}} \int_{\gamma k} f(z)\log(i+z)\,dz = -\pi i \operatorname*{Res}_{z=a_k} f(z)\log(i+z),$$

implies

(13) $\int_{\Gamma} f(z)\log(i+z)\,dz + \text{v.p.} \int_{-R}^{R} f(x)\log(i+x)\,dx$

$$= 2\pi i \sum_{k=1}^{n} \operatorname*{Res}_{z=z_k} f(z)\log(i+z) + \pi i \sum_{k=1}^{m} \operatorname*{Res}_{z=a_k} f(z)\log(i+z).$$

According to 4° and the inequality

$$|\log(i+z)| \leq |\log|i+z|| + \pi \leq \log(1+|z|) + \pi = O(\log|z|)$$

we have

$$\lim_{R\to +\infty} \int_{\Gamma} f(z)\log(i+z)\,dz = 0.$$

Therefore, if $R \to +\infty$ in (13), we get (9).

REMARK 5. This result is proved in [2].

REMARK 6. In book [4], p. 291, the following result is given:
Let R be an odd rational function, whose poles in the upper half-plane are denoted by $z_1, \ldots, z_n$, and which is regular on the real axis. Then

$$\int_{0}^{+\infty} R(x)\operatorname{arctg} dx = \pi \operatorname{Re} \sum_{k=1}^{n} \operatorname*{Res}_{z=z_k} R(z)\log(1-iz),$$

where log denotes the branch of the logarithm which is regular in the upper half-plane for which $|\arg(1-iz)| < \pi$.

EXAMPLE 3. Let $f(z) = z/(z^4 + 1)$. From (9) we obtain

$$\int_{0}^{+\infty} \frac{x}{x^4+1} \operatorname{arctg} x\,dx = -\pi \sum_{k=1}^{2} \operatorname*{Res}_{z=z_k} \frac{z}{z^4+1} \log(i+z) = \frac{\pi^2}{16},$$

where

$$z_k = \exp\left(\frac{2k-1}{4}\pi i\right).$$

THEOREM 6. *Suppose that the following conditions are satisfied:*

$1°$ *The function $z \mapsto f(z)$ is real and continuous on the real axis;*

$2°$ *For real x we have $f(-x) = -f(x)$;*

$3°$ *f is an analytic function in the upper half-plane, where it can have a finite number of poles $z_1, \ldots, z_n$;*

$4°$ *For some c we have*

$$f(z) = o\left(\frac{1}{z \log |z|}\right) + \frac{c}{z} \qquad (|z| \to +\infty).$$

Then

$$(14) \qquad I = \int_0^{+\infty} f(x) \log\left|\frac{x+1}{x-1}\right| dx = \frac{\pi^2 c}{2} - 2\pi \operatorname{Im} \sum_{k=1}^{n} \operatorname*{Res}_{z=z_k} f(z) \log(z+1).$$

Proof. We first prove that the integral I exists. When $x \to +\infty$ we have

$$\log\left|\frac{x+1}{x-1}\right| = \log\frac{1+\dfrac{1}{x}}{1-\dfrac{1}{x}} = \frac{2}{x} + O\left(\frac{1}{x^3}\right),$$

and by assumption $4°$ we conclude that I converges at $+\infty$.

Also the integral converges at $x = 1$ since by $1°$, the integrand is essentially $\log |x-1|$ in a neighbourhood of 1 and

$$\int_{1+\varepsilon}^{2} \log(x-1)\,dx = ((x-1)\log(x-1) - (x-1))\Big|_{1+\varepsilon}^{2} \to -1 \qquad (\varepsilon \to 0+).$$

Hence, putting $y = -x$ and using $2°$, we see that

$$I = \lim_{R\to+\infty}\left(\int_0^R f(x) \log\left|\frac{1+x}{1-x}\right| dx\right)$$

$$= \lim_{R\to+\infty}\left(\int_0^R f(x) \log|1+x|\,dx - \int_0^R f(x) \log|1-x|\,dx\right)$$

$$= \lim_{R\to+\infty}\left(\int_0^R f(x) \log|1+x|\,dx + \int_{-R}^0 f(y) \log|1+y|\,dy\right)$$

$$= \lim_{R\to+\infty}\left(\int_{-R}^R f(y) \log|1+y|\,dy\right) = \lim_{R\to+\infty}\left(\int_{-R}^R f(z-1) \log|z|\,dz\right.$$

$$\left. + \int_R^{R+1} f(z-1) \log|z|\,dz - \int_{-R}^{-R+1} f(z-1) \log|z|\,dz\right).$$

In view of 4°, we have

$$f(z-1) \log|z| = O\left(\frac{1}{R} \log R\right) \to 0, \qquad \text{as} \qquad R \to +\infty,$$

and therefore

$$I = \lim_{R \to +\infty} \int_{-R}^{R} f(z-1) \log|z| \, dz$$

$$= \lim_{R \to +\infty} \lim_{\varepsilon \to 0} \left(\int_{-R}^{-\varepsilon} f(z-1) \log|z| \, dz + \int_{\varepsilon}^{R} f(z-1) \log|z| \, dz \right),$$

inasmuch as we have already shown the convergence of the integral at $z = 0$.
  Now by 1°

$$f(z-1) \log|z| = f(z-1) \operatorname{Re} \log z = \operatorname{Re}\left(f(z-1) \log z\right),$$

so that

(15) $$I = \lim_{R \to +\infty} \lim_{\varepsilon \to 0} \operatorname{Re} J(R, \varepsilon),$$

where

$$J(R, \epsilon) = \int_{-R}^{-\varepsilon} f(z-1) \log z \, dz + \int_{\varepsilon}^{R} f(z-1) \log z \, dz.$$

Let

$$C_\varrho = \{z \,||\, z| = \varrho, \ \operatorname{Im} z \geq 0\},$$

and let

$$0 < \varepsilon < \min_{1 \leq k \leq n} |z_k + 1| \leq \max_{1 \leq k \leq n} |z_k + 1| < R.$$

By Cauchy's residue theorem, we have, by 1° and 3°,

(16) $$\int_{-R}^{-\varepsilon} f(z-1) \log z \, dz - \int_{C_\varepsilon} f(z-1) \log z \, dz + \int_{\varepsilon}^{R} f(z-1) \log z \, dz$$

$$+ \int_{C_R} f(z-1) \log z \, dz = 2\pi i \sum_{k=1}^{n} \operatorname*{Res}_{z=z_k+1} f(z-1) \log z$$

$$= 2\pi i \sum_{k=1}^{n} \operatorname*{Res}_{z=z_k} f(z) \log(z+1),$$

i.e.

$$(17) \qquad J(R, \varepsilon) = H(\varepsilon) - H(R) + 2 \pi i \sum_{k=1}^{n} \operatorname{Res}_{z=z_k} f(z) \log (z+1),$$

where

$$H(\varrho) = \int_{C_\varrho} f(z-1) \log z \, dz = \int_{0}^{\pi} f(\varrho \, e^{i\theta} - 1) (\log \varrho + i \, \theta) \varrho \, e^{i\theta} i \, d\theta$$

$$= i \log \varrho \int_{0}^{\pi} g(\varrho e^{i\theta}) \, d\theta - \int_{0}^{\pi} \theta g(\varrho e^{i\theta}) \, d\theta,$$

on putting $g(z) = z f(z-1)$.

As $z \to 0$, we have $g(z) = O(z)$. Hence, as $\varepsilon \to 0$, we find

$$H(\varepsilon) = i \int_{0}^{\pi} g(\varepsilon e^{i\theta}) \log (\varepsilon + i \, \theta) \, d\theta = O(\varepsilon \log \varepsilon) = o(1).$$

When $|z| \to +\infty$, by 4°, we have

$$g(z) = \frac{cz}{z-1} + o\left(\frac{z}{(z-1) \log |z-1|}\right) = c + o\left(\frac{1}{\log |z|}\right),$$

uniformly in the closed upper half-plane. Hence, we have

$$c = g(z) + o(1), \qquad c = \lim_{x \to +\infty} g(x) = \lim_{x \to +\infty} x f(x-1),$$

implying that c is real.

On the other hand, as $R \to +\infty$,

$$H(R) = i \log R \int_{0}^{\pi} \left(c + o\left(\frac{1}{\log R}\right)\right) d\theta - \int_{0}^{\pi} \left(c + o\left(\frac{1}{\log R}\right)\right) \theta \, d\theta$$

$$= \pi \, ic \log R + o(1) - \frac{\pi^2 c}{2}.$$

From (17), when $R \to +\infty$ follows

$$J(R, \varepsilon) = -\pi \, ic \log R + \frac{\pi^2 c}{2} + o(1) + 2 \pi i \sum_{k=1}^{n} \operatorname{Res}_{z=z_k} f(z) \log (z+1).$$

Since c is real, we have

$$\operatorname{Re} J(R, \ \varepsilon) = \frac{\pi^2 c}{2} + o(1) + 2\pi \operatorname{Im} \sum_{k=1}^{n} \operatorname*{Res}_{z=z_k} f(z) \log (z+1).$$

Finally, on using (15) and (16), we obtain (14).

REMARK 7. Theorem 6 was proved by Schoenfeld [5].

REMARK 8. As noted by Schoenfeld [5], the hypothesis $1°$ can be replaced by the weaker hypothesis:

$1_1°$ f is real on the real axis, and at each real ξ different from the real numbers $\xi_1, \ldots, \xi_m$, the function is continuous for approach to ξ through points in the closed upper half-plane. Furthermore, there are numbers $\alpha_1, \ldots, \alpha_m$ each exceeding -1 such that $f(z) = (z - \xi_j)^{\alpha_j} F_j(z)$, where F_j is continuous at ξ_j for approach to ξ_j through points in the closed upper half-plane.

REMARK 9. In addition [6] to [5], Schoenfeld states that a special case of this result was obtained by Mary Boas [7] who used the same method to evaluate the integral

$$\int_{0}^{+\infty} \frac{x}{3x^2 + 1} \log \left| \frac{x+1}{x-1} \right| dx.$$

REMARK 10. In connection with the integral

$$\int_{0}^{+\infty} f(x) \log \left| \frac{x+1}{x-1} \right| dx$$

see also [2].

EXAMPLE 4. The function

$$z \mapsto f(z) = \frac{z}{z^2 + a^2} \quad (a > 0)$$

satisfies the conditions $1° - 4°$. Since

$$\operatorname*{Res}_{z=ia} \frac{z}{z^2 + a^2} \log (z+1) = \frac{i}{2} \operatorname{arctg} \frac{1}{a},$$

we have

$$\int_{0}^{+\infty} \frac{x}{x^2 + a^2} \log \left| \frac{x+1}{x-1} \right| dx = \pi \operatorname{arctg} \frac{1}{a}.$$

Similarly we obtain the formula

$$\int\limits_{0}^{+\infty} \frac{x(1-x^2)}{(x^2+a^2)^2} \log\left|\frac{x+1}{x-1}\right| dx = \pi\left(\frac{1}{2a}-\mathrm{arctg}\,\frac{1}{a}\right).$$

EXAMPLE 5. Using the hypothesis 1°_1 instead of 1° for the function f, defined by

$$f(z) = \frac{1}{(4\log^2 z^{2/3}+\pi^2)\sqrt[3]{z(z^2-1)}},$$

we have

$$\xi_1 = -1,\ \xi_2 = 0,\ \xi_3 = 1,\ n = 1,\ z_1 = \exp\frac{3\pi i}{4},\ c = 0.$$

If we cut the plane along the rays $z = -1-iy$, $z = -iy$, $z = 1-iy$ ($y \geq 0$) and use principal values for the roots and the logarithm, we get

$$\int\limits_{0}^{+\infty} \frac{1}{(4\log^2 x^{2/3}+\pi^2)\sqrt[3]{x(x^2-1)}} \log\left|\frac{x+1}{x-1}\right| dx = -\frac{3}{8}\sqrt[3]{2}\left(\frac{1}{2}\log(2-\sqrt{2})+\mathrm{arctg}\,(1+\sqrt{2})\right).$$

REFERENCES

1. Copson, p. 154.
2. J. T. Baskin: 'Some General Theorems for Evaluating Definite Integrals by Cauchy's Residue Theorem', M. A. Thesis, The Pennsylvania State University, U.S.A., 1962.
3. M. L. Glasser: 'Some Recursive Formulas for Evaluation of a Class of Definite Integrals', *Am. Math. Monthly* 71 (1964), 75–76.
4. M. A. Evgrafov, Yu. V. Sidorov, M. V. Fedoryuk, M. I. Šabunin, and K. A. Bežanov: *A Collection of Exercices on the Theory of Analytic Functions*, (Russian), Moscow 1969, p. 291.
5. L. Schoenfeld: 'On Integrals of the Type $\int_0^\infty f(x) \log |(x+1)/(x-1)|\, dx$', *SIAM Review* 1 (1959), 154–157.
6. L. Schoenfeld: 'A Remark on a Previous Paper', *SIAM Review* 4 (1962), 147.
7. Mary Layne Boas: 'The Photodisintegration of H^3', Ph. D. Thesis in Physics, The Massachusetts Institute of Technology, U.S.A., 1948.

5.3.3. Integrals from a to + ∞

In paper [1] Neville considered integrals of the form $\int_a^{+\infty} f(x)\, dx$, where $a > 0$, and proved the following result.

THEOREM 1. *Let f be an analytic function inside the circle* $\Gamma = \{z\,|\,|z| = R\}$, *where R is large enough so that the region* int Γ *contains all the singular* points $z_1, \ldots, z_n$ of f. Furthermore, let f be a regular function on the half-line $\{z|\operatorname{Im} z = 0, \operatorname{Re} z \geq a\}$. If

$$\lim_{R \to +\infty} \int_\Gamma f(z)\log(a-z)\,dz = 0,$$

where $\log(a-z)$ takes real values for real values of z from $(-\infty, +\infty)$, then

$$\int_a^{+\infty} f(x)\,dx = -\sum_{k=1}^n \operatorname*{Res}_{z=z_k} f(z)\log(a-z).$$

Neville proved Theorem 1 by integrating the function $z \mapsto f(z)\log(a-z)$ round the contour shown on Figure 5.3.3. In fact, as noted by Neville himself, we have

$$\int_a^{+\infty} f(x)\,dx = \int_0^{+\infty} f(x+a)\,dx,$$

and Theorem 1 can be reduced to Theorem 1 from 5.3.1.

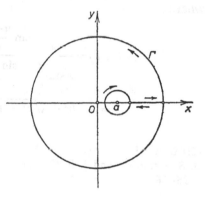

Fig. 5.3.3.

EXAMPLE 1. If $p > q > 0$, then

$$\int_0^{+\infty} \frac{1}{(x+p)^2 - q^2}\,dx = \frac{1}{2q}\log\frac{p+q}{p-q},$$

and hence

$$\int_a^{+\infty} \frac{1}{(x+p)^2 - q^2}\,dx = \int_0^{+\infty} \frac{1}{(x+p+a)^2 - q^2}\,dx = \frac{1}{2q}\log\frac{p+a+q}{p+a-q}.$$

REFERENCE

1. E. H. Neville: 'Indefinite Integration by Means of Residues', *Math. Student* 13 (1945), 16–25.

5.3.4. Miscellaneous Integrals

5.3.4.1. If $a, b, c \in \mathbf{R}, n \in \mathbf{N}$, then

$$\int_0^{+\infty} \frac{1}{a^2 + 2b^2 x^n + c^2 x^{2n}} \, dx = \frac{1}{a^2} \sqrt[n]{\frac{a}{c}} \, N\left(n, \arccos \frac{b^2}{ac}\right) \quad (b^2 < ac);$$

$$= \frac{1}{a^2} \sqrt[n]{\frac{a}{c}} \lim_{t \to 0} N(n, t) \quad (b^2 = ac);$$

$$= \frac{1}{a^2} \sqrt[n]{\frac{a}{c}} \, K(p, q) N\left(\frac{n}{2}, \frac{\pi}{2}\right) \quad (b^2 > ac),$$

where

$$N(n, t) = \frac{\pi}{n \sin \frac{\pi}{n}} \frac{\sin \frac{n-1}{n} t}{\sin t}, \quad K(p, q) = \frac{p^{(1-n)/n} - q^{(1-n)/n}}{q - p},$$

$$p = \frac{1}{ac}\left(b^2 + \sqrt{b^4 - a^2 c^2}\right), \quad q = \frac{1}{ac}\left(b^2 - \sqrt{b^4 - a^2 c^2}\right).$$

REFERENCE
D. S. Mitrinović and J. F. Georgatos: 'Problem 5533', *Am. Math. Monthly* **76** (1969). 95–96.

5.3.4.2. If $c < 0$ and $p < 0$ or $c \geq 0$ and $a > 0$, then

$$\int_0^{+\infty} e^{p \cos cx} \cos(ax + p \sin cx) \frac{1}{1 + x^2} \, dx = \pi e^{-a + pe^{-c}};$$

$$\text{v.p.} \int_0^{+\infty} e^{p \cos cx} \cos(ax + p \sin cx) \frac{1}{1 - x^2} \, dx = \frac{\pi}{2} e^{p \cos c} \sin(a + p \sin c).$$

REFERENCE
Hardy.

5.3.4.3. Let $a \in \mathbf{R}^+$, $m, n \in \mathbf{N}$ and $m + 1 < n$. Then

$$(1) \qquad \int_0^{+\infty} \frac{x^m}{x^n + a} \, dx = \frac{\pi}{n a^{(n-m-1)/n} \sin \frac{(m+1)\pi}{n}}.$$

Proof. Consider the integral

$$\oint_C \frac{z^m}{z^n+a}\,dz,$$

where

$$C=\{z\,|\,0\le\mathrm{Re}\,z\le R,\ \mathrm{Im}\,z=0\}\cup$$
$$\cup\left\{z\,\big|\,|z|=R,\ 0\le\arg z\le\frac{2\pi i}{n}\right\}\cup$$
$$\cup\left\{z\,\big|\,0\le|z|\le R,\ \arg z=\frac{2\pi i}{n}\right\}$$

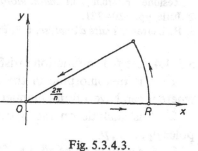

Fig. 5.3.4.3.

(see Figure 5.3.4.3) and where R is large enough. The region int C contains only one zero of $z\mapsto z^n+a$, namely $w=a^{1/n}e^{\pi i/n}$. According to Cauchy's theorem on residues we have

(2) $$\oint_C \frac{z^m}{z^n+a}\,dz=\frac{2\pi i w^m}{n w^{n-1}}=\frac{2\pi i}{n a^{(n-m-1)/n}\exp\dfrac{(n-m-1)\pi i}{n}},$$

where

(3) $$\oint_C \frac{z^m}{z^n+a}\,dz=\int_0^R \frac{x^m}{x^n+a}\,dx+\int_0^{2\pi/n}\frac{iR^{m+1}e^{i(m+1)\theta}}{R^ne^{in\theta}+a}\,d\theta$$
$$-\int_0^R \frac{x^m}{x^n+a}e^{2(m+1)\pi i/n}\,dx.$$

If $R\to+\infty$, from (2) and (3) follows (1).

REMARK 1. Formula (1) is proved in [1]. This result, however, is not new. Indeed, in book [2] the following formula is proved by means of Calculus of residues:

(4) $$\int_0^{+\infty} \frac{x^m}{1+x^n}\,dx=\frac{\pi}{n}\frac{1}{\sin\dfrac{(m+1)\pi}{n}},$$

where $n\in\mathbf{N}$ and $-1<m<n-1$. It is easily seen that (1) can be obtained from (4) by putting $x=t\sqrt[n]{a}$ in (1).

REMARK 2. We also mention the following formula which can be found in [3]:

$$\int_0^{+\infty} \frac{x^{2m}}{1+x^{2n}}\,dx=\frac{\pi}{2n\sin\dfrac{2m+1}{2n}\pi},\qquad (m,n\in\mathbf{N},\ m<n).$$

REFERENCES
1. O. J. Farrell and B. Ross: 'Note on Evaluating Certain Real Integrals by Cauchy's Residue Theorem', *Am. Math. Monthly* 68 (1961), 151–152.
2. Julia, pp. 220–221.
3. H. Laurent: *Traité d'analyse*, t. 3, Paris 1888, pp. 249–250.

5.3.4.4. Let f be a function satisfying the conditions:

1° f is meromorphic in the region $\{z \mid 0 < \arg z < 2\pi/n\}$, $n \in \mathbb{N}$, $n \geq 2$, where it can have a finite number of singular points $a_1, \ldots, a_m$;

2° f is analytic on the positive real axis where it can have only simple poles $p_1, \ldots, p_r$;

3° f is a function of z^n, i.e. $f(z) = g(z^n)$;

$$4° \qquad \lim_{R \to +\infty} \int_{C_R} f(z)\, dz = 0,$$

where

$$C_R = \left\{ z \mid |z| = R,\ 0 < \arg z < \frac{2\pi}{n} \right\}.$$

Then

$$\text{v.p.} \int_0^{+\infty} f(x)\, dx = \frac{-\pi e^{-\pi i/n}}{\sin \dfrac{\pi}{n}} \left(\sum_{k=1}^{m} \operatorname*{Res}_{z=a_k} f(z) + \frac{1}{2} \sum_{k=1}^{r} \operatorname*{Res}_{z=p_k} f(z) \right).$$

REMARK. Compare with Theorem 4 from 5.3.1.

EXAMPLES.

$$1. \int_0^{+\infty} \frac{1}{1+x^4}\, dx = \frac{\pi}{2\sqrt{2}}. \qquad 2. \int_0^{+\infty} \frac{1}{1+x^n}\, dx = \frac{\pi}{n \sin \dfrac{\pi}{n}} \quad \text{(see [2]).}$$

$$3. \text{ v.p.} \int_0^{+\infty} \frac{1}{a^3 - x^3}\, dx = \frac{\pi}{3\sqrt{3}\, a^2}.$$

REFERENCES
1. S. Melamed and H. Kaufman: 'Evaluation of Certain Improper Integrals by Residues', *Am. Math. Monthly* 72 (1965), 1111–1112.
2. H. Cartan: *Elementary Theory of Analytic Functions of One or Several Complex Variables*, Reading (Mass.) 1963, p. 117.

5.3.4.5. Let f be a function satisfying the following conditions:

1° f is meromorphic in the region $D = \{z \mid 0 < \arg z < 2\pi/n\}$, where $n \in \mathbb{N}$, $n \geq 2$, where it can have a finite number of singular points;

2° f is analytic on the positive real axis where it can have only simple poles $p_1, \ldots, p_r$;

3° f is a function of z^n, i.e. $f(z) = g(z^n)$;

4° $\displaystyle\lim_{z \to \infty} |z|^{a+1} |f(z)| = \lim_{z \to 0} |z|^{a+1} |f(z)| = 0,$

where $a \in \mathbb{R}$ and $a + 1 \neq kn$ $(k \in \mathbb{Z})$.

Then

$$(1) \quad \text{v.p.} \int_0^{+\infty} x^a f(x)\,dx = -\frac{\pi\, e^{-i\pi(a+1)/n}}{\sin \dfrac{(a+1)\pi}{n}} \left(\sum_{k=1}^{m} \operatorname*{Res}_{z=a_k} z^a f(z) \right.$$

$$\left. + \frac{1}{2} \sum_{k=1}^{r} \operatorname*{Res}_{z=p_k} z^a f(z) \right).$$

REMARK 1. Formula (1) is proved by integrating the function $z \mapsto z^a f(z)$ around the contour shown on Figure 5.3.4.5. The single valued branch of the multivalued function $z \mapsto z^a$ is defined by $z^a = e^{a \log z}$, where $\log z = \log|z| + i\theta$ $(\theta_0 < \theta < \theta_0 + 2\pi)$, where the ray θ_0 does not belong to the region D.

REMARK 2. Compare with Theorem 5 from 5.3.1.

EXAMPLE (see [2]).

$$\int_0^{+\infty} \frac{x^a}{1 + x^n}\,dx = \frac{\pi}{n \sin \dfrac{(a+1)\pi}{n}}.$$

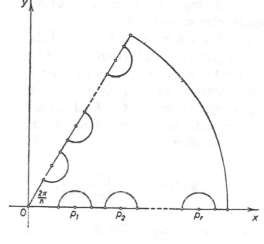

Fig.5.3.4.5.

REMARK 3. A similar result was again published in [3], and the following example was worked out:

$$\int_0^{+\infty} \frac{\sqrt{x}}{(1 + x^4)^3}\,dx = \frac{65\pi}{512 \sin 3\pi/8}$$

This equality can also be obtained from (1) by putting $a = \frac{1}{2}$, $n = 4$.

REFERENCES

1. H. Kaufman and S. Melamed: 'Integration of Multiple-valued Functions by Residues', *Elem. Math.* **21** (1966), 37–39.
2. H. Cartan: *Elementary Theory of Analytic Functions of One or Several Complex Variables*, Reading (Mass.) 1963.
3. B. K. Sachdeva and B. Ross: 'Evaluation of Certain Real Integrals by Contour Integration', *Am. Math. Monthly* **89** (1982), 246–249.

5.3.4.6. Let f be an analytic function, regular at $z = 0$, possessing a finite number of singular points $z_1, \ldots, z_n$. If $\lim_{|z| \to +\infty} |zf(z)| = 0$, then

$$(1) \qquad \int_0^{+\infty} \frac{f(x)}{(\log x + 2n\pi i)^2 + \pi^2} \, dx = \sum \operatorname{Res} \frac{f(z)}{\log z + (2n-1)\pi i},$$

where $z \mapsto \log z$ is defined as the branch whose value at $z = 1$ is 0.

Proof. Let $k = \{z \mid |z| = \epsilon\}$, $l = \{z \mid \epsilon \le \operatorname{Re} z \le R\}$, $K = \{z \mid |z| = R\}$, where ϵ is sufficiently small and R is sufficiently large positive number, such that the region int C, where $C = K \cup l^- \cup k \cup l$, contains all the singularities of the function g, where

$$g(z) = \frac{f(z)}{\log z + (2n-1)\pi i}.$$

Then, according to Cauchy's theorem on residues, we have

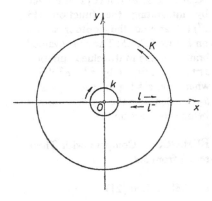

Fig. 5.3.4.6.

$$(2) \qquad \oint_C \frac{f(z)}{\log z + (2n-1)\pi i} \, dz = 2\pi i \sum \operatorname{Res} \frac{f(z)}{\log z + (2n-1)\pi i}$$

and clearly

$$(3) \qquad \oint_C g(z)\,dz = \int_l g(z)\,dz + \int_K g(z)\,dz + \int_{l^-} g(z)\,dz + \int_k g(z)\,dz.$$

If $R \to +\infty$ and $\epsilon \to 0$, we get

(4)
$$\lim_{\substack{\epsilon \to 0 \\ R \to +\infty}} \left(\int_l g(z)\,dz + \int_{l-} g(z)\,dz \right) =$$

$$= \int_0^{+\infty} \frac{f(x)}{\log x + (2n-1)\pi i}\,dx + \int_{+\infty}^0 \frac{f(x)}{\log x + (2n+1)\pi i}\,dx$$

$$= 2\pi i \int_0^{+\infty} \frac{f(x)}{(\log x + 2n\pi i)^2 + \pi^2}\,dx.$$

On the other hand,

(5)
$$\lim_{R \to +\infty} \int_K g(z)\,dz = \lim_{\epsilon \to 0} \int_k g(z)\,dz = 0.$$

From (2)–(5) follows (1).

EXAMPLE. Let $f(x) = 1/(x^2 + a^2)$ $(a > 0)$ and $n = 0$. The function g, defined by

$$g(z) = \frac{1}{(z^2 + a^2)(\log z - \pi i)},$$

has simple poles at $z_1 = ai$, $z_2 = -ai$, $z_3 = -1$. From (1) we get

$$\int_0^{+\infty} \frac{1}{(x^2 + a^2)(\log^2 x + \pi^2)}\,dx = \sum_{\nu=1}^3 \operatorname*{Res}_{z=z_\nu} \frac{1}{(z^2 + a^2)(\log z - \pi i)}.$$

Since

$$\operatorname*{Res}_{z=ai} g(z) = \frac{1}{2ai\left(\log a - \dfrac{\pi}{2}i\right)}, \quad \operatorname*{Res}_{z=-ai} g(z) = \frac{-1}{2ai\left(\log a + \dfrac{\pi}{2}i\right)}, \quad \operatorname*{Res}_{z=-1} g(z) = \frac{-1}{1+a^2},$$

we find

$$\int_0^{+\infty} \frac{1}{(x^2 + a^2)(\log^2 x + \pi^2)}\,dx = \frac{\pi}{2a} \frac{1}{\log^2 a + \pi^2/4} - \frac{1}{1+a^2}.$$

REFERENCE

Ş. Gheorghiu: 'On an Integral', (Romanian), Gaz. Mat. Fiz. 9 (62) (1957), 70–75.

5.3.4.7. Let $k > \frac{1}{2}$ be a real number. Then

(1)
$$\int_0^{+\infty} \cos(x^{2k})\,dx = \frac{1}{2k}\,\Gamma\left(\frac{1}{2k}\right)\cos\frac{\pi}{4k},$$

$$\int_0^{+\infty} \sin(x^{2k})\,dx = \frac{1}{2k}\,\Gamma\left(\frac{1}{2k}\right)\sin\frac{\pi}{4k},$$

where Γ is the gamma-function.

Proof. Let C be the contour shown on Figure 5.3.4.7, i.e. let

$$C = \{z\,|\,0 \le \operatorname{Re} z \le R,\ \operatorname{Im} z = 0\} \cup \left\{z\,\Big|\,|z| = R,\ 0 \le \arg z \le \frac{\pi}{4k}\right\} \cup$$

$$\cup \left\{z\,\Big|\,0 \le |z| \le R,\ \arg z = \frac{\pi}{4k}\right\}.$$

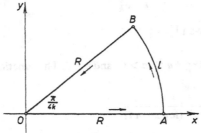

Fig. 5.3.4.7.

The function $z \mapsto e^{-z^{2k}}$ is regular in the region int C, and hence $\oint_C e^{-z^{2k}}\,dz = 0$.

However, clearly

$$\oint_C e^{-z^{2k}}\,dz = \int_0^R e^{-x^{2k}}\,dx + \int_l e^{-z^{2k}}\,dz$$

$$+ \int_{BO} e^{-z^{2k}}\,dz.$$

Hence, we have

$$\int_0^R e^{-x^{2k}}\,dx + \int_l e^{-z^{2k}}\,dz = \int_{OB} e^{-z^{2k}}\,dz,$$

which implies, if $R \to +\infty$, that

(2)
$$\int_0^{+\infty} e^{-x^{2k}}\,dx + \lim_{R \to +\infty}\int_l e^{-z^{2k}}\,dz = \lim_{R \to +\infty}\int_{OB} e^{-z^{2k}}\,dz.$$

We first show that

$$\lim_{R \to +\infty} \int_l e^{-z^{2k}} dz = 0.$$

Indeed, putting $z = Re^{i\theta/2k}$, for $k > \frac{1}{2}$ and $R \to +\infty$, we get

$$\left| \int_l e^{-z^{2k}} dz \right| = \left| \int_0^{\pi/2} e^{-R^{2k}(\cos\theta + i\sin\theta)} \frac{R}{2k} \left(-\sin\frac{\theta}{2k} + i\cos\frac{\theta}{2k} \right) d\theta \right|$$

$$\leq \frac{R}{2k} \int_0^{\pi/2} e^{-R^{2k}\cos\theta} d\theta = \frac{R}{2k} \int_0^{\pi/2} e^{-R^{2k}\sin\theta} d\theta$$

$$\leq \frac{R}{2k} \int_0^{\pi/2} e^{-\frac{2}{\pi}\theta R^{2k}} d\theta = \frac{\pi}{2k R^{2k-1}} (1 - e^{-R^{2k}}) \to 0.$$

Therefore, (2) becomes

$$(3) \qquad \int_0^{+\infty} e^{-x^{2k}} dx = \lim_{R \to +\infty} \int_{OB} e^{-z^{2k}} dz.$$

The integral appearing on the left-hand side of (3) can be expressed by means of the gamma-function. In fact, if we put $x^{2k} = y$, we get

$$\int_0^{+\infty} e^{-x^{2k}} dx = \frac{1}{2k} \int_0^{+\infty} y^{\frac{1}{2k}-1} e^{-y} dy = \frac{1}{2k} \Gamma\left(\frac{1}{2k}\right).$$

In order to evaluate the integral on the right-hand side of (3), put $z = re^{i\pi/4k}$. Then (3) becomes

$$\int_0^{+\infty} e^{-ir^{2k}} \left(\cos\frac{\pi}{4k} + i\sin\frac{\pi}{4k} \right) dr = \frac{1}{2k} \Gamma\left(\frac{1}{2k}\right),$$

i.e.

$$\int_0^{+\infty} (\cos r^{2k} - i\sin r^{2k}) dr = \frac{1}{2k} \Gamma\left(\frac{1}{2k}\right) \left(\cos\frac{\pi}{4k} - i\sin\frac{\pi}{4k} \right).$$

Hence, separating real and imaginary parts, we obtain (1).

REMARK. These results were obtained by Koutský [1]. Using a similar method, i.e. integrating the function $z \mapsto e^{iz^2}$ round the contour C, for the case $k = 1$, Olds [2] proved the well-known formulas for the Fresnel integrals

$$\int_0^{+\infty} \cos(x^2) dx = \int_0^{+\infty} \sin(x^2) dx = \frac{\sqrt{2\pi}}{4},$$

though this method of evaluating Fresnel integrals is not new (see, for instance, [3] and [4]).

REFERENCES
1. Z. Koutský: 'Generalized Fresnel Integrals', (Chech), *Časopis Mat. Fyz. Praha* 75 (1950), D257–D261.
2. C. D. Olds: 'The Fresnel Integrals', *Am. Math. Monthly* 75 (1968), 285–286.
3. H. Laurent: *Traité d'analyse*, t. 3, Paris 1888, pp. 258–259.
4. Tisserand, pp. 450–452.

5.3.4.8. Let $0 < a < 1$, let $a_1, \ldots, a_n, b_1, \ldots, b_m$ be nonnegative real numbers such that $a_1 + \cdots + a_n > 0$, and let the function g be defined by

$$g(z) = \sum_{k=1}^{n} a_k z^k - \sum_{k=1}^{m} b_k z^k.$$

If $0 \leq t \leq \pi/\max(m, n)$, then we have

(1) $\quad \displaystyle\int_0^{+\infty} x^{a-1} \cos g(x)\, dx = \int_0^{+\infty} x^{a-1} e^{-S(x)-s(x)} \cos\left(at + C(x) - c(x)\right) dx,$

(2) $\quad \displaystyle\int_0^{+\infty} x^{a-1} \sin g(x)\, dx = \int_0^{+\infty} x^{a-1} e^{-S(x)-s(x)} \sin\left(at + C(x) - c(x)\right) dx,$

where

$$S(x) = \sum_{k=1}^{n} a_k x^k \sin kt, \quad s(x) = \sum_{k=1}^{m} b_k x^{-k} \sin kt,$$

$$C(x) = \sum_{k=1}^{n} a_k x^k \cos kt, \quad c(x) = \sum_{k=1}^{m} b_k x^{-k} \cos kt.$$

The above result was proved by Tudor [1]. The proof follows the usual lines of integrating $z \mapsto z^{a-1} e^{ig(z)}$ around the contour given on Figure 5.3.4.8, and then letting $r \to 0, R \to +\infty$.

Formulas (1) and (2) can be used to evaluate various improper integrals. We give three examples.

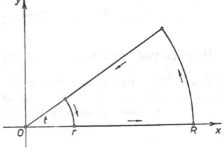

Fig. 5.3.4.8.

1° Let $a_1 = 1, a_2 = \cdots = a_n = b_1 = \cdots = b_m = 0, k = 1/a, t = \pi/2$. Putting $x = y^k$, we find

(3) $\quad \displaystyle\int_0^{+\infty} \cos y^k \, dy = \cos \frac{\pi}{2k} \int_0^{+\infty} e^{-y^k} \, dy = \frac{1}{k} \Gamma\left(\frac{1}{k}\right) \cos \frac{\pi}{2k},$

$$(k>1)$$

(4) $\quad \displaystyle\int_0^{+\infty} \sin y^k \, dy = \sin \frac{\pi}{2k} \int_0^{+\infty} e^{-y^k} \, dy = \frac{1}{k} \Gamma\left(\frac{1}{k}\right) \sin \frac{\pi}{2k}.$

Formulas (3) and (4) were proved by Koutský [2] (see 5.3.4.7).
2° For $a_1 = b_1 = 1, a_2 = \cdots = a_n = b_2 = \cdots = b_m = 0, a = \frac{1}{2}, t = \pi/2$, we get

$$\int_0^{+\infty} \cos\left(x^2 - \frac{1}{x^2}\right) dx = \int_0^{+\infty} \sin\left(x^2 - \frac{1}{x^2}\right) dx = \frac{1}{2e^2} \sqrt{\frac{\pi}{2}}.$$

3° The following equalities are valid:

$$\int_0^{+\infty} e^{-x^2 \sin t} \cos (x^2 \cos t) \, dx = \frac{1}{2} \sqrt{\frac{\pi}{2}} \left(\cos \frac{t}{2} + \sin \frac{t}{2} \right);$$

$$\int_0^{+\infty} e^{-x^2 \sin t} \sin (x^2 \cos t) \, dx = \frac{1}{2} \sqrt{\frac{\pi}{2}} \left(\cos \frac{t}{2} - \sin \frac{t}{2} \right).$$

REFERENCES
1. Gh. Tudor: 'On an Improper Integral', (Romanian), *Bul. Şti. Tehn. Inst. Politech. Timişoara* 15 (29) (1970), 51–54.
2. Z. Koutský: 'Generalised Fresnel Integrals', (Chech), *Časopys Mat. Fyz. Praha* 75 (1950), D257–D261.

5.3.4.9. If a and b are real numbers, then

$1°\quad \text{v.p.} \displaystyle\int_0^{+\infty} \frac{\cos ax}{\cos bx} \frac{1}{1-x^2} \, dx = 0; \qquad 2°\quad \text{v.p.} \int_0^{+\infty} \frac{\cos ax}{\cos bx} \frac{1}{1+x^2} \, dx = \frac{\pi}{2 \operatorname{ch} b},$

Let a, b, c, and d be real numbers such that $|c| > |a|$, $|d| > |b|$ and

$cd < 0$. Then

$$\text{v.p.} \int\limits_{0}^{+\infty} \frac{\sin\left(ax+\dfrac{b}{x}\right)}{\sin\left(cx+\dfrac{d}{x}\right)} \frac{1}{1+x^2}\, dx = \frac{\pi}{2}\frac{\text{sh}\,(a-b)}{\text{sh}\,(c-d)}.$$

REFERENCE
Hardy.

5.3.4.10. The value of the so-called probability integral, namely

(1) $$\int\limits_{0}^{+\infty} e^{-x^2}\, dx = \frac{1}{2}\sqrt{\pi}$$

is usually established by the methods or real analysis. Some books even state that its value cannot be found by contour methods (see, for example, [1] or [2]).

In fact, this is not so. We did our best to find the earliest proof of (1) by contour methods, but we did not arrive at a decisive conclusion. In a private communication dated 3 June, 1971, Professor Copson informed us that immediately after the publication of [2] (in 1935) somebody evaluated the integral (1) by the method of residues, but that he forgot who that was.

However, even before the publication of [2], Mordell [3] considered the integral

(2) $$\int\limits_{-\infty}^{+\infty} \frac{e^{ax^2+bx}}{e^{cx}+d}\, dx \qquad (a, b, c, d \text{ real or complex numbers})$$

and showed that it can be reduced to two standard forms

(3) $$\int\limits_{-\infty}^{+\infty} \frac{e^{\pi i\omega x^2 - 2\pi iu}}{e^{2\pi x}-1}\, dx \qquad \text{or} \qquad \int\limits_{-\infty}^{+\infty} \frac{e^{\pi i\omega x^2 - 2\pi iu}}{e^{2\pi\omega x}-1}\, dx$$

The integrals (3) are evaluated in [3] by contour integration and their values are expressed by means of some special functions. Though the integral (2) contains (1) as a special case, the methods used by Mordell are too complicated, and it is not really worthwhile applying them to (1).

The first direct published proofs of (1) we found are from the post-war period. At the Berkeley Symposium on Mathematical Statistics and Probability (held: 13–18 August 1945 and 27–29 January 1946) G. Pólya [4] gave the following proof.

Integrate the function $z \mapsto e^{\pi i z^2}$ tg πz counterclockwise around the parallelogram P with vertices $R + iR$, $-R-iR$, $-R+1-iR$, $R+1+iR$ where R is positive and large (see Figure 5.3.4.10 (1)). Since the integrand has

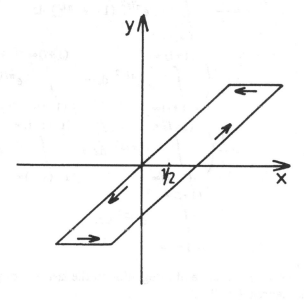

Fig. 5.3.4.10(1)

only a simple pole at $z = \frac{1}{2}$, with the residue $-(1/\pi)e^{\pi i/4}$, then

(4) $\oint e^{\pi i z^2}$ tg πz $dz = -2ie^{\pi i/4}$.

As $R \to +\infty$, the contribution of the two horizontal sides of P becomes negligible and the integral (4) reduces to two integrals along infinite parallel straight lines. The integral along the right-hand line can be transformed by a change of the variable of integration, and so (4) becomes

$$\int_{-(1+i)\infty}^{(1+i)\infty} (e^{\pi i(z + 1)^2} - e^{\pi i z^2})\ \text{tg}\ \pi z\ dz = -2ie^{\pi i/4}.$$

Hence, transforming the integrals, we find

$$-2ie^{\pi i/4} = \frac{1}{i} \int\limits_{-(1+i)\infty}^{(1+i)\infty} (-e^{-\pi iz^2 + 2\pi iz} - e^{\pi iz^2}) \frac{e^{2\pi iz} - 1}{e^{2\pi iz} + 1} \, dz$$

$$= \frac{1}{i} \int\limits_{-(1+i)\infty}^{(1+i)\infty} e^{\pi iz^2} (1 - e^{2\pi iz}) \, dz$$

$$= \frac{1}{i} \left(\int\limits_{-(1+i)\infty}^{(1+i)\infty} e^{\pi iz^2} \, dz + \int\limits_{-(1+i)\infty}^{(1+i)\infty} e^{\pi i(z+1)^2} \, dz \right)$$

$$= \frac{1}{i} \left(\int\limits_{-(1+i)\infty}^{(1+i)\infty} e^{\pi iz^2} \, dz + \int\limits_{1-(1+i)\infty}^{1+(1+i)\infty} e^{\pi iz^2} \, dz \right)$$

$$= \frac{1}{i} \int\limits_{-(1+i)\infty}^{(1+i)\infty} e^{\pi iz^2} \, dz.$$

The changes of the variable of integration in the last two steps are justified by Cauchy's theorem (see 3.1.5).

Finally, we change the variable of integration in the last integral, setting $z = e^{\pi i/4} t$ $(t \in \mathbf{R})$. We then obtain

$$-2ie^{\pi i/4} = \frac{2}{i} e^{\pi i/4} \int\limits_{-\infty}^{+\infty} e^{-\pi t^2} \, dt,$$

i.e.

$$\int\limits_{-\infty}^{+\infty} e^{-\pi t^2} \, dt = 1,$$

and (1) follows after a trivial transformation.

A different approach was adopted by Cadwell [5]. He used the fact that

$$(5) \quad \int_0^{+\infty} \cos x^2 \, dx = \int_0^{+\infty} \sin x^2 \, dx = \frac{1}{\sqrt{2}} \int_0^{+\infty} e^{-x^2} \, dx,$$

which is established by integrating $z \mapsto e^{-z^2}$ round the
contour of Figure 5.3.4.10(2).
This procedure is standard (see,
for example, [6]), but the
value of $\int_0^{+\infty} e^{-x^2} \, dx$ is usually assumed and is used to
obtain the values of the first
two integrals in (5). Cadwell
went the other way round;
he evaluated the first two
integrals in (5) by the method
of residues, and hence obtained
(1).

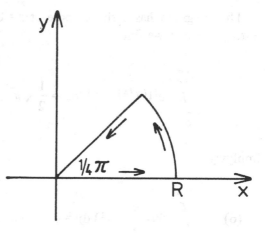

Fig. 5.3.4.10(2).

In order to evaluate those
integrals, integrate the function

$$z \mapsto \frac{e^{iz^2}}{\sin \sqrt{\pi} \, z}$$

round the contour of Figure
5.3.4.10(3), where $P = (\frac{1}{2}\sqrt{\pi}, -R)$, $Q = (\frac{1}{2}\sqrt{\pi}, R)$, $S = (-\frac{1}{2}\sqrt{\pi}, R)$, $T = (-\frac{1}{2}\sqrt{\pi}, -R)$. The sum
of the contributions of PQ
and ST is

$$2 \int_{-R}^{R} i \exp(i(\frac{\pi}{4} - y^2)) \, dy.$$

On QS we have $|\sin \sqrt{\pi} \, z| >$
sh $\sqrt{\pi} R$ and $|e^{iz^2}| = e^{-2Rx}$;

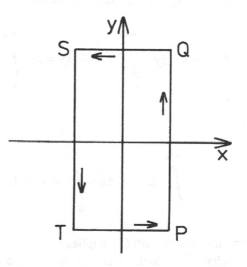

Fig. 5.3.4.10(3).

thus the contribution of QS is in modulus less than

$$\int_{-\sqrt{\pi}/2}^{\sqrt{\pi}/2} \frac{e^{-2Rx}}{\text{sh}\,\sqrt{\pi}\,R}\,dx = \frac{1}{R}.$$

The integrand has a simple pole at $z = 0$ with residue $1/\sqrt{\pi}$, and hence, letting $R \to +\infty$, we find

$$\int_0^{+\infty} e^{i[(\pi/4)-y^2]}\,dy = \frac{1}{2}\sqrt{\pi},$$

implying

$$(6) \qquad \int_0^{+\infty} \sin\left(\frac{\pi}{4}-y^2\right)dy = 0 \quad \text{and} \quad \int_0^{+\infty} \cos\left(\frac{\pi}{4}-y^2\right)dy = \frac{1}{2}\sqrt{\pi}.$$

Expanding the integrals (6), we find

$$\int_0^{+\infty} \cos y^2\,dy - \int_0^{+\infty} \sin y^2\,dy = 0; \quad \int_0^{+\infty} \cos y^2\,dy + \int_0^{+\infty} \sin y^2\,dy = \sqrt{\frac{\pi}{2}},$$

i.e.

$$\int_0^{+\infty} \cos x^2\,dx = \int_0^{+\infty} \sin x^2\,dx = \frac{1}{2}\sqrt{\frac{\pi}{2}},$$

which together with (5) implies (1).

Mirsky [7] arrived at (1) by a single contour integration. He integrated the function F, defined by $F(z) = e^{i\pi z^2}/\sin \pi z$ round the parallelogram having vertices at $R \pm \frac{1}{2} + iR$ and $-R \pm \frac{1}{2} - iR$ where $R > 0$ is sufficiently large. The integrand has a simple pole of residue $1/\pi$ at $z = 0$, and it is easily verified that the integrals along the horizontal lines of the parallelogram tend to zero as $R \to + \infty$. Furthermore, the sum of the integrals along the sides inclined to the real axis is

$$\int_{-R\sqrt{2}}^{R\sqrt{2}} [F(te^{\pi i/4} + \tfrac{1}{2}) - F(te^{\pi i/4} - \tfrac{1}{2})] e^{\pi i/4} \, dt$$

$$= 2i \int_{-R\sqrt{2}}^{R\sqrt{2}} e^{-\pi t^2} \, dt.$$

Using Cauchy's residue theorem, and proceeding to the limit as $R \to + \infty$, we obtain

$$\int_{-\infty}^{+\infty} e^{-\pi t^2} \, dt = 1,$$

i.e. (1).

REMARK 1. Notice that the Proceedings of the Berkeley Symposium which contain the paper [4], appeared as late as 1949, and so the authors of [5] and [7] were unaware of Pólya's proof.

None of the proofs given so far are really proofs of (1), but proofs of an equality which has further to be transformed (though trivially) into (1). A direct proof of (1) was given by Srinivasa Rao [8]. It runs as follows.

Let f be defined by

$$f(z) = \frac{\exp \dfrac{iz^2}{2}}{e^{-z\sqrt{\pi}} - 1},$$

and let Γ be the contour $ABCDEFGA$ shown on Figure 5.3.4.10(4), i.e. parallelogram $ADEG$, indented at the origin. In fact, if R and r are positive numbers and $r < R$, then $A = -R - iR$, $B = -r - ir$, $C = r + ir$, $D = R + iR$, $E = R + i(R + \sqrt{\pi})$, $F = i\sqrt{\pi}$, $G = -R - i(R - \sqrt{\pi})$.

We have

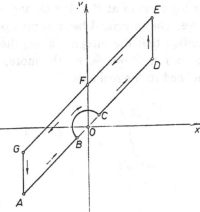

Fig. 5.3.4.10(4).

(7)

$$\left| \int_{DE} f(z)\, dz \right| = \left| \int_0^{\sqrt{\pi}} \frac{e^{i(R+iR+it)^2}}{e^{-\sqrt{\pi}(R+iR+it)} - 1}\, dt \right|$$

$$\leqq \frac{e^{-R^2}}{1 - e^{-\sqrt{\pi}R}} \int_0^{\sqrt{\pi}} e^{-Rt}\, dt = \frac{e^{-R^2}}{R} \to 0$$

$$(R \to +\infty).$$

Similarly,

(8) $$\lim_{R \to +\infty} \int_{GA} f(z)\, dz = 0.$$

On the other hand,

$$\int_{CD} f(z)\, dz = (1+i) \int_{-R}^{-r} \frac{e^{-x^2}}{e^{-(1+i)\sqrt{\pi}x} - 1}\, dx,$$

$$\int_{AB} f(z)\, dz = (1+i) \int_{-R}^{-r} \frac{e^{-x^2}}{e^{-(1+i)\sqrt{\pi}x} - 1}\, dx = (1+i) \int_{r}^{R} \frac{e^{-x^2}}{e^{(1+i)\sqrt{\pi}x} - 1}\, dx,$$

which implies

$$\int_{AB} f(z)\, dz + \int_{CD} f(z)\, dz = -(1+i) \int_{r}^{R} e^{-x^2}\, dx.$$

Similarly,

$$\int_{GF} f(z)\,dz + \int_{FE} f(z)\,dz = i(1+i)\int_0^K e^{-x^2}\,dx.$$

Finally, for the integral along the semicircle BC we find

$$\int_{BC} f(z)\,dz = i\sqrt{\pi}.$$

The function f has no singularities in the region int Γ, and therefore

$$\oint_\Gamma f(z)\,dz = 0.$$

However, clearly

$$\oint_\Gamma f(z)\,dz = \int_{AB} f(z)\,dz + \int_{BC} f(z)\,dz + \int_{CD} f(z)\,dz + \int_{DE} f(z)\,dz$$

$$+ \int_{EF} f(z)\,dz + \int_{FG} f(z)\,dz + \int_{GA} f(z)\,dz.$$

Therefore, we get

$$i\sqrt{\pi} - (1+i)\int_r^R e^{-x^2}\,dx + \int_{DE} f(z)\,dz - i(1+i)\int_0^R e^{-x^2}\,dx + \int_{GA} f(z)\,dz = 0,$$

and hence, if $R \to +\infty$, $r \to 0$, having in mind (7) and (8), we obtain

$$i\sqrt{\pi} - (1+i)\int_0^{+\infty} e^{-x^2}\,dx - i(1+i)\int_0^{+\infty} e^{-x^2}\,dx = 0,$$

i.e. (1).

REMARK 2. More generally, by integrating the function

$$z \mapsto \frac{\exp(iz^2\cos^2\alpha)}{e^{-\sqrt{2\pi}z\cos\alpha} - 1}$$

along a parallelogram inclined at α to the real axis ($0 \le \alpha < \pi/2$), we obtain in exactly the same manner

(9)
$$\int_0^{+\infty} e^{-x^2\sin 2\alpha + ix^2\cos 2\alpha}\,dx = \frac{1+i}{2}\sqrt{\frac{\pi}{2}}\,e^{-i\alpha}.$$

Formula (9) includes for $\alpha = 0$ the Fresnel integrals (see 5.3.4.7) and for $\alpha = \pi/4$ the Poisson integral (1).

REMARK 3. The integral $\int_0^{+\infty} e^{-x^2}\,dx$ was evaluated by residues by Dimitrovski [9], but indirectly. We sketch briefly his idea. First, applying the standard residue procedure it is shown that

$$I(n) = \int\limits_{-\infty}^{+\infty} \left(1 + \frac{x^2}{n}\right)^{-n} dx = \pi \frac{(2n-3)!!}{(2n-2)!!}\sqrt{n},$$

which implies

$$\int\limits_0^{+\infty} e^{-x^2}\,dx = \lim_{n\to+\infty} \frac{1}{2} I(n) = \lim_{n\to+\infty} \pi \frac{(2n-3)!!}{(2n-2)!!}\sqrt{n},$$

and hence, using Wallis' formula

$$\lim_{n\to+\infty} \frac{(2n-3)!!}{(2n-2)!!}\sqrt{n} = \frac{1}{\sqrt{\pi}},$$

we get (1).

We might mention that Wallis' formula can be proved by the method of residues (see 4.5.4) which means that Dimitrovski actually evaluated (1) by residues, though indirectly.

REMARK 4. The Gauss integrals

$$I_{2k} = \int\limits_0^{+\infty} e^{-x^2} x^{2k}\,dx \qquad\qquad (k \in \mathbf{N})$$

were evaluated by Dekker [10] by a method which is similar to the method applied by Dimitrovski. Namely, using real analysis methods, Dekker proves that

$$I_{2k} = \int\limits_0^{+\infty} e^{-1/x} x^{-k-(3/2)}\,dx = \int\limits_0^{+\infty} \lim_{n\to+\infty}\left(1 + \frac{1}{nx}\right)^{-n} x^{-k-(3/2)}\,dx$$

$$= \lim_{n\to+\infty} \int\limits_0^{+\infty} \left(1 + \frac{1}{nx}\right)^{-n} x^{-k-(3/2)}\,dx,$$

and then integrating the function

$$z \mapsto f(z) = \left(1 - \frac{1}{nz}\right)^{-n} z^{-k-(3/2)}$$

along the contour shown on Figure 5.3.4.10(5), in the limiting case $r \to 0$, $R \to + \infty$, he obtains

(10) $$I_{2k} = \lim_{n \to + \infty} (-1)^k \pi \operatorname*{Res}_{z=1/n} f(z).$$

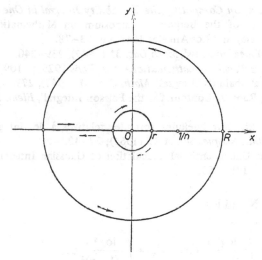

Fig. 5.3.4.10(5).

On the other hand

$$\operatorname*{Res}_{z=1/n} f(z) = \frac{1}{(n-1)!} \lim_{z \to 1/n} \frac{d^{n-1}}{dz^{n-1}} z^{n-k-(3/2)}$$

$$= \binom{n-k-(3/2)}{n-1} n^{k+(1/2)}$$

$$= \frac{(-1)^k}{\pi} n^{k+\frac{1}{2}} \frac{\Gamma\left(n-k-\frac{1}{2}\right)}{\Gamma(n)} \Gamma\left(k+\frac{1}{2}\right),$$

which together with (10) yields

$$I_{2k} = \lim_{n \to + \infty} n^{k+\frac{1}{2}} \frac{\Gamma\left(n-k-\frac{1}{2}\right)}{\Gamma(n)} \Gamma\left(k+\frac{1}{2}\right)$$

$$= \Gamma\left(k+\frac{1}{2}\right).$$

Dekker emphasizes in [10] the importance of the integral I_{2k} in theoretical physics, and believes that many integrals of that form can be evaluated by the method of residues.

REFERENCES
1. Watson, p. 79.
2. Copson, p. 125.
3. L. J. Mordell: 'The Value of the Definite Integral $\int_{-\infty}^{+\infty} (e^{at^2+bt} \, dt)/(e^{ct}+d)$, *Quarterly J. Math.* 48 (1920), 329–342.
4. G. Pólya: *Remarks on Computing the Probability Integral in One and Two Dimensions*, Proceedings of the Berkeley Symposium on Mathematical Statistics and Probability, Berkeley and Los Angeles 1949, pp. 63–78.
5. J. H. Cadwell: 'Three Integrals', *Math Gaz.* 31 (1947), 239–240.
6. É. Goursat: *Cours d'analyse mathématique*, t. 2, Paris 1929, p. 109.
7. L. Mirsky: 'The Probability Integral', *Math. Gaz.* 31 (1947), 279.
8. K. N. Srinivasa Rao: 'A Contour for the Poisson Integral', *Elem. Math.* 27 (1972), 88–90.
9. D. S. Dimitrovski: 'Sur quelques formules relatives à des intégrales impropres', *Ann. Fac. Sci. Univ. Skopje. Sec. A* 15 (1964), 27–42.
10. H. Dekker: 'An Unconventional Calculation of Gaussian Integrals', *Simon Stevin* 50 (1976–1977), 145–153.

5.3.4.11. Let $n \in \mathbf{N}$ and let

$$H_n = \int\limits_0^{+\infty} \frac{\log^n x}{(1+x)^2} \, dx, \quad J_n = \int\limits_0^{+\infty} \frac{\log^n x}{(1-x)^2} \, dx.$$

The following equalities are valid

$$H_{2n+1} = J_{2n+1} = 0 \quad (n \in \mathbf{N}),$$

$$H_{2n} = (-1)^{n-1} \, 2 \, (2^{2n-1}-1) \, \pi^{2n} B_{2n}, \quad J_{2n} = (-1)^{n-1} \, 2^{2n} \, \pi^{2n} B_{2n} \quad (n \in \mathbf{N}_0),$$

where $B_0, B_1, \ldots$ are Bernoulli's numbers. In particular, we have

$$H_2 = \frac{\pi^2}{3}, \quad H_4 = \frac{7\pi^4}{15}, \quad H_6 = \frac{31\pi^6}{21}, \ldots;$$

$$J_2 = \frac{2\pi^2}{3}, \quad J_4 = \frac{8\pi^4}{15}, \quad J_6 = \frac{32\pi^6}{21}, \ldots$$

REFERENCE
Th. Angheluţă: *On a Class of Integrals*, (Romanian), Lucrări ştiinţifice, Institutul Politehnic Cluj 1959, 21–28.

5.3.4.12. If the function f is defined by

$$f(x) = \int\limits_0^{\pi/2} \left(1 - \exp(x - x \operatorname{cosec} t)\right) \sec^2 t \, dt,$$

then

$$\int\limits_{0}^{+\infty} xe^{-2x} f(x)^2 \, dx = \frac{1}{3} \, .$$

REFERENCES
1. W. L. Bade: 'Problem 61–9', *SIAM Review* 3 (1962), 329.
2. L. Schoenfeld: 'The Evaluation of Certain Definite Integrals', *SIAM Review* 5 (1963), 358–369.

5.3.4.13. If a and b are positive numbers, then

$$1° \quad \int\limits_{0}^{+\infty} \frac{1}{1+x^2} \cos\left(ax - \frac{b}{x}\right) dx = \frac{\pi}{2} e^{-a-b};$$

$$2° \quad \text{v.p.} \int\limits_{0}^{+\infty} \frac{1}{1-x^2} \cos\left(ax - \frac{b}{x}\right) dx = \frac{\pi}{2} \sin(a-b);$$

$$3° \quad \int\limits_{0}^{+\infty} \frac{x}{1+x^2} \sin\left(ax - \frac{b}{x}\right) dx = \frac{\pi}{2} e^{-a-b};$$

$$4° \quad \text{v.p.} \int\limits_{0}^{+\infty} \frac{x}{1-x^2} \sin\left(ax - \frac{b}{x}\right) dx = \frac{\pi}{2} \cos(a-b);$$

$$5° \quad \int\limits_{0}^{+\infty} \frac{1}{x} \sin\left(ax - \frac{b}{x}\right) dx = 0.$$

REFERENCE
Hardy

5.3.4.14. Let f and g be differentiable functions on $[0, +\infty)$ and suppose that $f'(0) \neq 0$ and $g'(0) \neq 0$ have the same sign. Furthermore, let $\lim_{x \to 0} xF(x) = K \in \mathbf{R}$. Then

$$(1) \quad \int\limits_{0}^{+\infty} \left(g'(x) F(g(x)) - f'(x) F(f(x))\right) dx = K \log \frac{f'(0)}{g'(0)} \, .$$

The following formulas are consequences of (1):

$$1° \quad \int\limits_{0}^{+\infty} \left(\frac{g'(x)}{g(x)} e^{-g(x)} - \frac{f'(x)}{f(x)} e^{-f(x)} \right) dx = \log \frac{f'(0)}{g'(0)} ;$$

$$2° \quad \int\limits_{0}^{+\infty} \left(\frac{e^{-x}}{x} - \frac{1}{(1+x)^2 \log(1+x)} \right) dx = 0;$$

$$3° \quad \int\limits_{0}^{+\infty} \frac{e^{-bx} - e^{-ax}}{x} dx = \log \frac{a}{b} .$$

REFERENCES

1. A.-L. Cauchy: 'Mémoire sur la théorie des intégrales définies singulières appliquée généralement à la détermination des intégrales définies, et en particulier à l'évaluation des intégrales eulériennes', *Exercices d'analyse et de physique mathématique*, Paris 1841. ≡ *Oeuvres* (2) 12; 409–469.
2. A.-L. Cauchy: 'Mémoire sur la théorie des intégrales définies singulières appliquée généralement à la détermination des intégrales définies, et en particulier à l'évaluation des intégrales eulériennes', *Compt. Rend. Acad. Sci. Paris* 16 (1843), 422–433. ≡ *Oeuvres* (1) 7, 271–283.

5.3.4.15. If a, b, and c are positive numbers, then

$$(1) \qquad \int\limits_{0}^{+\infty} \frac{c x^{a-1} \sin\left(\frac{a\pi}{2} - bx \right)}{x^2 + c^2} dx = \frac{\pi}{2} c^{a-1} e^{-bc}.$$

Formula (1) was proved by Cauchy. It contains, as particular cases, the following Laplace integrals

$$\int\limits_{0}^{+\infty} \frac{b \cos ax}{x^2 + b^2} dx = \frac{\pi}{2} e^{-ab}, \qquad \int\limits_{0}^{+\infty} \frac{x \sin ax}{x^2 + b^2} dx = \frac{\pi}{2} e^{-ab},$$

as well as the Euler integral

$$\int\limits_{0}^{+\infty} \frac{x^{a-1}}{1+x^2} dx = \frac{\pi}{2 \sin \dfrac{a\pi}{2}} .$$

REFERENCE

A.-L. Cauchy: 'Sur diverses relations qui existent entre les résidus des fonctions et les intégrales définies', *Exercices de mathématiques*, Paris 1826. ≡ *Oeuvres* (2) 6, 139.

5.3.4.16. If a, b, and c are positive numbers, then

$$1° \quad \int_0^{+\infty} \frac{\sin ax}{x}\, dx = \frac{\pi}{2};$$

$$2° \quad \int_0^{+\infty} \frac{e^{a\cos bx} \sin(a\sin bx)}{x}\, dx = \frac{\pi}{2}(e^a - 1);$$

$$3° \quad \int_0^{+\infty} \frac{c\, e^{a\cos bx} \cos(a\sin bx)}{x^2 + c^2}\, dx = \frac{\pi}{2} e^{ae^{-bc}};$$

$$4° \quad \int_0^{+\infty} \frac{x\, e^{a\cos bx} \sin(a\sin bx)}{x^2 + c^2}\, dx = \frac{\pi}{2}\left(e^{ae^{-bc}} - 1\right);$$

$$5° \quad \int_0^{+\infty} \frac{\cos ax - e^{-ax}}{x}\, dx = 0;$$

$$6° \quad \int_0^{+\infty} \frac{\cos ax - e^{-ax}}{x(x^4 + b^4)}\, dx = \frac{\pi}{2b^4} e^{-ab} \sin \frac{ab}{\sqrt{2}}.$$

REFERENCES

A.-L. Cauchy: 'Sur diverses relations qui existent entre les résidus des fonctions et les intégrales définies', *Exercices de mathématiques*, Paris 1826. ≡ *Oeuvres* (2) 6, 139.

5.3.4.17. The following equalities are valid:

$$1° \quad \int_0^{+\infty} \frac{1}{x} \log \left| \frac{x+1}{x-1} \right|\, dx = \frac{1}{2}\pi^2;$$

$$2° \quad \int_0^{+\infty} \frac{x}{x^2 + a^2} \log \left| \frac{x+1}{x-1} \right|\, dx = \pi\, \text{arctg}\, \frac{1}{a} \qquad (a > 0).$$

REMARK. The first integral is given in [1], and the second in [2]. If $a \to 0 +$ in the second integral, we obtain the first.

REFERENCES

1. L. Schoenfeld: 'On Integrals of the Type $\int_0^{+\infty} f(x) \log |(x + 1)/(x - 1)| dx$', *SIAM Review* 1 (1959), 154–157.

2. J. T. Baskin: 'Some General Theorems for Evaluating Definite Integrals by Cauchy's Residue Theorem', M. A. Thesis, The Pennsylvania State University, U.S.A, 1962, 84 pp.

5.3.4.18. If $p < 1, a > 0, b > 0$, then

1°
$$\int\limits_0^{+\infty} \frac{1}{1 - 2p\cos\left(ax - \dfrac{b}{x}\right) + p^2} \frac{1}{1 + x^2} dx = \frac{\pi}{2} \frac{1}{1 - p^2} \frac{1 + pe^{-a-b}}{1 - pe^{-a-b}};$$

2° v.p.
$$\int\limits_0^{+\infty} \frac{1}{1 - 2p\cos\left(ax - \dfrac{b}{x}\right) + p^2} \frac{1}{1 - x^2} dx = \frac{\pi}{1 - p^2} \frac{p\sin(a - b)}{1 - 2p\cos(a - b) + p^2}.$$

REFERENCE
Hardy.

5.3.4.19. Let $a \neq 1, 0 < \operatorname{Re} a < 2, b > 0$. Then

1° v.p.
$$\int\limits_0^{+\infty} \frac{x^{a-1}}{x^2 - b^2} dx = -\frac{\pi b^{a-2}}{2} (\operatorname{cosec} \pi a + \operatorname{cotg} \pi a);$$

2° v.p.
$$\int\limits_0^{+\infty} \frac{x^{a-1}}{x^2 + b^2} dx = \frac{\pi b^{a-2}}{2} \operatorname{cosec} \frac{\pi a}{2}.$$

REFERENCE
J. T. Baskin: 'Some General Theorems for Evaluating Definite Integrals by Cauchy's Residue Theorem', M. A. Thesis, The Pennsylvania State University, U.S.A., 1962.

5.3.4.20. The following formulas are valid:

1°
$$\int\limits_0^{+\infty} \frac{x^{a-1}}{1 + x} dx = \pi \operatorname{cosec} \pi a \qquad (0 < a < 1);$$

2°
$$\int\limits_0^{+\infty} \frac{x^{a-1}}{1 + x^2 + x^4} dx = \frac{\pi}{\sqrt{3}} \operatorname{cosec} \frac{\pi a}{2} \sin \frac{2 - a}{6} \pi \qquad (0 < a < 4);$$

3°
$$\int\limits_0^{+\infty} \frac{\log px}{q^2 + x^2} dx = \frac{\pi}{2q} \log pq \qquad (p, q > 0);$$

4° $\displaystyle\int\limits_{0}^{+\infty}\frac{\log x}{(x-1)x^a}\,dx=\frac{\pi^2}{\sin^2 a\pi}$ $(0<a<1)$;

5° $\displaystyle\int\limits_{0}^{+\infty}\frac{\text{arctg}\,px}{x(1+x^2)}\,dx=\frac{\pi}{2}\log(1+p)$ $(p>0)$;

6° $\displaystyle\int\limits_{0}^{+\infty}x^{a-2}\log(1+x^2)\,dx=\frac{\pi}{1-a}\sec\frac{\pi a}{2}$ $(0<a<1)$;

7° $\displaystyle\int\limits_{0}^{+\infty}x^{2a}e^{-x^2}\,dx=\frac{1}{2}\Gamma\left(\frac{2a+1}{2}\right)$ $\left(-\frac{1}{2}<a<\frac{1}{2}\right)$;

8° $\displaystyle\int\limits_{0}^{+\infty}e^{-kx}x^{a-1}\frac{\sin}{\cos}\,\theta x\,dx=\frac{\Gamma(a)}{(k^2+\theta^2)^{a/2}}\frac{\sin}{\cos}\left(a\,\text{arctg}\,\frac{\theta}{k}\right)$ $(k\geq 0;\ 0<a<1)$.

REFERENCE

E. Schmid: *Die Cauchy'sche Methode der Auswertung bestimmter Integrale zwischen reellen Grenzen*, Stuttgart 1903.

5.3.4.21. Let $a\geq 0,\,b\geq 0,\,a+b>0,\,m>0,\,n>0,\,p\in\mathbf{R},\,q\in\mathbf{R}$. Then

1° $\displaystyle\int\limits_{0}^{+\infty}\frac{\cos(ax^2+bx)-e^{-bx}\cos(ax^2)}{x}\,\frac{p+qx^4}{m^4+n^4x^4}\,dx$

$$=\frac{\pi}{2}\frac{pn^4-qm^4}{m^4n^4}\exp\left(-\frac{m}{n}\left(a\frac{m}{n}+\frac{b}{\sqrt{2}}\right)\right)\sin\frac{bm}{n\sqrt{2}};$$

2° $\displaystyle\int\limits_{0}^{+\infty}\frac{\sin(ax^2+bx)+e^{-bx}\sin(ax^2)}{x}\,\frac{p+qx^4}{m^4+n^4x^4}\,dx$

$$=\frac{\pi p}{2m^4}-\frac{\pi}{2}\frac{pn^4-qm^4}{m^4n^4}\exp\left(-\frac{m}{n}\left(a\frac{m}{n}+\frac{b}{\sqrt{2}}\right)\right)\cos\frac{bm}{n\sqrt{2}}.$$

The above integrals were published in [1], and they are generalizations of the integrals evaluated in [2]. Further, rather complicated, generalizations can be found in [3].

REFERENCES

1. D. Dimitrovski: 'Application of Calculus of Residues to the Evaluation of some Definite Integrals', (Serbian), *Matematička biblioteka* 22 (1962), 61–70.

2. F. Guglielmino: 'Calcolo di integrali singolari mediante il teorema dei residui', *Matematiche Catania* 6 (1951), 97–112.
3. Gh. Tudor: 'The Evaluation of an Improper Integral by the Theorem of Residues (Romanian), *Bul. Stiin. Tehnic Inst. Politehn. Timişoara. Serie noua* 12 (26) (1967), 403–407.

5.3.4.22. We have

$$1° \int_0^{+\infty} \frac{\log x}{(1+x^2)^2}\, \mathrm{d}x = -\frac{\pi}{4}; \qquad 2° \int_0^{+\infty} \frac{x\log x}{(1+x^2)^3}\, \mathrm{d}x = -\frac{1}{8};$$

$$3° \int_0^{+\infty} \frac{\log x}{\sqrt{x}\,(x^2+a^2)^2}\, \mathrm{d}x = \frac{3\,\pi}{8\,a^3}\sqrt{\frac{2}{a}} \quad (a>0);$$

$$4° \int_0^{+\infty} \frac{\log x}{(1+x)^n}\, \mathrm{d}x = -\frac{1}{n-1}\left(1+\frac{1}{2}+\cdots+\frac{1}{n-2}\right) \qquad (n\in\mathbf{N},\ n\neq 1).$$

REFERENCE
Julia, pp. 196, 199, 221, 191.

5.3.4.23. We have

$$1° \int_0^{+\infty} \frac{\sqrt{x}\,\log x}{(1+x)^2}\, \mathrm{d}x = \pi; \qquad 2° \int_0^{+\infty} \log\left(\frac{e^x+1}{e^x-1}\right)\mathrm{d}x = \frac{\pi^2}{4}.$$

REFERENCE
Tisserand, pp. 487, 492.

5.3.4.24. If a is a positive number, then

$$1° \int_a^{+\infty} \frac{1}{x^2-b^2}\, \mathrm{d}x = \frac{1}{2b}\log\frac{a+b}{a-b} \qquad (0<b<a);$$

$$2° \int_a^{+\infty} \frac{1}{x^4+1}\, \mathrm{d}x = \frac{1}{2\sqrt{2}}\operatorname{arctg}\frac{a\sqrt{2}}{a^2-1} - \operatorname{arth}\frac{a\sqrt{2}}{a^2+1};$$

$$3° \int_a^{+\infty} \frac{x^2+1}{x^4-2x^2\cos 2p+1}\, \mathrm{d}x = \frac{1}{2\sin p}\operatorname{arctg}\frac{2a\sin p}{3a^2-1} \qquad \left(0<p<\frac{\pi}{2}\right);$$

$$4° \int_a^{+\infty} \frac{1}{x^2 - 2x \cos p + 1} \, dx = \frac{1}{\sin p} \, \text{arctg} \, \frac{\sin p}{a - \cos p} \qquad \left(0 < p < \frac{\pi}{2}\right);$$

$$5° \int_a^{+\infty} \frac{1}{(x^4 - 2x^2 \cos 2p + 1)^2} \, dx = \frac{1 + 4\cos^2 p}{32 \cos^2 p} \, \text{arth} \, \frac{2a \cos p}{a^2 + 1}$$

$$+ \frac{1 + 4\sin^2 p}{32 \sin^2 p} \, \text{arctg} \, \frac{2a \sin p}{a^2 - 1} - \frac{a(a^2 \cos 2p - \cos 4p)}{4(a^4 - 2a^2 \cos 2p + 1) \sin^2 2p} \qquad \left(0 < p < \frac{\pi}{2}\right).$$

REFERENCE

E. H. Neville: 'Indefinite Integration by Means of Residues', *Math. Student* 13 (1945), 16–25.

5.4. INTEGRALS WITH FINITE LIMITS

5.4.1. *Integrals of Periodic Functions*

Integrals of the form $\int_0^{2\pi} f(\cos x, \sin x) \, dx$ are very important and they often appear in various applications. Hence our first result is devoted to them.

THEOREM 1. *Let the function*

$$z \mapsto f\left(\frac{z + z^{-1}}{2}, \frac{z - z^{-1}}{2i}\right)$$

be regular on the circle $\{z \mid |z| = 1\}$, *and let it have a finite number of singularities in the disc* $\{z \mid |z| < 1\}$, *which are denoted by* $z_1, \ldots, z_n$. *Then*

$$(1) \qquad \int_0^{2\pi} f(\cos x, \sin x) \, dx = 2\pi \sum_{k=1}^{n} \text{Res}_{z=z_k} \frac{1}{z} f\left(\frac{z + z^{-1}}{2}, \frac{z - z^{-1}}{2i}\right).$$

Proof. If we put $z = e^{ix}$, the integral $\int_0^{2\pi} f(\cos x, \sin x) \, dx$ is transformed into the integral

$$(2) \qquad \oint_{|z|=1} f\left(\frac{z + z^{-1}}{2}, \frac{z - z^{-1}}{2i}\right) \frac{1}{iz} \, dz,$$

since when x varies from 0 to 2π, $z = e^{ix}$ describes the circle $\{z \mid |z| = 1\}$ in the positive direction.

Applying Cauchy's theorem on residues to (2), we immediately get (1).

REMARK 1. Notice that

$$\int_{a}^{a+2\pi} f(\cos x, \ \sin x)\,dx = \int_{0}^{2\pi} f(\cos x, \ \sin x)\,dx.$$

REMARK 2. This method is particularly useful for the evaluation of integrals of form $\int_{0}^{2\pi} R(\cos x, \sin x)\,dx$, where $R(u, v) = [P(u, v)]/[Q(u, v)]$ (P and Q are polynomials in u and v). The hypothesis of Theorem 1 for the function f is, in this case, reduced to the condition that $Q(x, y) \neq 0$ on the circle $\{(x, y)|x^2 + y^2 = 1\}$.

REMARK 3. The above result can be found in almost all books in which calculus of residues is applied to the evaluation of real definite integrals.

REMARK 4. If the integrand contains expressions $\cos nx$ or $\sin nx$, where n is an integer, then the substitution $z = e^{ix}$ gives

$$\cos nx = \frac{z^n + z^{-n}}{2}, \qquad \sin nx = \frac{z^n - z^{-n}}{2i}.$$

REMARK 5. If the integral has the form

$$\int_{0}^{2\pi} f(\cos x, \sin x) \cos nx\,dx \quad \text{or} \quad \int_{0}^{2\pi} f(\cos x, \sin x) \sin nx\,dx,$$

where n is an integer, then it is more convenient to start with the integral

$$\int_{0}^{2\pi} f(\cos x, \sin x)\,e^{inx}\,dx,$$

to replace z by e^{ix}, and after the evaluation, to separate the real and imaginary parts.

REMARK 6. For the integral

$$\int_{0}^{2\pi} R(\cos x, \sin x)\,dx = \int_{-\pi}^{\pi} R(\cos x, \sin x)\,dx$$

Angheluţă [1, p. 202] introduces the substitution

(3) $x = 2\,\mathrm{arctg}\,t, \quad dx = \dfrac{2\,dt}{1 + t^2}, \quad \sin x = \dfrac{2t}{1 + t^2}, \quad \cos x = \dfrac{1 - t^2}{1 + t^2},$

and reduces it to the form

$$\int_{-\infty}^{+\infty} R\left(\frac{2t}{1+t^2}, \frac{1-t^2}{1+t^2}\right) \frac{2\,dt}{1+t^2} = \int_{-\infty}^{+\infty} f(t)\,dt,$$

which we considered in 5.2.1.

REMARK 7. The same method can be applied to the integrals of the form

$$\int_0^{2\pi} f(\cos x, \sin x)\, g\,(e^{ix})\,dx,$$

where f satisfies the conditions of Theorem 1, and g is a regular function in the disc $\{z \,|\,|z| \le 1\}$. We then find

$$\int_0^{2\pi} f(\cos x, \sin x)\, g\,(e^{ix})\,dx = 2\pi \sum_{k=1}^{n} \operatorname*{Res}_{z=z_k} \frac{g(z)}{z}\, f\left(\frac{z+z^{-1}}{2}, \frac{z-z^{-1}}{2i}\right).$$

EXAMPLE 1. Applying the exposed method to the integral

$$\int_0^{2\pi} \frac{1}{3 + \cos x + 2 \sin x}\,dx,$$

we see that the function f, defined by

$$f(z) = \frac{-2i}{(1-2i)z^2 + 6z + 1 + 2i},$$

has only one simple pole $z_0 = (-1-2i)/5$ in the disc $\{z \,|\,|z| < 1\}$, and also $\operatorname*{Res}_{z=z_0} f(z) = -(i/2)$. Hence, the value of the given integral is $(2\pi i)\,(-i/2) = \pi$.

EXAMPLE 2. If we introduce the substitution (3) into the integral

$$\int_{-\pi}^{\pi} \frac{1}{\cos x + 2}\,dx$$

we find

$$\int_{-\pi}^{\pi} \frac{1}{\cos x + 2}\,dx = \int_{-\infty}^{+\infty} \frac{1}{t^2 + 3}\,dt = \frac{2\pi}{\sqrt{3}}.$$

THEOREM 2. *Suppose that the following conditions are fulfilled:*

$1°$ *The function f is analytic in the extended plane, where it can have a finite number of singularities;*

$2°$ *On the arc* $l = \{z \,|\, |z| = 1, \alpha < \arg z < \beta\}$ *the function f can only have simple poles;*

$3°$ *f is regular at* $z = e^{i\alpha}$ *and* $z = e^{i\beta}$.

Then

$$(4) \qquad \int_{\alpha}^{\beta} f(e^{ix})\, dx = i\left(\sum_{k=1}^{n} \operatorname*{Res}_{z=z_k} \frac{f(z)}{z} \log e^{i\mu} \frac{z - e^{i\alpha}}{z - e^{i\beta}} \right.$$

$$\left. + \sum_{k=1}^{m} \operatorname*{Res}_{z=a_k} \frac{f(z)}{z} \log\left(- e^{i\mu} \frac{z - e^{i\alpha}}{z - e^{i\beta}} \right) \right),$$

where $\mu = (\beta - \alpha)/2$, *and where* $z_1, \ldots, z_n$ *are singularities of the function*

$$z \mapsto \frac{f(z)}{z} \log e^{i\mu} \frac{z - e^{i\alpha}}{z - e^{i\beta}}$$

which do not belong to l, and $a_1, \ldots, a_m$ *are the singularities of the function*

$$z \mapsto \frac{f(z)}{z} \log\left(- e^{i\mu} \frac{z - e^{i\alpha}}{z - e^{i\beta}} \right)$$

belonging to l.

Proof. Since $\alpha < \beta < \alpha + 2\pi$ and $\mu = (\beta - \alpha)/2$, we have $0 < \mu < \pi$. Put

$$(5) \qquad z = \frac{\sin \dfrac{x - \alpha}{2}}{\sin \dfrac{\beta - x}{2}} = e^{i\mu} \frac{e^{ix} - e^{i\alpha}}{e^{i\beta} - e^{ix}}.$$

Then

$$\frac{dz}{dx} = \frac{\sin \mu}{2 \sin^2 \dfrac{\beta - x}{2}} > 0,$$

so that the arc l is bijectively mapped onto the interval $(0, +\infty)$. We also get

$$e^{ix} = e^{i\beta} \frac{z + e^{-i\mu}}{z + e^{i\mu}}, \qquad \frac{dx}{dz} = \frac{2 \sin \mu}{(z + e^{i\mu})(z + e^{-i\mu})},$$

and so

$$(6) \qquad \int_{\alpha}^{\beta} f(e^{ix})\, dx = \int_{0}^{+\infty} F(z)\, dz,$$

where

$$(7) \qquad F(z) = \frac{2 \sin \mu}{(z + e^{i\mu})(z + e^{-i\mu})} f\left(e^{i\beta} \frac{z + e^{-i\mu}}{z + e^{i\mu}}\right).$$

We have

$$\lim_{z \to 0} e^{i\beta} \frac{z + e^{-i\mu}}{z + e^{i\mu}} = e^{i\alpha}, \qquad \lim_{|z| \to +\infty} e^{i\beta} \frac{z + e^{-i\mu}}{z + e^{i\mu}} = e^{i\beta}.$$

Since the function f is regular at $z = e^{i\alpha}$ and $z = e^{i\beta}$ from (7) follows that F is regular at $z = 0$ and that $F(z) = O(|z|^{-2})$ as $|z| \to +\infty$. Hence, we may apply Theorem 4 from 5.3.1 to the function F to obtain

$$(8) \qquad \int_0^{+\infty} F(z)\, dz = -\left(\sum_{k=1}^{r} \operatorname*{Res}_{z=\zeta_k} F(z) \log(-z) + \sum_{k=1}^{s} \operatorname*{Res}_{z=a_k} F(z) \log z\right),$$

where $\zeta_1, \ldots, \zeta_r$ are singularities of F which are not on the positive real axis, while $\alpha_1, \ldots, \alpha_s$ are simple poles of F which belong to the positive real axis.
Putting

$$H(z) = e^{i\beta} \frac{z + e^{-i\mu}}{z + e^{i\mu}} \quad \text{i} \quad G(w) = -\frac{e^{-i\beta}}{2 \sin \mu} \frac{(e^{i\beta} - w)^2}{w} F(w) \log e^{i\mu} \frac{w - e^{i\alpha}}{w - e^{i\beta}}.$$

into the formula

$$\operatorname*{Res}_{z=\zeta} G(H(z)) = \operatorname*{Res}_{w=H(\zeta)} G(w)\, h'(w) \qquad (h = H^{-1}),$$

(see Theorem 7 from 2.1.2) we get

$$(9) \qquad \sum_{k=1}^{r} \operatorname*{Res}_{z=\zeta_k} F(z) \log(-z) = -i \sum_{k=1}^{n} \operatorname*{Res}_{z=z_k} \frac{f(z)}{z} \log\left(e^{i\mu} \frac{z - e^{i\alpha}}{z - e^{i\beta}}\right).$$

Similarly, we obtain the formula

$$(10) \qquad \sum_{k=1}^{s} \operatorname*{Res}_{z=a_k} F(z) \log z = -i \sum_{k=1}^{m} \operatorname*{Res}_{z=a_k} \frac{f(z)}{z} \log e^{i\mu} \frac{z - e^{i\alpha}}{e^{i\beta} - z}.$$

From (6), (8), (9) and (10) follows (4).

REMARK 8. This result is due to Boas and Schoenfeld [2].
The integral $\int_\alpha^\beta f(e^{ix})\, dx$ was considered by Boas [3], where he made a mistake, later corrected in [4].

REMARK 9. If $\alpha = 0, \beta = \pi$, the transformation (5) becomes the standard transformation $z = \mathrm{tg}(x/2)$, which reduces the given integral to an integral of a rational function.

REMARK 10. The case $\alpha + 2\pi = \beta$ was considered in Theorem 1.

REMARK 11. The integral $\int_\alpha^\beta f(e^{ix})\,dx$ is connected to the integral $\int_\alpha^\beta g(\cos x, \sin x)$ dx, since we have

$$f(e^{ix}) = f(\cos x + i \sin x); \quad g(\cos x, \ \sin x) = g\left(\frac{e^{ix}+e^{-ix}}{2}, \ \frac{e^{ix}-e^{-ix}}{2i}\right).$$

REMARK 12. In connection with the considered class of definite integrals, we also mention the following result of Dimitrovski [5]:

Let $0 \le \alpha < \beta \le 2\pi$ and let R be a rational function in two variables which has no singularities on the segment $[\alpha, \beta]$ nor on half-lines $\{z\mid \mathrm{Re}\,z = \alpha, 0 \le \mathrm{Im}\,z < +\infty\}$, $\{z\mid \mathrm{Re}\,z = \beta, 0 \le \mathrm{Im}\,z < +\infty\}$. Then

$$\int\limits_\alpha^\beta e^{ix} R(\sin x, \ \cos x)\,dx = 2\pi e^{i\alpha} \sum \mathrm{Res}\,R(\alpha, \ z)$$

$$+ \sum \mathrm{Res}\,\frac{i\log z}{(1+z)^2}\left(e^{i\beta} R\left(\beta, \ \frac{1}{1+z}\right) - e^{i\alpha} R\left(\alpha, \ \frac{1}{1+z}\right)\right),$$

where the first sum is taken over all the singularities of $z \mapsto R(\alpha, z)$ in the region $\{z\mid |z| < 1, 0 \le \arg z \le \beta - \alpha\}$, and the second over all the singularities of the mentioned function which do not belong to the half-line $\{z\mid \mathrm{Im}\,z = 0, 0 \le \mathrm{Re}\,z < +\infty\}$.

EXAMPLE 3. Applying this method to the integral

$$I = \int\limits_0^{\pi/2} e^{\cos x}(\cos \sin x)\,dx = \mathrm{Re}\int\limits_0^{\pi/2} e^{\cos x}\,e^{i\sin x}\,dx$$

we get

$$I = \frac{\pi}{2} + \sum_{m=1}^{+\infty}\frac{(-1)^{m-1}}{(2m-1)!\,(2m-1)}.$$

EXAMPLE 4. If $0 < a < 1$, then

$$\int\limits_0^\pi \frac{1}{(a+\sin x)^2}\,dx = \frac{2}{a(1-a^2)} - \frac{a}{(1-a^2)^{3/2}}\log\frac{1+\sqrt{1-a^2}}{1-\sqrt{1-a^2}}.$$

THEOREM 3. *Suppose that the following conditions are satisfied:*

$1°$ *The function f is analytic in the region* $\{z \mid 0 \leq \operatorname{Re} z < 2\pi, \operatorname{Im} z > 0\}$ *where it can have only a finite number of singularities;*

$2°$ *The function f is regular at $z = 0$;*

$3°$ *On the interval $(0, 2\pi)$ of the real axis, f can have only simple poles* $a_1, \ldots, a_m$;

$4°$ $(e^{-iz})^s f(z) = ic + o(1)$ *as* $|z| \to +\infty$;

$5°$ $f(z + 2\pi) = f(z)$ *when* $\operatorname{Im} z \geq 0$.

Then

$$(11) \quad \sum_{j=0}^{s} \left\{ \binom{s}{j} \left(\frac{i}{2}\right)^{s-j} \text{v.p.} \int_0^{2\pi} f(x)(x-\pi)^{s-j} \log^j \left(2 \sin \frac{x}{2}\right) dx \right\}$$

$$= 2\pi i \sum_{k=1}^{n} \operatorname*{Res}_{z=z_k} \left(f(z) \log^s(1 - e^{iz})\right) + i \sum_{k=1}^{m} \operatorname*{Res}_{z=a_k} \left(f(z) \log^s(1 - e^{iz})\right) + 2\pi ci.$$

The proof of this theorem, given in [6], is rather long, and we shall leave it out. In essence, it consists of an application of Cauchy's theorem on residues to the integral of the function $z \mapsto f(z) \log^s(1 - e^{iz})$ along the contour given on Figure 5.4.1, where log denotes the principal value of the logarithm.

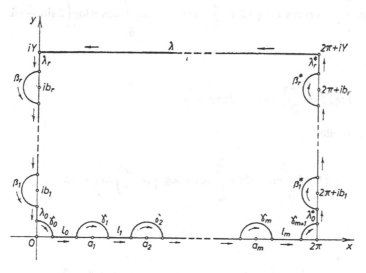

Fig. 5.4.1.

REMARK 13. Theorem 3 is useful for evaluating integrals of the form

$$\int_0^{2\pi} f(x) \log \sin \frac{x}{2}\, dx.$$

EXAMPLE 5. In order to evaluate the integral

$$\int_0^{2\pi} \cos x \cdot \log \sin \frac{x}{2}\, dx,$$

take $s = 1$, $f(z) = \cos z$ in Theorem 3. Then

$$-e^{iz} f(z) = -\frac{1}{2}\,(e^{2iz} + 1) = -\frac{1}{2} + o\,(1),$$

and hence $c = i/2$.

The function $z \mapsto \cos z$ has no singularities in the finite plane. Therefore, from (11) follows

$$\frac{i}{2}\,\text{v.p.} \int_0^{2\pi} (x - \pi)\cos x \cdot \log \left(2 \sin \frac{x}{2}\right) dx + \text{v.p.} \int_0^{2\pi} \cos x \cdot \log \left(2 \sin \frac{x}{2}\right) dx = -\pi,$$

which implies

$$(12)\ \ \text{v.p.} \int_0^{2\pi} (x - \pi)\cos x \cdot \log \left(2 \sin \frac{x}{2}\right) dx = 0, \ \ \text{v.p.} \int_0^{2\pi} \cos x \cdot \log \left(2 \sin \frac{x}{2}\right) dx = -\pi.$$

However,

$$\log \left(2 \sin \frac{x}{2}\right) = \log 2 + \log \sin \frac{x}{2}\,,$$

and from (12) follows

$$\text{v.p.} \left(\log 2 \int_0^{2\pi} \cos x\, dx + \int_0^{2\pi} \cos x \cdot \log \left(\sin \frac{x}{2}\right) dx\right) = -\pi,$$

i.e.

$$(13) \qquad \int_0^{2\pi} \cos x \cdot \log \sin \frac{x}{2}\, dx = -\pi, \quad \text{since} \quad \int_0^{2\pi} \cos x\, dx = 0.$$

The sign for the principal value is omitted, since the integral (13) exists.

THEOREM 4. *Suppose that the following conditions are fulfilled:*

1° *The function f is analytic in the region* $G = \{z \mid 0 \leq \operatorname{Re} z \leq 2\pi, \operatorname{Im} z > 0\}$ *where it can have a finite number of singularities;*

2° *On the interval* $(0, 2\pi)$ *of the real axis the function f can have only simple poles* $a_1, \ldots, a_n$;

3° *On the half-line* $\{z \mid \operatorname{Re} z = 0, \operatorname{Im} z > 0\}$ *the function f can have only simple poles* $ib_1, \ldots, ib_r$;

4° $zf(z) = ic + o(1)$ *as* $|z| \to +\infty$ *in* G;

5° $f(z + 2\pi) = f(z)$ *when* $\operatorname{Im} z \geq 0$.

If the integral v.p. $\int_0^{+\infty} f(iy)\, dy$ *exists, then the integral* v.p. $\int_0^{2\pi} xf(x)\, dx$ *also exists, and*

$$(14) \quad \text{v.p.} \int_0^{2\pi} xf(x)\, dx = 2\pi i \sum_{k=1}^{n} \operatorname*{Res}_{z=z_k} zf(z) + 2\pi^2 i \sum_{k=1}^{r} \operatorname*{Res}_{z=ib_k} f(z)$$

$$+ \pi i \sum_{k=1}^{m} \operatorname*{Res}_{z=a_k} zf(z) + 2\pi ci + \frac{1}{2}\pi i \operatorname*{Res}_{z=2\pi} zf(z) - 2\pi i \ \text{v.p.} \int_0^{+\infty} f(iy)\, dy.$$

The proof is carried out by applying Cauchy's theorem on residues to the integral $\oint zf(z)\, dz$, where the notations are the same as in Theorem 3.

REMARK 14. Theorem 4 is proved in [6].

EXAMPLE 6. Let $f(z) = 1/(2 + \cos z)$. Then in the region G we have $\lim_{|z| \to +\infty} zf(z) = 0$, and so $c = 0$. The function f is regular in all points of the segment $[0, 2\pi]$. Furthermore, if $z = x + iy$, for $x = 0$ f becomes $1/(2 + \cos iy)$, and $2 + \cos iy > 0$ for $y \in \mathbb{R}$. This means that f has no singularities on the imaginary axis.

Therefore, according to (14) we get

$$\text{v. p.} \int_0^{2\pi} \frac{x}{2 + \cos x}\, dx = 2\pi i \, \Sigma \operatorname{Res} zf(z) - 2\pi i \left(\text{v. p.} \int_0^{+\infty} \frac{1}{2 + \cos iy}\, dy \right),$$

where the summation is taken over all the singularities of the function $z \mapsto zf(z)$ in the region

$$D = \{z \mid 0 < \operatorname{Re} z < 2\pi, \ \operatorname{Im} z > 0\}.$$

The expression $2 + \cos z$ is equal to 0 if and only if

$$z = z_k = -i \log\left(2 \pm \sqrt{3}\right) + (2k+1)\pi \qquad (k \in \mathbb{Z}).$$

In the region D there is only one of those singularities, namely z_0. Hence, we have

$$\int_0^{2\pi} \frac{x}{2+\cos x}\,dx = 2\pi i \operatorname*{Res}_{z=z_0} zf(z) - 2\pi i \int_0^{+\infty} \frac{1}{2+\cos iy}\,dy$$

$$= 2\pi i \lim_{z\to z_0} (z-z_0)\,zf(z) - 2\pi i \int_0^{+\infty} \frac{1}{2+\operatorname{ch} y}\,dy$$

$$= -\frac{2\pi i}{\sqrt{3}}\left(i\pi + \log(2-\sqrt{3})\right) - 2\pi i \int_0^{+\infty} \frac{1}{2+\operatorname{ch} y}\,dy,$$

and, separating the real and imaginary parts, we find

$$\int_0^{2\pi} \frac{x}{2+\cos x}\,dx = \frac{2\pi^2}{\sqrt{3}}, \quad \int_0^{+\infty} \frac{1}{2+\operatorname{ch} y}\,dy = \frac{1}{\sqrt{3}}\log(2+\sqrt{3}).$$

REFERENCES
1. Th. Angheluță: *A Course of the Theory of Functions of a Complex Variable*, (Romanian), București 1957, p. 202.
2. R. P. Boas, Jr. and L. Schoenfeld: 'Indefinite Integration by Residues', *SIAM Review* 8 (1966), 173–183.
3. R. P. Boas, Jr.: 'Indefinite Integration by Residues', *Am. Math. Monthly* 71 (1964), 298–300.
4. R. P. Boas, Jr.: 'Correction for "Indefinite Integration by Residues" ', *Am. Math. Monthly* 71 (1964), 906.
5. D. S. Dimitrovski: 'Note sur les intégrales trigonométriques', *Bull. Soc. Math. Phys. Macédoine* 24 (1973), 91–94.
6. J. T. Baskin: 'Some General Theorems for Evaluating Definite Integrals by Cauchy's Residue Theorem', M. A. Thesis, The Pennsylvania State University, U.S.A., 1962.

5.4.2. *Integrals of Nonperiodic Functions*

The evaluation of the definite integral $\int_a^b f(x)\,dx$, where $-\infty < a < b < +\infty$ by means of residues is effected by introducing a suitable transformation which transforms $\int_a^b f(x)\,dx$ into an integral of the form $\int_0^{+\infty} F(t)\,dt$.

THEOREM 1. *Suppose that the following conditions are satisfied:*

   1° *The function f is analytic in the extended plane, where it can have a finite number of singularities;*

$2°$ On the interval (a, b) of the real axis f may have only simple poles;
$3°$ f is regular at $z = a$ and $z = b$.
Then

$$(1) \qquad \int_a^b f(x)\,dx = -\left(\sum_{k=1}^{n} \operatorname*{Res}_{z=z_k} f(z) \log \frac{z-a}{z-b} + \sum_{k=1}^{m} \operatorname*{Res}_{z=a_k} f(z) \log \frac{z-a}{b-z} \right),$$

where $z_1, \ldots, z_n$ are the singularities of f which do not belong to (a, b), and $a_1, \ldots, a_m$ are the poles from (a, b).

Proof. Introduce the transformation $t = (x - a)/(b - x)$. Then

$$(2) \qquad \int_a^b f(x)\,dx = \int_0^{+\infty} \frac{b-a}{(t+1)^2} f\left(\frac{bt+a}{t+1}\right) dt = \int_0^{+\infty} F(t)\,dt.$$

From (2) we conclude that F is an analytic function in the finite plane, where it can have a finite number of singularities, while in the interval $(0, +\infty)$ it can only have simple poles. Besides, F is regular at $t = 0$. Furthermore, since

$$\lim_{z \to \infty} \frac{bz + a}{z+1} = b,$$

and since f is regular at $z = b$, we conclude that $F(z) = O(|z|^{-2})$, as $|z| \to +\infty$. Hence, we may apply the result given in Theorem 4 from 5.3.1 to the function F, and we obtain

$$(3) \qquad \int_0^{+\infty} \frac{b-a}{(t+1)^2} f\left(\frac{bt+a}{t+1}\right) dt$$

$$= -\left(\sum_1 \operatorname{Res} \frac{b-a}{(1+z)^2} f\left(\frac{bz+a}{1+z}\right) \log(-z) + \sum_2 \operatorname{Res} \frac{b-a}{(1+z)^2} f\left(\frac{bz+a}{1+z}\right) \log z \right),$$

where the first sum refers to the singularities of F which do not belong to $(0, +\infty)$, and the second to the poles of F from $(0, +\infty)$.

However, setting

$$H(z) = \frac{bz+a}{1+z}, \qquad G(w) = \frac{(b-w)^2}{b-a} f(w) \log \frac{w-a}{w-b},$$

into

$$\operatorname*{Res}_{z=\zeta} G\big(H(z)\big) = \operatorname*{Res}_{w=H(\zeta)} G(w)\, h'(w) \qquad (h = H^{-1})$$

(see Theorem 7 from 2.1.2) we find

$$(4) \qquad \sum_1 \operatorname*{Res} \frac{b-a}{(1+z)^2}\, f\!\left(\frac{bz+a}{1+z}\right) \log(-z) = \sum_{k=1}^{n} \operatorname*{Res}_{z=z_k} f(z)\log\frac{z-a}{z-b}.$$

Similarly we prove

$$(5) \qquad \sum_2 \operatorname*{Res} \frac{b-a}{(1+z)^2}\, f\!\left(\frac{bz+a}{1+z}\right) \log z = \sum_{k=1}^{m} \operatorname*{Res}_{z=a_k} f(z)\log\frac{z-a}{b-z}.$$

From (2)–(5) follows (1).

REMARK 1. This result is due to Boas and Schoenfeld [1].

REMARK 2. Formulas of the form (1) existed before the publication of [1], but under the supposition that f is regular on (a, b). Hence, (1) is a generalization of those results, though the authors of [1] do not mention that fact.

The following equality can be found in [2] and [3]:

$$(6) \qquad \int_a^b f(x)\,dx = \sum_{k=1}^{n} \operatorname*{Res}_{z=z_k} f(z)\log\frac{z-a}{z-b}.$$

It is obtained by the same method as (1), but as we said before, f is supposed to be regular on (a, b).

Prešić [4, problem 235, pp. 474–476] evaluated the integral $\int_1^2 (1/(1+x^3))dx$ by the method of residues, and posed as a problem the possibility of evaluating integrals $\int_a^b R(x)\,dx$, where R is a rational function and $\int_a^b (\sin x/x)\,dx$, by a similar method.

Dimitrovski [5] proved the following result:

Let R be a rational function, continuous on $[a, b]$. Then

$$\int_a^b R(x)\,dx = -4(b-a)^2 \sum \operatorname*{Res} \frac{z\log z}{(1+(b-a)z^2)^2}\, R\!\left(\frac{a(b-a)z^2+b}{1+(b-a)z^2}\right),$$

where the summation is taken over all the singularities of the function

$$z \mapsto \frac{z\log z}{(1+(b-a)z^2)^2}\, R\!\left(\frac{a(b-a)z^2+b}{1+(b-a)z^2}\right)$$

in the upper half-plane.

Also, in [6] we find the formula

$$\int_a^b R(x)\,dx = -\sum_{k=1}^n \operatorname*{Res}_{z=z_k} R(z) \log \frac{z-a}{z-b},$$

where R is a rational function, namely $R(z) = P(z)/Q(z)$, where P and Q are polynomials such that $\deg P \le \deg Q - 1$, and $Q(x) \ne 0$ on $[a, b]$, and where $z_1, \ldots, z_n$ are the zeros of Q.

REMARK 3. As far as we could find, calculus of residues was first applied to the integrals of the form $\int_a^b f(x)\,dx$ in book [2], first edition from 1955. No references regarding that integral are given there. We asked the authors of [2], Behnke and Sommer, to inform us about the origins of the method. They replied that the result was not theirs, and that they could not remember where they took it from.

REMARK 4. In view of (2), we can apply any one of the formulas given in 5.3.1 to the integral on the right-hand side of (2), which may lead to different formulas for the integral $\int_a^b f(x)\,dx$. So, for example, Dimitrovski and Adamović [7], starting with (5) from 5.3.1, proved the following result:

If f is an analytic function with a finite number of singularities which do not belong to the segment $[a, b]$, then

(7) $$\int_a^b f(x)\,dx = (a-b)\sum_{k=1}^n \operatorname*{Res}_{z=z_k} f\left(\frac{az+b}{z+1}\right)\frac{\log z}{(1+z)^2},$$

where $z_1, \ldots, z_n$ are the singularities which do not belong to the half-line $\{z\mid \operatorname{Im} z = 0, \operatorname{Re} z \ge 0\}$, of that branch of the function

$$z \mapsto f\left(\frac{az+b}{z+1}\right)\frac{\log z}{(1+z)^2}$$

which corresponds to the cut $[0, +\infty)$ and for which $\arg z \in [0, 2\pi)$.

EXAMPLE 1. Boas and Schoenfeld [1] applied (1) to the evaluation of the integral

$$I = \int_a^b \frac{1}{x^4+1}\,dx.$$

After a lengthy calculation they found that

(8) $$I = \frac{1}{4\sqrt{2}}\left(\log\frac{1-a\sqrt{2}+a^2}{1+a\sqrt{2}+a^2} + \log\frac{1+b\sqrt{2}+b^2}{1-b\sqrt{2}+b^2}\right)$$

$$+ \frac{1}{2\sqrt{2}}\left(\operatorname{arctg}\frac{b-a}{\sqrt{2}-a-b+ab\sqrt{2}} + \operatorname{arctg}\frac{b-a}{\sqrt{2}+a+b+ab\sqrt{2}}\right).$$

In particular, for $x > 0$ we find

$$(9) \quad \int_0^x \frac{1}{x^4+1}\,dx = \frac{1}{4\sqrt{2}}\log\frac{x^2+x\sqrt{2}+1}{x^2-x\sqrt{2}+1}$$

$$+\frac{1}{2\sqrt{2}}\left(\operatorname{arctg}\frac{x}{\sqrt{2}-x}+\operatorname{arctg}\frac{x}{\sqrt{2}+x}\right).$$

which, at the same time, corrects the wrong formulas present in [8], formula 170, and [9], formula 61.

Notice, however, that the function f defined by $f(z) = 1/(z^4 + 1)$, has no singularities on any real interval of the form (a, b), and hence Boas and Schoenfeld obtained (8) and (9) as a result of the simpler formula (6).

EXAMPLE 2. Applying (7) to the integral $\int_1^2 e^{1/x}\,dx$, the authors of [7] obtained the equality

$$\int_1^2 e^{1/x}\,dx = \log 2 + e\sum_{k=1}^{+\infty}\frac{1}{2^k k}\sum_{n=k+1}^{+\infty}(-1)^n\frac{n-k}{n!}.$$

We give one more class of real definite integrals with finite limits which are evaluated directly, i.e. without reduction to previous cases.

THEOREM 2. *Suppose that the following conditions are satisfied:*

$1°$ *The function f is analytic in the complex plane; its singularities are denoted by $z_1,\ldots,z_n$;*

$2°$ *The function f is regular on the segment $\{z\,|\,z\in[a,b]\}$;*

$3°$ $\lim_{z\to\infty} z^{r+s+1}\,f(z) = A \neq \infty$;

$4°$ $r,s\in(-1,1), r+s=-1, 0$ *or* 1.

Then

$$(10)\int_a^b (x-a)^r(b-x)^s f(x)\,dx = \frac{\pi}{\sin\pi s}\left(\sum_{k=1}^n \operatorname*{Res}_{z=z_k}(z-a)^r(z-b)^s f(z) - A\right).$$

Proof. Let $K = \{z\,|\,|z|=R\}$, where R is large enough so that the circle K contains all the singularities of f, let $k_1 = \{z\,|\,|z-a|=\epsilon\}$, $k_2 = \{z\,|\,|z-b|=\epsilon\}$, and let $\Gamma = K\cup C$, where C is the contour shown on Figure 5.4.2.

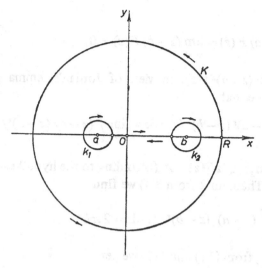

Fig. 5.4.2.

According to Cauchy's theorem on residues we have

(11) $$\oint_\Gamma (z-a)^r (z-b)^s f(z)\, dz$$

$$= 2\pi i \sum_{k=1}^{n} \operatorname*{Res}_{z=z_k} (z-a)^r (z-b)^s f(z).$$

However,

(12) $$\oint_\Gamma (z-a)^r (z-b)^s f(z)\, dz = e^{i\pi s} \int_{a+\varepsilon}^{b-\varepsilon} (x-a)^r (b-x)^s f(x)\, dx$$

$$- e^{-i\pi s} \int_{a+\varepsilon}^{b-\varepsilon} (x-a)^r (b-x)^s\, dx + \int_{k_1} (z-a)^r (z-b)^s f(z)\, dz$$

$$+ \int_{k_2} (z-a)^r (z-b)^s f(z)\, dz + \int_{K} (z-a)^r (z-b)^s f(z)\, dz$$

$$= 2i \sin \pi s \int_{a+\varepsilon}^{b-\varepsilon} (x-a)^r (b-x)^s f(x)\, dx + \int_{k_1} (z-a)^r (z-b)^s f(z)\, dz$$

$$+ \int_{k_2} (z-a)^r (z-b)^s f(z)\, dz + \int_{K} (z-a)^r (z-b)^s f(z)\, dz.$$

Since

$$\lim_{z \to a}(z-a)F(z) = \lim_{z \to b}(z-b)F(z) = 0,$$

where $F(z) = (z-a)^r(z-b)^s f(z)$, in view of Jordan's lemma (Theorem 1 from 3.4) we conclude that

$$\lim_{\varepsilon \to 0}\int_{k_1}(z-a)^r(z-b)^s f(z)\,dz = \lim_{\varepsilon \to 0}\int_{k_2}(z-a)^r(z-b)^s f(z)\,dz = 0.$$

Further, since $\lim_{z \to \infty} zF(z) = A$ (according to the hypothesis $3°$) in view of Jordan's lemma (Theorem 2 from 3.4) we find

$$\lim_{R \to +\infty}\int_K (z-a)^r(z-b)^s f(z)\,dz = 2\pi iA.$$

If $\varepsilon \to 0, R \to +\infty$, from (11) and (12) we get

$$2i\sin \pi s \int_a^b (x-a)^r(b-x)^s f(x)\,dx + 2\pi iA = 2\pi i\sum_{k=1}^n \operatorname*{Res}_{z=z_k}(z-a)^r(z-b)^s f(z),$$

which implies (10).

REMARK 5. Formula (10) is given in the book [6], but under the condition that f is a rational function without poles in (a, b).

REMARK 6. If $\lim_{z \to \infty} z^{r+s+1} f(z) = \infty$, then start with the integral

$$\int_a^b \frac{(x-a)^r(b-x)^s f(x)}{(1-\lambda x)^k}\,dx,$$

where $k \in \mathbf{N}$ is sufficiently large to ensure that

$$\lim_{z \to \infty} \frac{z(z-a)^r(b-z)^s f(z)}{(1-\lambda z)^k}$$

is finite, and then

$$\int_a^b (x-a)^r(b-x)^s f(x)\,dx = \lim_{\lambda \to 0}\int_a^b \frac{x(x-a)^r(b-x)^s f(x)}{(1-\lambda x)^k}\,dx.$$

REMARK 7. If r, or s, is positive, f can be allowed to have a simple pole at $z = a$, or $z = b$, respectively.

REMARK 8. If f has a finite numbers of simple poles $a_1, \ldots, a_m$ in the interval (a, b), then we have

$$
\text{v.p.} \int_a^b (x-a)^r (b-x)^s f(x) \, dx =
$$

$$
= \frac{\pi}{\sin \pi s} \left(\sum_{k=1}^n \operatorname*{Res}_{z=z_k} (z-a)^r (z-b)^s f(z) - A \right) + \pi \cotg \pi s \sum_{k=1}^m \operatorname*{Res}_{z=a_k} (z-a)^r (b-z)^s f(z).
$$

REMARK 9. The following result is given in [10]:

Suppose that the rational function R fulfils the conditions:

$1°$ $R(z) = O((z-a)^{-p})$ as $z \to a$, and $R(z) = O((z-b)^q)$ as $z \to b$, where p and q are integers such that $p < q$;

$2°$ The function R has only simple poles in the interval (a, b);

$3°$ $p < \operatorname{Re} \alpha < q$.

Then

$$
\text{v.p.} \int_a^b \left(\frac{x-a}{b-x} \right)^{\alpha-1} R(x) \, dx
$$

$$
= \frac{\pi}{\sin \pi \alpha} \sum \operatorname{Res} R(z) \left(\frac{z-a}{z-b} \right)^{\alpha-1} + \pi \cotg \pi \alpha \sum_{k=1}^m \operatorname*{Res}_{z=a_k} R(z) \left(\frac{z-a}{b-z} \right)^{\alpha-1},
$$

where the first summation is taken over all the singularities of the function f which do not belong to the segment $[a, b]$, including the point ∞.

It is easily seen that this result is a special case of Theorem 2.

EXAMPLE 3. For $r = s = -\frac{1}{2}$, $f(x) = 1$, we have $\lim_{z \to \infty} z^{r+s+1} f(z) = 1$, and so

$$
\int_a^b \frac{1}{\sqrt{(x-a)(b-x)}} \, dx = \pi.
$$

EXAMPLE 4. In order to evaluate the integral

$$
I = \int_a^b \sqrt{(x-a)(b-x)} \, dx,
$$

start with the integral

$$
I(\lambda) = \int_a^b \frac{\sqrt{(x-a)(b-x)}}{(1-\lambda x)^2} \, dx.
$$

Since

$$
\lim_{z \to \infty} \frac{z}{(1-\lambda z)^2} \sqrt{(z-a)(z-b)} = \frac{1}{\lambda^2},
$$

we get

$$I(\lambda) = \frac{\pi}{\lambda^2} \frac{d}{dz} \left(\sqrt{(z-a)(z-b)} \right) \Bigg|_{z=1/\lambda} - \frac{\pi}{\lambda^2} = \frac{\pi}{2} \frac{2-\lambda(a+b)-2\sqrt{(1-a\lambda)(1-b\lambda)}}{\lambda^2 \sqrt{(1-a\lambda)(1-b\lambda)}}.$$

However, $I = \lim_{\lambda \to 0} I(\lambda)$. By a repeated application of L'Hospital's rule, we finally obtain

$$\int_a^b \sqrt{(x-a)(b-x)}\, dx = \frac{\pi}{8}(b-a)^2.$$

REFERENCES

1. R. P. Boas, Jr. and L. Schoenfeld: 'Indefinite Integration by Residues', *SIAM Review* 8 (1966), 173–183.
2. H. Behnke und F. Sommer: *Theorie der analytischen Funktionen einer komplexen Veränderlichen*, First edition, Berlin-Göttingen-Heidelberg 1955, pp. 198–202; Second edition, Berlin-Göttingen-Heidelberg 1962, pp. 212–215.
3. L. I. Volkovyskiĭ, G. L. Lunc, and I. G. Aramanovič: *A Collection of Exercices from theory of Functions of a Complex Variable*', (Russian), Moscow 1962, Problem 906.
4. D. S. Mitrinović: *A Collection of Mathematical Problems I*', (Serbian), Third edition, Belgrade 1962.
5. D. Dimitrovski: 'On a Method of Evaluating Definite Integrals of Rational Functions by Means of Finite Sums', (Macedonian), *Bull. Soc. Math. Phys. Macédoine* 13 (1962), 21–32.
6. Garnir–Gobert, pp. 102–107.
7. D. S. Dimitrovski and D. D. Adamović: 'Sur quelques formules du calcul des résidus', *Mat. Vesnik* 1 (16) (1964), 113–117.
8. H. B. Dwight: *Tables of Integrals and Other Mathematical Data*, New York 1961.
9. B. O. Pierce: *A Short Table of Integrals*, Boston 1956.
10. M. A. Evgrafov, Ju. V. Sidorov, M. V. Fedoryuk, M. I. Šabunin, and K. A. Bežanov: *A Collection of Exercices on the Theory of Analytic Functions*', (Russian), Moscow 1969, p. 279.

5.4.3. *Miscellaneous Integrals*

5.4.3.1. Let r (Re $r > -1$) and t be complex numbers. Then

$$(1) \qquad \int_{-\pi/2}^{\pi/2} \left(\frac{\cos x}{t \cos x + i \sin x} \right)^r dx = \begin{cases} \dfrac{\pi}{(t+1)^r} & (\operatorname{Re} t > 0); \\[2ex] \dfrac{\pi}{(t-1)^r} & (\operatorname{Re} t < 0). \end{cases}$$

Proof. Put

$$z = \frac{\cos x}{t \cos x + i \sin x}$$

Then

$$(2) \qquad \int\limits_{-\pi/2}^{\pi/2} \left(\frac{\cos x}{t \cos x + i \sin x}\right)^r dx = \int\limits_C \frac{iz^r}{(1-t^2)z^2 + 2tz - 1} \, dz,$$

where

$$C = \left\{ z \,\middle|\, \left|z - \frac{1}{2t}\right| = \frac{1}{2|t|} \right\}$$

is traversed in the clockwise or counterclockwise directions according as $\mathrm{Re}\, t > 0$, or $\mathrm{Re}\, t < 0$, respectively. The function

$$z \mapsto \frac{iz^r}{(1-t^2)z^2 + 2tz - 1}$$

has simple poles $1/(t+1)$ and $1/(t-1)$. If $\mathrm{Re}\, t > 0$, the first is inside C, and the second outside C, and if $\mathrm{Re}\, t < 0$, the second is inside C, and the first outside C. Applying Cauchy's theorem on residues, we immediately obtain (1).

REMARK. The integral appearing in (2) has a singularity at $z = 0$. Hence, C has to be indented, so that $z = 0$ lies outside C. The standard limiting process, using the restriction $\mathrm{Re}\, r > -1$ is then applied.

REFERENCE
L. Carlitz and M. R. Spiegel: 'Problem 5110', *Am. Math. Monthly* **71** (1964), 570–571.

5.4.3.2. If $a > -1$, then

$$\int\limits_0^{\pi/2} \cos^a x \, \frac{\sin(a+1)x}{\sin x} \, dx = \frac{\pi}{2}.$$

REFERENCE
B. Crstici, and R. Meynieux: 'Compléments au traité de D. S. Mitrinović (IV): Sur une intégrale dépendant d'un paramètre réel', *Univ. Beograd. Publ. Elektrotehn. Fak. Ser. Mat. Fiz.* **498–541** (1975), 155–158.

5.4.3.3. If $p > 0, a < 1$, then:

$$1° \quad \int\limits_0^\pi \frac{\cos(1-a)x}{\sin^a x} \, \frac{\sin x}{\mathrm{ch}\, 2p - \cos 2x} \, dx = \frac{\pi}{2} \frac{e^{-(1-a)p}}{\mathrm{ch}\, p \,\mathrm{sh}^a p} \sin \frac{a\pi}{2};$$

$$2° \int_0^\pi \frac{\sin(1-a)x}{\sin^a x} \frac{\sin x}{\text{ch}\, 2p - \cos 2x} dx = \frac{\pi}{2} \frac{e^{-(1-a)p}}{\text{ch}\, p\, \text{sh}^a p} \cos \frac{a\pi}{2};$$

$$3° \int_0^\pi \frac{\cos(1-a)x}{\sin^a x} \frac{\cos x}{\text{ch}\, 2p - \cos 2x} dx = \frac{\pi}{2} \frac{e^{-(1-a)p}}{\text{sh}^{1+a} p} \cos \frac{a\pi}{2};$$

$$4° \int_0^\pi \frac{\sin(1-a)x}{\sin^a x} \frac{\cos x}{\text{ch}\, 2p - \cos 2x} dx = -\frac{\pi}{2} \frac{e^{-(1-a)p}}{\text{sh}^{1+a} p} \sin \frac{a\pi}{2}.$$

REFERENCE
Hardy.

5.4.3.4. We have

$$\int_0^{\pi/2} \frac{1}{a^2 \sin^6 x + b^2 \cos^6 x} dx = \frac{\pi}{3(ab)^{5/3}} (a^{4/3} + (ab)^{2/3} + b^{4/3}).$$

REFERENCE
E.-N. Barisien: 'Question 352', *Intermédiaire des Math.* 1 (1894), 'Réponse 211'. *Ibidem*
2 (1895), 173.

5.4.3.5. If $0 < a < 1$ and $0 < c < \pi$, then:

$$1° \text{ v.p. } \int_0^\pi \frac{\cos(1-a)(c-x)}{\sin^a x \sin(c-x)} dx = 0; \qquad 2° \int_0^\pi \frac{\sin(1-a)(c-x)}{\sin^a x \sin(c-x)} dx = \frac{\pi}{\sin^a c}.$$

If $0 < a < 1$ and $-(\pi/2) < c < \pi/2$, we have:

$$3° \text{ v.p. } \int_{-\pi/2}^{\pi/2} \frac{\cos(1-a)(c-x)}{\cos^a x \sin(c-x)} dx = 0; \qquad 4° \int_{-\pi/2}^{\pi/2} \frac{\sin(1-a)(c-x)}{\cos^a x \sin(c-x)} dx = \frac{\pi}{\cos^a c}.$$

REFERENCE
Hardy.

5.4.3.6. If

$$F(x) = \frac{\sin 2x \sin v + (1 - \cos 2x) \cos v}{(1 - 2c \cos n(x-a) + c^2)^{1/n}} \quad \text{and} \quad \text{tg}\, \frac{nv}{2} = \frac{c \sin n(x-a)}{1 - c \cos n(x-a)},$$

where $n \in N$ and $n > 2, -1 \leq c < 1$, then

(1) $\int_0^{2\pi} F(x)\, dx = 2\pi.$

Proof. The integrand is the real part of the function

$$z \mapsto \frac{e^{-iv}(1-z^2 e^{-2ia})}{(1-cz^n)^{2/n}}, \quad \text{where} \quad z = e^{i(a-x)}.$$

Therefore

$$\int_0^{2\pi} F(x)\, dx = \mathrm{Re}\left(-i \oint_C \frac{1-z^2 e^{-2ia}}{z(1-cz^n)^{2/n}}\, dz\right), \quad \text{where} \quad C = \{z \mid |z| = 1\}.$$

The integrand has no poles in the region int C, except for $z = 0$. Hence, the contour integral is equal to $2\pi i$, and so

$$\int_0^{2\pi} F(x)\, dx = 2\pi.$$

REMARK 1. If $c = 1$, it can be proved that $\int_0^{2\pi} F(x)\, dx = 0$.

REMARK 2. The integral (1) has arisen in the analysis of the elastic stresses around polygonal openings with straight lines.

REFERENCE
S. K. Dhir and J. S. Brock-H. E. Fettis: 'Problem 70–17', *SIAM Review* **13** (1971), 401–404.

5.4.3.7. If $a > |b| > 0; m, n \in N_0$, then

$$\int_0^{2\pi} \frac{\cos mt}{(a+b\cos t)^{n+1}}\, dt = \frac{2\pi (\sqrt{a^2-b^2}-a)^m}{b^m (a^2-b^2)^{(n+1)/2}} \sum_{v=0}^n \left(-\frac{1}{2}\right)^v \binom{n+v}{v}\binom{m+n}{n-v}\left(1-\frac{a}{\sqrt{a^2-b^2}}\right)^v$$

REMARK. This formula was proved by Hähnel [1]. For $m = 0$ we get

$$\int_0^{2\pi} \frac{1}{(a+b\cos t)^{n+1}}\, dt = \frac{2\pi}{(a^2-b^2)^{(n+1)/2}} \sum_{v=0}^n \left(-\frac{1}{2}\right)^v \frac{(n+v)\cdots(n-v+1)}{(v!)^2}\left(1-\frac{a}{\sqrt{a^2+b^2}}\right)^v$$

$$= \frac{2\pi}{(a^2-b^2)^{(n+1)/2}} P_n\left(\frac{a}{\sqrt{a^2-b^2}}\right),$$

where P_n is the Legendre polynomial of order n.

REFERENCE
1. E. Hähnel: 'Geschlossene Dartstellung einiger bestimmter Integrale', *Z. Angew. Math. Mech.* **45** (1965), 259.

5.4.3.8. We have

$$1° \int_0^{\pi/2} e^{\cos x} \cos (\sin x)\, dx = \frac{\pi}{2} + \sum_{m=1}^{+\infty} \frac{(-1)^{m-1}}{(2m-1)!\,(2m-1)} \; ;$$

$$2° \int_0^{\pi/2} e^{\cos x} \cos (x + \sin x)\, dx = \sin 1;$$

$$3° \int_0^{\pi} \frac{1}{(a+\sin x)^2}\, dx = \frac{2}{a\,(1-a^2)} - \frac{a}{(1-a^2)^{3/2}} \log \frac{1+\sqrt{1-a^2}}{1-\sqrt{1-a^2}} \qquad (0 < a < 1).$$

REFERENCE
R. P. Boas, Jr. and L. Schoenfeld: 'Indefinite Integration by Residues', *SIAM Review* **8** (1966), 173–183.

5.4.3.9. Let $a < 1$ and $p > 0$. Then

$$1° \int_0^{\pi/2} \frac{\sin (1-a)\, x}{\cos^a x} \; \frac{\sin x}{\operatorname{ch} 2p - \cos 2x}\, dx = \frac{\pi}{4} \frac{e^{-(1-a)p}}{\operatorname{ch}^{1+a} p} \; ;$$

$$2° \int_0^{\pi/2} \frac{\cos (1-a)\, x}{\cos^a x} \; \frac{\cos x}{\operatorname{ch} 2p - \cos 2x}\, dx = \frac{\pi}{4} \frac{e^{-(1-a)p}}{\operatorname{ch}^a p \; \operatorname{sh} p} \; .$$

REFERENCE
Hardy.

5.4.3.10. Let $a\ (|a| < 1)$ be a real, and n a natural number. Then

$$1° \int_0^{2\pi} \frac{1}{1 - 2a\cos x + a^2}\, dx = \frac{2\pi}{1-a^2} \; ;$$

$$2° \int_0^{2\pi} e^{\cos x} \cos (nx - \sin x)\, dx = \frac{2\pi}{n!} \; ;$$

$3°$ $\displaystyle\int_0^{2\pi} e^{\cos x} \sin(nx - \sin x)\,dx = 0;$

$4°$ $\displaystyle\int_0^{\pi} \cos 2x \log \sin x\,dx = -\frac{\pi}{2};$

$5°$ $\displaystyle\int_0^{2\pi} \cos x \log^2 \sin\frac{x}{2}\,dx = \pi(2\log 2 + 1);$

$6°$ $\displaystyle\int_0^{2\pi} e^{r\cos x}\,dx = 2\pi I_0(r),$

where r is a complex number, and I_0 the Bessel function.

REFERENCE

J. T. Baskin: 'Some General Theorems for Evaluating Definite Integrals by Cauchy's Residue Theorem', M. A. Thesis, The Pennsylvania State University, U.S.A., 1962.

5.4.3.11. Let $a > 0$. The following integrals were evaluated by Cauchy, who used his residue theorem:

$1°$ $\displaystyle\int_0^{\pi/2} \cos^a x \cos ax\,dx = \frac{\pi}{2^{a+1}};$ $2°$ $\displaystyle\int_0^{\pi/2} \sin^a x \cos a\left(\frac{\pi}{2} - x\right)dx = \frac{\pi}{2^{a+1}};$

$3°$ $\displaystyle\int_0^{\pi/2} \mathrm{tg}^a x\,dx = \frac{\pi}{2\cos\dfrac{a\pi}{2}};$ $4°$ $\displaystyle\int_0^{\pi/2} \log(\cos x)\,dx = \frac{\pi}{2}\log\frac{1}{2};$

$5°$ $\displaystyle\int_0^{\pi/2} \log(\sin x)\,dx = \frac{\pi}{2}\log\frac{1}{2};$ $6°$ $\displaystyle\int_0^{\pi/2} \log(\mathrm{tg}\,x)\,dx = 0;$

$7°$ $\displaystyle\int_0^{\pi/2} \cos^{a-1}x\,\frac{\sin ax}{\sin x}\,dx = \frac{\pi}{2};$ $8°$ $\displaystyle\int_0^{\pi/2} \frac{x}{\sin 2x}\,dx = +\infty;$

$9°$ $\displaystyle\int_0^{\pi/2} \frac{\log(\cos x)}{x^2 + \log^2 \cos x}\,dx = \frac{\pi}{2}\left(1 - \frac{1}{\log 2}\right);$

$10°$ $\displaystyle\int_0^{\pi/2} \frac{x\,\mathrm{tg}\,x}{x^2 + \log^2 \cos x}\,dx = \frac{\pi}{2\log 2}.$

Cauchy mentions that 5° was earlier obtained by Euler, and 1°, 9° and 10° by Poisson.

REFERENCE
A.-L. Cauchy: 'Sur les limites placées à droite et à gauche du signe E dans le calcul des résidus', *Exercices de mathématiques*, Paris 1826. ≡ *Oeuvres* (2) 6, 256–285.

5.4.3.12. Let f be a rational function and a a positive number. Then

$$(1) \qquad \int_{\xi}^{x} \frac{f(u)+f(-u)}{\sqrt{a+u}-\sqrt{a-u}}\, du = \sqrt{a}\, f(0) \log \frac{2\sqrt{a}-t}{2\sqrt{a}+t} \cdot \frac{2\sqrt{a}+\tau}{2\sqrt{a}-\tau}$$

$$+ \sum \mathrm{Res}\big(F(z,\, t) - F(z,\, \tau)\big) f(z)$$

$$- \mathrm{Res}_{z=0} \frac{1}{z^2} \left(F\Big(\frac{1}{z},\, t\Big) - F\Big(\frac{1}{z},\, \tau\Big) \right) f\Big(\frac{1}{z}\Big),$$

where

$$t = \sqrt{a+x} - \sqrt{a-x}, \quad \tau = \sqrt{a+\xi} - \sqrt{a-\xi}$$

and

$$F(z,\, t) = \frac{1}{\sqrt{a+z}+\sqrt{a-z}} \log \frac{\sqrt{a+z}-\sqrt{a-z}-t}{\sqrt{a+z}-\sqrt{a-z}+t}$$

$$+ \frac{1}{\sqrt{a+z}-\sqrt{a-z}} \log \frac{\sqrt{a+z}+\sqrt{a-z}-t}{\sqrt{a+z}+\sqrt{a-z}+t}.$$

The summation on the right-hand side of (1) is taken over all the singularities of f.

In particular, if $f(x) = 1$, we find

$$\int_{\xi}^{x} \frac{1}{\sqrt{a+u}+\sqrt{a-u}}\, du = t - \tau + \frac{\sqrt{a}}{2} \log \frac{2\sqrt{a}-t}{2\sqrt{a}+t} \cdot \frac{2\sqrt{a}+\tau}{2\sqrt{a}-\tau}.$$

REFERENCE
A.-L. Cauchy: 'Sur la détermination et la réduction des intégrales dont les dérivées renferment une ou plusieurs fonctions implicites d'une même variable', *Compt. Rend. Acad. Sci. Paris* 12 (1841), 1029–1045. ≡ *Oeuvres* (1) 6, 159–175.

5.4.3.13. For a positive integer n, let

$$K_n = \int_0^1 \frac{\log^n x}{1+x}\, dx.$$

Then

$$K_n = (-1)^n n! \sum_{\nu=1}^{+\infty} (-1)^{\nu-1} \frac{1}{\nu^{n+1}}; \quad K_{2n-1} = (-1)^n \frac{(2^{2n-1}-1)\pi^{2n}}{2n} B_{2n},$$

where $B_1, B_2, \ldots$ are Bernoulli's numbers.

REFERENCE
Th. Angheluță: *On a Class of Integrals*, (Romanian), Lucrări Ştiinţifice, Institutul Politehnic Cluj 1959, pp. 21−28.

5.4.3.14. If $0 < p < 1$, then

$$1° \quad \int_0^1 \frac{x^{p-1}}{(1-x)^p(x+a)}\, dx = \left(\frac{a}{1+a}\right)^p \frac{\pi}{a\sin p\pi} \qquad (a \in \mathbf{R};\ a \notin (-1, 0));$$

$$2° \quad \int_0^1 \frac{x^{p-1}}{(1-x)^p(x+a)}\, dx = \left(\frac{a}{1+a}\right)^p \frac{\pi}{a\sin p\pi} \qquad (a \in \mathbf{C};\ \text{Im}\, a \neq 0).$$

REFERENCE
E. Lainé: 'Sur une classe d'intégrales définies qui se ramènent aux intégrales eulériennes', *Enseignement Math.* 29 (1930), 238−244.

5.4.3.15. Let f be a regular function in the disc $\{z \mid |z| < 1\}$. Then

$$\frac{1}{2\pi} \int_0^{2\pi} f(re^{i\theta})\, d\theta = f(0).$$

In particular, putting $f(z) = 1/(1-z)$, for $r < 1$, we get

$$\frac{1}{2\pi} \int_0^{2\pi} \frac{1}{1 - r\cos\theta - ir\sin\theta}\, d\theta = 1,$$

and, separating the real and imaginary parts:

$$\int_0^{2\pi} \frac{1-r\cos\theta}{1-2r\cos\theta+r^2}\,d\theta=2\,\pi, \qquad \int_0^{2\pi} \frac{\sin\theta}{1-2r\cos\theta+r^2}\,d\theta=0.$$

REFERENCE
H. Laurent: *Traité d'analyse*, t. 3, Paris 1888, pp. 56–57.

5.4.3.16. Applying the calculus of residues to some special classes of non-analytic functions, Dimitrovski [1] proved, among other things, the following results:

1° Let $a>1$, and define the functions C and S by:

$$C(x)=\sum_{n=0}^{+\infty}\frac{(-1)^n}{(2n)!!(2n-1+k)!!}\,x^{2n} \quad S(x)=\sum_{n=0}^{+\infty}\frac{(-1)^n}{(2n+1)!!(2n-k)!!}\,x^{2n+1}.$$

Then

$$\int_1^a xC(x)\,dx=S(a)-S(1).$$

2° The following equalities are valid:

$$\int_0^{2\pi} e^{\cos x}\left(\cos(2x+\sin x)+\cos(x+\sin x)\right)dx=0,$$

$$\int_0^{2\pi} e^{\cos x}\left(\sin(2x+\sin x)+\sin(x+\sin x)\right)dx=0.$$

REFERENCE
1. D. S. Dimitrovski: 'A Contribution to the Theory of Generalized Analytic Functions',
(Macedonian), *Ann. Fac. Sci. Univ. Skopje. Sec.* A20 (1970), 25–214.

5.4.3.17. If $a<1$, then

$$1° \text{ v.p. } \int_{-\pi/2}^{\pi/2} \frac{\cos(1-a)x}{\cos^a x \sin(c-x)}\,dx=\pi\,\frac{\sin(1-a)c}{\cos^a c} \qquad \left(-\frac{\pi}{2}<c<\frac{\pi}{2}\right);$$

$$2° \text{ v.p. } \int_{-\pi/2}^{\pi/2} \frac{\sin(1-a)x}{\cos^a x \sin(c-x)}\,dx=-\pi\,\frac{\cos(1-a)c}{\cos^a c} \qquad \left(-\frac{\pi}{2}<c<\frac{\pi}{2}\right);$$

$3°$ v.p. $\displaystyle\int_0^\pi \frac{\cos(1-a)x}{\sin^a x \sin(c-x)} dx = \pi \frac{\sin(1-a)c}{\sin^a c}$ $(0<c<\pi)$;

$4°$ v.p. $\displaystyle\int_0^\pi \frac{\sin(1-a)x}{\sin^a x \sin(c-x)} dx = -\pi \frac{\cos(1-a)c}{\sin^a c}$ $(0<c<\pi)$;

$5°$ v.p. $\displaystyle\int_0^{\pi/2} \frac{\sin(1-a)x}{\cos^a x} \frac{\sin x}{\cos 2x - \cos 2c} dx = -\frac{\pi}{4} \frac{\cos(1-a)c}{\cos^{1+a}c}$ $\left(0<c<\frac{\pi}{2}\right)$;

$6°$ v.p. $\displaystyle\int_0^{\pi/2} \frac{\cos(1-a)x}{\cos^a x} \frac{\cos x}{\cos 2x - \cos 2c} dx = \frac{\pi}{4} \frac{\sin(1-a)c}{\sin c \cos^a c}$ $\left(0<c<\frac{\pi}{2}\right)$;

$7°$ v.p. $\displaystyle\int_0^\pi \frac{\cos(1-a)x}{\sin^a x} \frac{\cos x}{\cos 2x - \cos 2c} dx = \frac{\pi}{2} \frac{\sin(1-a)c}{\sin^{1+a}c}$ $(0<c<\pi)$.

REFERENCE
Hardy.

5.4.3.18. For a positive integer n we have

$$(1) \qquad \int_0^\pi \frac{1}{(a+\sqrt{a^2-1}\cos t)^{n+1}} dt = \pm \frac{\pi}{2^n n!} \frac{d^n(a^2-1)^n}{da^n}.$$

Proof. Set $z = a + \sqrt{a^2-1}\,\cos t$, to obtain

$$(2) \qquad \int_0^\pi \frac{1}{(a+\sqrt{a^2-1}\cos t)^{n+1}} dt = \pm \frac{1}{i} \int_{c_1}^{c_2} \frac{1}{z^{n+1}\sqrt{1-2az+z^2}} dz,$$

where $c_1 = a + \sqrt{a^2-1}$, $c_2 = a - \sqrt{a^2-1}$. The function f, defined by

$$f(z) = \frac{1}{z^{n+1}\sqrt{1-2az+z^2}},$$

has a pole of order $n+1$ at $z = 0$, and critical singularities at $z = c_1$ and $z = c_2$. Let

$$K = \{z \,||z| = R\}, \quad k_1 = \{z \,||z-c_1| = r_1\},$$

$$k_2 = \{z \,||z-c_2| = r_2\}, \quad k_3 = \{z \,||z| = r_3\},$$

where r_1, r_2, r_3 are sufficiently small positive numbers so that the circles k_1, k_2, k_3 are disjoint, and R is a sufficiently large positive number so that the circle K contains the circles k_1, k_2, k_3. Furthermore, let l_1 and l_2 be segments joining the points c_1 and c_2 (see Figure 5.4.3.18). Finally, let $\Gamma = k_1 \cup l_1 \cup k_2 \cup l_2$. Then

$$(3) \qquad \oint_K f(z)\,dz = \oint_\Gamma f(z)\,dz + \oint_{k_3} f(z)\,dz.$$

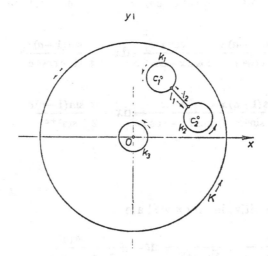

Fig. 5.4.3.18.

However, we have

$$\oint_{k_3} f(z)\,dz = 2\pi i \operatorname*{Res}_{z=0} f(z).$$

Since

$$\frac{1}{\sqrt{1-2az+z^2}} = 1 + \sum_{\nu=1}^{+\infty} A_\nu z^\nu,$$

where

$$A_n = \frac{1}{2^n\,n!}\frac{d^n(a^2-1)^n}{da^n} \qquad (n=1, 2, \ldots),$$

we conclude that

$$\operatorname*{Res}_{z=0} f(z) = A_n = \frac{1}{2^n n!} \frac{d^n (a^2-1)^n}{da^n},$$

which implies

$$\oint_{k_3} f(z)\, dz = \frac{2\pi i}{2^n n!} \frac{d^n (a^2-1)^n}{da^n}.$$

On the other hand, since $\lim_{z \to \infty} zf(z) = 0$, we conclude that (see Theorem 2 in 3.1.4) $\oint_K f(z)\, dz \to 0$, as $R \to +\infty$.

It remains to evaluate the integral along Γ. We have

$$\oint_{\Gamma} f(z)\, dz = \int_{c_1}^{c_2} f(z)\, dz + \oint_{k_2} f(z)\, dz + \int_{c_2}^{c_1} f(z)\, dz + \oint_{k_1} f(z)\, dz.$$

However, since $\lim_{z \to c_1} (z - c_1)f(z) = \lim_{z \to c_2} (z - c_2)f(z) = 0$, in view of Theorem 1 from 3.1.4 we deduce that the integrals along the circles k_1 and k_2 tend to zero, as their radii r_1 and r_2 tend to zero.

It is also easily established that

$$\int_{c_1}^{c_2} f(z)\, dz = \int_{c_2}^{c_1} f(z)\, dz,$$

and therefore letting in (3) $R \to +\infty, r_1 \to 0, r_2 \to 0$, we find

$$0 = \oint_{k_3} f(z)\, dz + 2 \int_{c_1}^{c_2} f(z)\, dz, \quad \text{i.e.} \quad \int_{c_1}^{c_2} f(z)\, dz = -\frac{\pi i}{2^n n!} \frac{d^n (a^2-1)^n}{da^n},$$

which together with (2) implies (1).

REFERENCE

E. Schmid: *Die Cauchy'sche Methode der Auswertung bestimmter Integrale zwischen reellen Grenzen*, Stuttgart 1903.

5.4.3.19. If $n \in N$ and $r \in R^+$, then:

$$1° \int_0^{\pi} \frac{\cos nt}{1-2r\cos t+r^2}\, dt = \begin{cases} \dfrac{\pi r^n}{1-r^2} & (r<1), \\[2ex] \dfrac{-\pi r^{-n}}{1-r^2} & (r>1); \end{cases}$$

$2°$ $\displaystyle\int_0^{\pi} \frac{\sin t \sin nt}{1-2r\cos t+r^2}\,dt = \begin{cases} \dfrac{\pi}{2}\,r^{n-1} & (r<1), \\[2ex] \dfrac{\pi}{2}\,r^{-n-1} & (r>1); \end{cases}$

$3°$ $\displaystyle\int_0^{\pi} \frac{\cos nt}{1\pm r\cos t}\,dt = \pm\frac{\pi}{\sqrt{1-r^2}}\left(\frac{-1+\sqrt{1-r^2}}{r}\right)^n \quad (r<1);$

$4°$ $\displaystyle\int_0^{\pi} \frac{\sin t\,\operatorname{tg}\dfrac{t}{2}}{1-2r\cos t+r^2}\,dt = \begin{cases} \dfrac{\pi}{1+r} & (r<1), \\[2ex] \dfrac{\pi}{r\,(r+1)} & (r>1); \end{cases}$

$5°$ $\displaystyle\int_0^{2\pi} \frac{1}{a+b\cos t+c\sin t}\,dt = \frac{2\pi}{\sqrt{a^2-b^2-c^2}} \quad (a^2>b^2+c^2);$

$6°$ v.p. $\displaystyle\int_0^{2\pi} \frac{1}{a+b\cos t+c\sin t}\,dt = 0 \quad (a^2<b^2+c^2);$

$7°$ v.p. $\displaystyle\int_0^{\pi} \frac{e^{p\cos t}\cos(p\sin t)}{\cos t}\,dt = \pi\sin p \quad (p\in\mathbf{R});$

$8°$ $\displaystyle\int_0^{\pi} \log(1+2r\cos t+r^2)\,dt = \begin{cases} 0 & (r<1), \\ 2\pi\log r & (r>1); \end{cases}$

$9°$ $\displaystyle\int_0^{\pi} \log(1+2r\cos t+r^2)\cos nt\,dt = \begin{cases} \dfrac{1}{n}\,\pi(-1)^{n-1}r^n & (r<1), \\[2ex] \dfrac{1}{n}\,\pi(-1)^{n-1}r^{-n} & (r>1); \end{cases}$

10 v.p. $\displaystyle\int_0^{\pi} \frac{\log(1-2r\cos t+r^2)}{\cos t}\,dt = 2\pi\operatorname{arctg} r \quad (r^2<1).$

REFERENCE

E. Schmid: *Die Cauchy'sche Methode der Auswertung bestimmter Integrale zwischen reellen Grenzen*, Stuttgart 1903.

5.4.3.20. Let n be a positive integer. Then

$$1° \int_0^1 \frac{x^{2n}}{\sqrt[3]{x(1-x^2)}}\,dx = \frac{\pi}{\sqrt{3}}\,\frac{1\cdot 4\cdots(3\,n-2)}{3\cdot 6\cdots(3\,n)};$$

$$2° \int_0^1 \frac{x^{2n}}{(1+x^2)\sqrt{1-x^2}}\,dx = \begin{cases} \dfrac{\pi}{2}\left(\dfrac{1}{\sqrt 2} - S_{2k}\right) & (n = 2\,k), \\[2ex] -\dfrac{\pi}{2}\left(\dfrac{1}{\sqrt 2} - S_{2k+1}\right) & (n = 2\,k+1), \end{cases}$$

where

$$S_m = 1 + \sum_{v=1}^{m-1}(-1)^v\,\frac{(2\,v-1)!!}{(2\,v)!!}.$$

REFERENCE
Tisserand, pp. 474, 470.

5.4.3.21. We have

$$1° \int_0^1 \frac{1}{1+x^2}\log\left(x+\frac{1}{x}\right)dx - \frac{\pi}{2}\log 2; \qquad 2° \int_{-1}^1 \frac{1}{\sqrt[3]{(1+x)^2(1-x)}}\,dx = \frac{2\,\pi}{\sqrt 3}.$$

REFERENCE
Tisserand, pp. 474, 487.

5.4.3.22. We have

$$1° \int_0^1 \frac{1}{(1+x)\sqrt[3]{x^2(1-x)}}\,dx = \pi\sqrt[3]{\frac{4}{3}}; \qquad 2° \int_0^1 \frac{\sqrt[3]{x^2(1-x)}}{(1+x)^3}\,dx = \frac{\pi\sqrt[3]{2}}{18\sqrt 3}.$$

REFERENCE
Julia, pp. 171, 177.

5.4.3.23. If n is a positive integer, then

$$\int_0^1 \frac{1}{\sqrt{1-\sqrt[n]{x}}}\,dx = \frac{2^{2n}(n!)^2}{(2\,n)!}.$$

REFERENCE

G. Oltramare: 'Mémoire sur quelques propositions du calcul des résidus', *Mémoires de l'Institut Genevois* **3** (1855), 15 pp.

5.4.3.24. Suppose that P is a real polynomial of degree $2r$, and that n is a positive integer. For the integral

$$J = \int_{-\pi}^{\pi} (\cos nx)\, \sqrt{P(\cos x)}\, dx$$

we have

1° For $r = 1$, J can be reduced to an elliptic integral;
2° For $r > 1$, J can be reduced to an hyperelliptic integral;
3° For $n \geq r + 1$, we have $J = 0$.

REFERENCE

D. S. Mitrinović and Z. R. Pop-Stojanović: 'About Integrals Expressible in Terms of Hyperelliptic Integrals', *Glasnik Mat.-Fiz. Astr.* (2) **18** (1963), 235–239.

5.4.3.25. Let f be an analytic function with a finite number of poles, which is regular on the unit circle. Then

$$(1) \qquad \int_{0}^{\pi/2} \left(f(e^{2it}) \left(\log\mathrm{tg}\, t - i\frac{\pi}{2} \right)^m + f(e^{-2it}) \left(\log\mathrm{tg}\, t + i\frac{\pi}{2} \right)^m \right) dt$$

$$= \pi \sum \mathrm{Res}\, \frac{f(z)}{z} \log^m \frac{1-z}{1+z},$$

where $m \in \mathbf{N}$ and where the summation is taken over all the singularities of the function $z \mapsto f(z)/z$ which lie inside the unit circle.

The formula (1) is obtained by integrating the function $z \mapsto f(z) \log^m (1-z)/(1+z)$ round the unit circle indented at $z = \pm 1$. It contains a number of interesting real integrals. We give two examples:

$$1° \int_{0}^{\pi/2} (\log\mathrm{tg}\, t)^{2p+1}\, dt = 0;$$

$$2° \int_{0}^{\pi/2} (\cos 2\,(2p+1)\,t)\,(\log\mathrm{tg}\, t)\, dt = -\frac{\pi}{2\,(2p+1)},$$

where in both cases $p \in \mathbf{N}$.

REFERENCE
W. Kapteijn: 'On a Definite Integral', (Dutch), *Amsterdam Versl.* 6 (1898), 329–335.

5.4.3.26. Let $a, b \in \mathbf{R}$ and consider the integral

$$J = \int_0^{\pi/2} \frac{\log \sin x}{a - b \sin^2 x} \, dx = \int_0^{\pi/2} \frac{\log \cos x}{a - b \cos^2 x} \, dx.$$

If $a > b > 0$, then

$$J = \frac{-\pi}{4\sqrt{a(a-b)}} \log \frac{2a - b + 2\sqrt{a(a-b)}}{a};$$

if $a = b > 0$, then $J = -(\pi/2)$;
if $b > a > 0$, then J does not exist.

This result, as well as some other results related to the integral (1) are proved in [1].

REFERENCE
1. G. Huber: 'Auswertung einiger bestimmter Integrale mit Anwendung des freien Integrationsweges', *Monatsh. Math.* 16 (1905), 141–160.

5.5. A SUMMARY OF MORE IMPORTANT TYPES OF INTEGRALS

$\int_{-\infty}^{+\infty} f(x) \, dx$	Theorem 1 in 5.2.1. Theorem 2 in 5.2.2.
$\int_{-\infty}^{+\infty} e^{iax} f(x) \, dx$	Theorem 2 in 5.2.1.
$\int_{-\infty}^{+\infty} f(x) \cos ax \, dx, \quad \int_{-\infty}^{+\infty} f(x) \sin ax \, dx$	Remark 5 in 5.2.1.
$\int_{-\infty}^{+\infty} f(x) \left(g(x) - g(x + bi) \right) dx$	Theorem 1 in 5.2.2.

$\int\limits_{-\infty}^{+\infty} e^{ax} f(e^{bx}) \, \mathrm{d}x$	Remark 1 in 5.3.1.
$\int\limits_{-\infty}^{+\infty} x^n e^{ax} f(e^{bx}) \, \mathrm{d}x$	Remark 3 in 5.3.1.
$\int\limits_{0}^{+\infty} f(x) \, \mathrm{d}x$	Theorem 1 in 5.3.1. Theorem 4 in 5.3.1. Theorem 1 in 5.3.2. Remark 3 in 5.3.2. Remark 4 in 5.3.2.
$\int\limits_{0}^{+\infty} x^a f(x) \, \mathrm{d}x$	Theorem 2 in 5.3.1. Theorem 5 in 5.3.1.
$\int\limits_{0}^{+\infty} x^a f(x) \log^m x \, \mathrm{d}x$	Remark 2 in 5.3.1.
$\int\limits_{0}^{+\infty} f(x) \cos ax \, \mathrm{d}x, \quad \int\limits_{0}^{+\infty} f(x) \sin ax \, \mathrm{d}x$	Theorems 2 and 3 in 5.3.2.
$\int\limits_{0}^{+\infty} f(x) \log x \, \mathrm{d}x$	Remark 4 in 5.3.2.
$\int\limits_{0}^{+\infty} f(x) \operatorname{arctg} x \, \mathrm{d}x$	Theorem 5 in 5.3.2.
$\int\limits_{0}^{+\infty} f(x) \log \left| \dfrac{x+1}{x-1} \right| \, \mathrm{d}x$	Theorem 6 in 5.3.2.

$\displaystyle\int\limits_{0}^{+\infty} f(iy)\,dy$	Theorem 4 in 5.4.1.
$\displaystyle\int\limits_{a}^{+\infty} f(x)\,dx \qquad (a>0)$	Theorem 1 in 5.3.3.
$\displaystyle\int\limits_{0}^{2\pi} f(\cos x,\ \sin x)\,dx$	Theorem 1 in 5.4.1.
$\displaystyle\int\limits_{0}^{2\pi} f(\cos x,\ \sin x)\cos nx\,dx$ $\displaystyle\int\limits_{0}^{2\pi} f(\cos x,\ \sin x)\sin nx\,dx$	Remark 5 in 5.4.1.
$\displaystyle\int\limits_{0}^{2\pi} f(\cos x,\ \sin x)\,g\,(e^{ix})\,dx$	Remark 7 in 5.4.1.
$\displaystyle\int\limits_{\alpha}^{\beta} f(e^{ix})\,dx$	Theorem 2 in 5.4.1.
$\displaystyle\int\limits_{0}^{2\pi} f(x)\log\sin\frac{x}{2}\,dx$	Remark 12 in 5.4.1.
$\displaystyle\int\limits_{0}^{2\pi} xf(x)\,dx$	Theorem 4 in 5.4.1.
$\displaystyle\int\limits_{a}^{b} f(x)\,dx$	Theorem 1 in 5.4.2.

$\displaystyle\int_a^b (x-a)^r (b-x)^s f(x)\,dx$	Theorem 2 in 5.4.2.
$\displaystyle\int_a^b \left(\frac{x-a}{b-x}\right)^a f(x)\,dx$	Remark 9 in 5.4.2.

Chapter 6

Evaluation of Finite and Infinite Sums by Residues

6.1. INTRODUCTION

Suppose that a_1, a_2, ... are singularities of the function f, and that b_1, b_2, ... are singularities of the function g in a region $G = \operatorname{int} \Gamma$, where Γ is a closed contour in the complex z-plane. Then, by Cauchy's theorem on residues we have

(1) $\qquad \dfrac{1}{2\pi i} \oint_{\Gamma} f(z) g(z) \, dz = \sum_{\nu} \operatorname*{Res}_{z=a_\nu} f(z) g(z) + \sum_{\nu} \operatorname*{Res}_{z=b_\nu} f(z) g(z).$

Formula (1), under certain conditions, can be used for evaluating various types of sums, finite or infinite.

For evaluating sums of the form $\sum_{n \in D} f(n)$, where $D \subset \mathbf{Z}$, it is convenient to use a function g which has simple poles at all the points of the set D, with residues equal to 1. It is also useful, in case of infinite sums, to make the integral on the left-hand side of (1) vanish.

For example, the functions g_1, g_2, g_3, defined by

(2) $\qquad g_1(z) = \pi \cotg \pi z,$

(3) $\qquad g_2(z) = \dfrac{2 i \pi}{e^{2i\pi z} - 1},$

(4) $\qquad g_3(z) = \dfrac{-2 i \pi}{e^{-2i\pi z} - 1},$

possess the mentioned properties, i.e. they have simple poles at $z = m \in \mathbf{Z}$. Besides,

$$\operatorname*{Res}_{z=m} g_1(z) = \operatorname*{Res}_{z=m} g_2(z) = \operatorname*{Res}_{z=m} g_3(z) = 1 \quad (m \in \mathbf{Z}).$$

The functions g_1, g_2, g_3 are bounded when $|y| \to +\infty$, where $y = \operatorname{Im} z$.

211

This follows from the inequalities

$$\left|\pi \cotg \pi z\right| \leq \pi \frac{e^{\pi y}+e^{-\pi y}}{\left|e^{-\pi y}-e^{\pi y}\right|}, \quad \left|\frac{2i\pi}{e^{2i\pi z}-1}\right| \leq \frac{2\pi}{\left|e^{-2\pi y}-1\right|},$$

$$\left|\frac{2i\pi}{e^{-2i\pi z}-1}\right| \leq \frac{2\pi}{\left|e^{2\pi y}-1\right|},$$

which hold for $y = \operatorname{Im} z \neq 0$.

The function g_4, defined by $g_4(z) = \pi \operatorname{cosec} \pi z$, will also prove useful. That function has simple poles at $z = m \in \mathbf{Z}$, and

$$\operatorname*{Res}_{z=m} g_4(z) = (-1)^m \cdot \quad (m \in \mathbf{Z}).$$

It is also bounded when $|\operatorname{Im} z| \to +\infty$, because

$$\left|\pi \operatorname{cosec} \pi z\right| \leq \frac{2\pi}{\left|e^{-\pi y}-e^{\pi y}\right|} \quad (y = \operatorname{Im} z \neq 0).$$

In further text we shall mostly be using the functions g_1 and g_4. In many cases it is possible to replace the function g_1 by any one of the functions g_2 or g_3.

6.2. BASIC THEOREMS

THEOREM 1. *Let f be an analytic function in the region G defined by*

$$G = \{z \mid a \leq \operatorname{Re} z \leq p, \quad |\operatorname{Im} z| \leq \delta\},$$

where

$$m-1 < a < m, \quad n < \beta < n+1 \ (m, n \in \mathbf{Z}),$$

in which it can have a finite number of singularities $a_1, \ldots, a_p$. *If* $\Gamma = \partial G$, *then*

(1) $$\sum_{v=m}^{n} f(v) = \frac{1}{2\pi i} \oint_{\Gamma} f(z)\,\pi \cotg \pi z \, dz - \sum_{k=1}^{p} \operatorname*{Res}_{z=a_k} (f(z)\pi \cotg \pi z),$$

(2) $$\sum_{v=m}^{n} (-1)^v f(v) = \frac{1}{2\pi i} \oint_{\Gamma} f(z)\,\pi \operatorname{cosec} \pi z \, dz - \sum_{k=1}^{p} \operatorname*{Res}_{z=a_k} (f(z)\pi \operatorname{cosec} \pi z).$$

Formulas (1) and (2) are direct consequences of the equality (1) from 6.1, where in the first case $g(z) = \pi \cot \pi z$, and in the second $g(z) = \pi \operatorname{cosec} \pi z$.

THEOREM 2. *Let f be an analytic function in the extended complex plane where it can have a finite number of singularities* $a_1, \ldots, a_p$, *and let* $f(z) = O(z^{-2})$, *as* $z \to \infty$. *Then*

$$(3) \qquad \sum_{\nu=-\infty}^{+\infty} f(\nu) = \sum_{k=1}^{p} \operatorname*{Res}_{z=a_k} (f(z)\,\pi \cot \pi z),$$

$$(4) \qquad \sum_{\nu=-\infty}^{+\infty} (-1)^\nu f(\nu) = \sum_{k=1}^{p} \operatorname*{Res}_{z=a_k} (f(z)\,\pi \operatorname{cosec} \pi z).$$

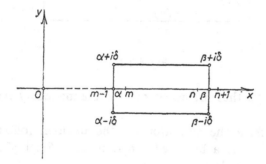

Fig. 6.2.1.

Proof. In the previous theorem let $\alpha = -n - \frac{1}{2}, \beta = \delta = n + \frac{1}{2}$. This defines a contour Γ_n (see Figure 6.2.2).
On the lines $\{z \mid \operatorname{Re} z = n + \frac{1}{2}\}$ and $\{z \mid \operatorname{Re} z = -n - \frac{1}{2}\}$ we have

$$|\cot \pi z|^2 = \frac{\operatorname{sh}^2 \pi y}{1 + \operatorname{sh}^2 \pi y} < 1 \quad (y = \operatorname{Im} z),$$

and on the lines $\{z \mid \operatorname{Im} z = n + \frac{1}{2}\}$ and $\{z \mid \operatorname{Im} z = -n - \frac{1}{2}\}$ we have

$$|\cot \pi z|^2 \le \frac{\operatorname{sh}^2\left(n+\dfrac{1}{2}\right)\pi + 1}{\operatorname{sh}^2\left(n+\dfrac{1}{2}\right)\pi} \le 1 +$$

$$+ \frac{1}{\operatorname{sh}^2\left(n+\dfrac{1}{2}\right)\pi} \le 1 + \frac{1}{\operatorname{sh}^2 \dfrac{\pi}{2}} < 4.$$

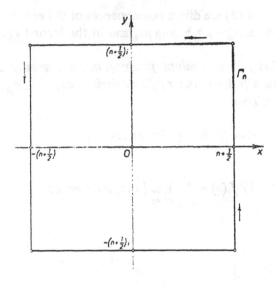

Fig. 6.2.2.

Hence, we conclude that for every $z \in \Gamma_n$ the inequality $|\cot g\, \pi z| < 2$ is valid.

Furthermore, from the conditions of the theorem follows that $f(z) = (1/z^2)F(z)$, where F is a bounded function, i.e. $|F(z)| \leq M$ (M positive number). Therefore, for large enough n,

$$|f(z)| = \left| \frac{1}{z^2} F(z) \right| \leq \frac{M}{|z|^2},$$

and so

(5) $$\left| \frac{1}{2\pi i} \oint_{\Gamma_n} \pi \cot g\, \pi z f(z)\, dz \right| \leq \oint_{\Gamma_n} \frac{M}{|z|^2} |dz|.$$

From (5) follows that the integral along Γ_n tends to zero, as $n \to +\infty$. If $n \to +\infty$ in (1), we get (3).

Formula (4) is proved in a similar manner. Namely, start with (2) and use the inequality $|\cosec\, \pi z| < 1$ ($z \in \Gamma_n$).

REMARK 1. If the function f has a countable number of singularities, then on the right-hand sides of (3) and (4), we get infinite sums, which means that the given series is transformed into another series. In certain cases this transformation may be useful, since it speeds up the convergence. For instance, the function f defined by

$$f(z) = \frac{z}{e^{\pi a z} - e^{-\pi a z}} \qquad (a \text{ real}),$$

has a countable number of singularities. Applying (4), with $p = +\infty$, we obtain

$$\sum_{\nu=1}^{+\infty} (-1)^\nu \frac{\nu}{e^{\pi a \nu} - e^{-\pi a \nu}} = -\frac{1}{4\pi a} - \frac{1}{a^2} \sum_{\nu=1}^{+\infty} (-1)^\nu \frac{\nu}{e^{\pi \nu/a} - e^{-\pi \nu/a}}.$$

For $a < 1$, the series on the right-hand side of the last equality converges faster than the series on the left-hand side.

REFERENCE

E. Lindelöf: *Le calcul des résidus et ses applications à la théorie des fonctions*, Paris 1905, reprinted 1947, pp. 52–55.

6.3. EVALUATION OF SOME SPECIAL CLASSES OF INFINITE SERIES

In this section we apply Theorem 2 from 6.2 to some special series.

6.3.1. *Rational and Trigonometric Functions*

THEOREM 1. *Let P and Q be polynomials of degree p and q respectively, and let $a_1, \dots, a_k$ be mutually distinct zeros of Q. Then*

$$(1) \qquad \sum_{\nu=-\infty}^{+\infty} \frac{P(\nu)}{Q(\nu)} e^{i\nu\theta} = -\sum_{\nu=1}^{k} \operatorname*{Res}_{z=a_\nu} \left(\frac{P(z)}{Q(z)} e^{i\theta z} \pi \cotg \pi z \right),$$

$$(2) \qquad \sum_{\nu=-\infty}^{+\infty} (-1)^\nu \frac{P(\nu)}{Q(\nu)} e^{i\nu\theta} = -\sum_{\nu=1}^{k} \operatorname*{Res}_{z=a_\nu} \left(\frac{P(z)}{Q(z)} e^{i\theta z} \pi \operatorname{cosec} \pi z \right),$$

where $-\pi \leq \theta \leq \pi$.

This formula is a direct consequence of Theorem 2 from 6.2.

REMARK 1. Formulas (1) and (2) hold if the series on their left-hand sides are convergent. The convergence is secured if, for example, $p \leq q - 2$. If $p = q - 1$, it may happen that the series

$$\sum_{\nu=-\infty}^{+\infty} \frac{P(\nu)}{Q(\nu)}$$

is divergent, but that there exists a finite limit

$$(3) \qquad \lim_{n \to +\infty} \sum_{\nu=-n}^{n} \frac{P(\nu)}{Q(\nu)}.$$

In that case formulas (1) and (2) remain valid if the series which appears in term is replaced by (3).

EXAMPLE 1. From (1) for $\theta = 0$ and $a \notin Z$, we get

$$\lim_{n \to +\infty} \sum_{\nu=-n}^{n} \frac{1}{\nu-a} = -\frac{\pi}{\operatorname{tg} \pi a} ; \sum_{\nu=-\infty}^{+\infty} \frac{1}{(\nu-a)^2} = \frac{\pi^2}{\sin^2 a\pi} ; \sum_{\nu=-\infty}^{+\infty} \frac{1}{(\nu-a)^3} = -\frac{\pi^3 \cos a\pi}{\sin^3 a\pi}.$$

EXAMPLE 2. If we put $P(z) = 1, Q(z) = z^{2m}$ $(m \in N)$, $\theta = 0$ in (1) and (2), we obtain

$$\sum_{\nu=1}^{+\infty} \frac{1}{\nu^{2m}} = -\frac{1}{2} \frac{1}{(2m)!} \lim_{z \to 0} \frac{\mathrm{d}^{2m}}{\mathrm{d} z^{2m}} (\pi \cot g \, \pi z),$$

$$\sum_{\nu=1}^{+\infty} (-1)^\nu \frac{1}{\nu^{2m}} = -\frac{1}{2} \frac{1}{(2m)!} \lim_{z \to 0} \frac{\mathrm{d}^{2m}}{\mathrm{d} z^{2m}} (\pi \operatorname{cosec} \pi z).$$

EXAMPLE 3. If we put in (2):

$$P(z) = bt, \qquad Q(z) = \pi^2 z^2 + b^2 t^2, \qquad \theta = \frac{\pi a}{b},$$

where a, b, and t are real numbers such that $|a| < |b|$, we find

$$\sum_{\nu=-\infty}^{+\infty} (-1)^\nu \frac{bt}{\pi^2 \nu^2 + b^2 t^2} \exp \frac{i\pi a\nu}{b} =$$

$$= -\operatorname*{Res}_{z=\frac{ibt}{\pi}} \frac{bt}{\pi^2 z^2 + b^2 t^2} \exp\left(\frac{i\pi az}{b}\right) \pi \operatorname{cosec} \pi z - \operatorname*{Res}_{z=-\frac{ibt}{\pi}} \frac{bt}{\pi^2 z^2 + b^2 t^2} \exp\left(\frac{i\pi az}{b}\right) \pi \operatorname{cosec} \pi z =$$

$$= -\frac{e^{-at}}{2i \sin ibt} - \frac{e^{at}}{2i \sin ibt} = \frac{\operatorname{ch} at}{\operatorname{sh} at}$$

and hence, separating the real and imaginary parts,

$$\sum_{\nu=-\infty}^{+\infty} (-1)^\nu \frac{bt}{\pi^2 \nu^2 + b^2 t^2} \cos \frac{\pi a\nu}{b} = \frac{\operatorname{ch} at}{\operatorname{sh} at}, \qquad \sum_{\nu=-\infty}^{+\infty} (-1)^\nu \frac{bt}{\pi^2 \nu^2 + b^2 t^2} \sin \frac{\pi a\nu}{b} = 0.$$

6.3.2. Irrational Functions

Calculus of residues may be applied to the summation of series whose general term is an irrational algebraic function. For example, in order to sum the series

$$\sum_{\nu=-\infty}^{+\infty} f(\nu, \sqrt{\nu^2+a^2}) \quad \text{or} \quad \sum_{\nu=-\infty}^{+\infty} (-1)^\nu f(\nu, \sqrt{\nu^2+a^2}),$$

where f is a rational function and $a \notin \mathbf{Z}$, start with the integral

$$\lim_{R\to+\infty} \frac{1}{2i} \oint_{C_R} f(z, \sqrt{z^2+a^2})\,\mathrm{cotg}\,\pi z\,dz, \quad \text{or}$$

$$\lim_{R\to+\infty} \frac{1}{2i} \oint_{C_R} f(z, \sqrt{z^2+a^2})\,\mathrm{cosec}\,\pi z\,dz,$$

respectively, where $C_R = \{z \,|\, |z| = R + \tfrac{1}{2}\}$ $(R = 1, 2, \ldots)$, with cuts along the imaginary axis, so that the critical singularities ia and $-ia$ are eliminated (see Figure 6.3.2).

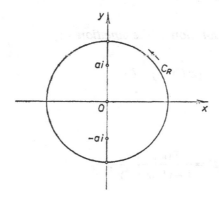

Fig. 6.3.2.

EXAMPLE 1. Let us evaluate the sum

$$\sum_{\nu=-\infty}^{+\infty} \frac{(-1)^\nu}{\sqrt{\nu^2+a^2}}.$$

Since the function

$$z \mapsto \frac{1}{\sqrt{z^2+a^2}}$$

is regular inside the described contour C_R, we have

$$\sum_{\nu=-\infty}^{+\infty} \frac{(-1)^\nu}{\sqrt{\nu^2+a^2}} = \lim_{R\to+\infty} \frac{1}{2i} \oint_{C_R} \frac{1}{\sqrt{z^2+a^2}} \operatorname{cosec} \pi z \, dz.$$

It is easily established that the integral along the circular part of the contour C_R tends to zero, as $R \to +\infty$. Hence, we obtain

$$\sum_{\nu=-\infty}^{+\infty} \frac{(-1)^\nu}{\sqrt{\nu^2+a^2}} = 2 \int_{ia}^{i\infty} \frac{1}{\sqrt{z^2+a^2}} \frac{1}{\operatorname{sh}\pi z}\, dz = 2 \int_a^\infty \frac{1}{\operatorname{sh}\pi u \sqrt{u^2-a^2}}\, du.$$

6.3.3. Potential Series with Functional Coefficients

THEOREM 1. *Let F and f be regular functions in a simply connected region G, and let* $f(\zeta) \ne 0$ *for* $\zeta \in G$. *Define* G_n $(n \in N_0)$ *by*

(1) $$G_n(\zeta) = \frac{1}{n!} \frac{d^n}{d\zeta^n} (F(\zeta)f(\zeta)^n) \quad (n \in N_0).$$

If $w = w(z, \zeta)$ *is a solution of the equation*

$$z = \frac{w-\zeta}{f(w)} \quad (w(0,\zeta)=\zeta),$$

then

(2) $$\sum_{n=0}^{+\infty} G_n(\zeta)z^n = \frac{F(w(z,\zeta))}{1-zf'(w(z,\zeta))}.$$

Proof. We shall determine the functions H_n $(n \in N_0)$ defined by

(3) $$\frac{F(w(z,\zeta))}{1-zf'(w(z,\zeta))} = \sum_{n=0}^{+\infty} H_n(\zeta)z^n,$$

and we shall prove that

(4) $$H_n(\zeta) = G_n(\zeta) \quad (n \in N_0).$$

If both sides of (3) are multiplied by z^{-n-1} and then integrated along a contour Γ which surrounds the points $z = 0$ and $z = \zeta$, we get

$$(5) \qquad H_n(\zeta) = \oint_\Gamma \frac{1}{z^{n+1}} \frac{F(w(z,\zeta))}{1-zf'(w(z,\zeta))} \, dz.$$

On the other hand, from (1) follows

$$(6) \qquad G_n(\zeta) = \oint_\Gamma \frac{F(w) f(w)^n}{(w-\zeta)^{n+1}} \, dw.$$

The substitution

$$\frac{f(w)}{w-\zeta} = \frac{1}{z},$$

reduces the integral on the right-hand side of (6) to the integral on the right-hand side of (5), implying (4).

EXAMPLE 1. Laguerre's polynomials L_n ($n \in \mathbb{N}_0$) are defined by

$$L_n(x) = e^x \frac{d^n}{dx^n} (e^{-x} x^n) \quad (x \in \mathbb{R}^+).$$

By an application of the formula (2) we obtain

$$\sum_{n=0}^{+\infty} L_n(\zeta) \frac{z^n}{n!} = \frac{1}{1-z} \exp\left(-\frac{z\zeta}{1-z}\right).$$

6.4. TRANSFORMATION OF FINITE AND INFINITE SUMS INTO INTEGRALS

6.4.1. Integration Along a Rectangle

THEOREM 1. *Suppose that f is regular in the region $G_1 = \{z \mid \alpha \leq \mathrm{Re}\, z \leq \beta\}$ and suppose that the equality*

$$(1) \qquad \lim_{|y| \to +\infty} e^{-2\pi|y|} f(x+iy) = 0 \qquad (z = x+iy)$$

holds uniformly in G_1. If $m - 1 < \alpha < m$, $n < \beta < n + 1$ ($m, n \in \mathbb{Z}$), then

$$(2) \qquad \sum_{v=m}^{n} f(v) = \int_a^\beta f(x)\,dx + \lim_{\delta \to +\infty} (J(\alpha,\delta) - J(\beta,\delta)),$$

where

$$J(x, \delta) = \int\limits_{x}^{x+i\delta} \frac{f(z)}{e^{-2\pi iz}-1}\, dz + \int\limits_{x}^{x-i\delta} \frac{f(z)}{e^{2\pi iz}-1}\, dz.$$

Proof. We shall need the following equalities:

$$\frac{1}{2i}\cotg \pi z = -\frac{1}{2} - \frac{1}{e^{-2\pi iz}-1} \quad (\text{Im}\, z>0),$$

(3)

$$\frac{1}{2i}\cotg \pi z = \frac{1}{2} + \frac{1}{e^{2\pi iz}-1} \quad (\text{Im}\, z<0).$$

Using Theorem 1 from 6.2, and having in mind that f has no singularites in G, we find

$$\sum_{\nu=m}^{n} f(\nu) = \frac{1}{2i} \oint\limits_{\Gamma} f(z) \cotg \pi z\, dz.$$

Let $\Gamma_1 = \Gamma \cap \{z \mid \text{Im}\, z > 0\}$ and $\Gamma_2 = \Gamma \cap \{z \mid \text{Im}\, z < 0\}$. Then

$$\sum_{\nu=m}^{n} f(\nu) = \frac{1}{2i} \int\limits_{\Gamma_1} f(z) \cotg \pi z\, dz + \frac{1}{2i} \int\limits_{\Gamma_2} f(z) \cotg \pi z\, dz,$$

and using (3) we get

$$\sum_{\nu=m}^{n} f(\nu) = \int\limits_{\Gamma_1} f(z)\left(-\frac{1}{2} - \frac{1}{e^{-2\pi iz}-1}\right) dz + \int\limits_{\Gamma_2} f(z)\left(\frac{1}{2} + \frac{1}{e^{2\pi iz}-1}\right) dz,$$

i.e.

(4) $$\sum_{\nu=m}^{n} f(\nu) = \int\limits_{a}^{\beta} f(x)\, dx + \int\limits_{a}^{a+i\delta} \frac{f(z)}{e^{-2\pi iz}-1}\, dz + \int\limits_{a}^{a-i\delta} \frac{f(z)}{e^{2\pi iz}-1}\, dz$$

$$-\int\limits_{\beta}^{\beta+i\delta} \frac{f(z)}{e^{-2\pi iz}-1}\, dz - \int\limits_{\beta}^{\beta-i\delta} \frac{f(z)}{e^{2\pi iz}-1}\, dz + \int\limits_{a}^{\beta} \frac{f(x+i\delta)}{e^{-2\pi ix}e^{2\pi\delta}-1}\, dx$$

$$+\int\limits_{a}^{\beta} \frac{f(x-i\delta)}{e^{2\pi ix}e^{2\pi\delta}-1}\, dx.$$

If $\delta \to +\infty$, using (1), we see that from (4) follows (2).

THEOREM 2. *Let f be an analytic function in the region $G_1 = \{z \,|\, \alpha < \operatorname{Re} z < \beta\}$, where it can have a finite number of singularities $a_1, \ldots, a_p$. If the condition (1) is fulfilled, and if m and n are defined as in Theorem 1, then*

$$\sum_{\nu=m}^{n} f(\nu) = \int_{\alpha}^{\beta} f(x)\,dx + \lim_{\delta \to +\infty} (J(\alpha, \beta) - J(\beta, \delta))$$

$$+ \pi i \sum_{\nu=1}^{p} \operatorname*{Res}_{z=a_\nu} (\pi \cot \pi z\, f(z))$$

$$+ \pi i \sum_{\operatorname{Im} a_\nu > 0} \operatorname*{Res}_{z=a_\nu} f(z) - \pi i \sum_{\operatorname{Im} a_\nu < 0} \operatorname*{Res}_{z=a_\nu} f(z).$$

THEOREM 3. *Suppose that the following conditions are satisfied:*

$1°$ *f is regular in the region $\{z \,|\, \operatorname{Re} z \geq a\}\, (m - 1 < a < m)$;*

$2°$ $\lim\limits_{|y| \to +\infty} e^{-2\pi|y|} f(x+iy) = 0 \ (z = x+iy), \quad$ *uniformly for $x \geq \alpha$;*

$3°$ $\lim\limits_{x \to +\infty} \int\limits_{-\infty}^{+\infty} e^{-2\pi|y|}|f(x+iy)|\,dy = 0.$

Then, if the series $\sum_{\nu=m}^{+\infty} f(\nu)$ is convergent, we have

$$(5) \qquad \sum_{\nu=m}^{+\infty} f(\nu) = \int_{a}^{+\infty} f(x)\,dx + \int_{a}^{a+i\infty} \frac{f(z)}{e^{-2\pi iz}-1}\,dz + \int_{a}^{a-i\infty} \frac{f(z)}{e^{2\pi iz}-1}\,dz,$$

and if it is divergent, we have

$$\lim_{n \to +\infty} \left(\sum_{\nu=m}^{n} f(\nu) - \int_{a}^{n+\frac{1}{2}} f(x)\,dx \right) = \int_{a}^{a+i\infty} \frac{f(z)}{e^{-2\pi iz}-1}\,dz + \int_{a}^{a+i\infty} \frac{f(z)}{e^{2\pi iz}-1}\,dz.$$

Proof. Put $\beta = n + \frac{1}{2}$ into (2). Using the inequality

$$\left| \int_{n+\frac{1}{2}}^{n+\frac{1}{2}+i\delta} \frac{f(z)}{e^{-2\pi iz}-1}\,dz + \int_{n+\frac{1}{2}}^{n+\frac{1}{2}-i\delta} \frac{f(z)}{e^{2\pi iz}-1}\,dz \right| < \int_{-\delta}^{\delta} e^{-2\pi|y|}|f(\beta+iy)|\,dy,$$

if $\delta \to +\infty$, we get (5).

THEOREM 4. *Let f be a regular function in the region* $\{z \,|\, |\mathrm{Im}\, z| \leq \delta\,\}$ *and let*

$$\int\limits_{-\infty}^{+\infty} \max_{|y|\leq\delta} |f(x+iy)|\, dx < +\infty \qquad (z=x+iy).$$

Then

(6) $$\sum_{\nu=-\infty}^{+\infty} f(\nu) = \sum_{\nu=-\infty}^{+\infty} \int\limits_{-\infty}^{+\infty} f(x)\, e^{2\nu\pi ix}\, dx.$$

Proof. The following formulas are valid

(7)
$$\operatorname{cotg} \pi z = -i - 2i \sum_{\nu=1}^{+\infty} e^{2\nu\pi iz} \qquad (\mathrm{Im}\, z > 0),$$

$$\operatorname{cotg} \pi z = i + 2i \sum_{\nu=1}^{+\infty} e^{-2\nu\pi iz} \qquad (\mathrm{Im}\, z < 0).$$

Let

$$G_n = \left\{ z \,\Big|\, |\mathrm{Re}\, z| \leq n + \frac{1}{2},\ |\mathrm{Im}\, z| < \delta \right\}$$

and let $\Gamma_n = \partial G_n$. Then, using the expansions (7), we have

$$\sum_{\nu=-n}^{n} f(\nu) = \frac{1}{2\pi i} \oint\limits_{\Gamma_n} f(z)\, \pi \operatorname{cotg} \pi z\, dz$$

$$= i \int\limits_{\delta}^{0} f\left(-n-\frac{1}{2}+iy\right)\left(-\frac{1}{2}-\sum_{\nu=1}^{+\infty} e^{2\nu\pi i\left(-n-\frac{1}{2}+iy\right)}\right) dy$$

$$+ i \int\limits_{0}^{-\delta} f\left(-n-\frac{1}{2}+iy\right)\left(\frac{1}{2}+\sum_{\nu=1}^{+\infty} e^{-2\nu\pi i\left(-n-\frac{1}{2}+iy\right)}\right) dy$$

$$+ \int\limits_{-n-\frac{1}{2}}^{n+\frac{1}{2}} f(x-i\delta)\left(\frac{1}{2}+\sum_{\nu=1}^{+\infty} e^{-2\nu\pi i(x-i\delta)}\right) dx$$

$$+i \int_0^\delta f\left(n+\frac{1}{2}+iy\right)\left(\frac{1}{2}+\sum_{\nu=1}^{+\infty} e^{-2\pi i\left(n+\frac{1}{2}+iy\right)}\right) dy$$

$$+i \int_0^\delta f\left(n+\frac{1}{2}+iy\right)\left(-\frac{1}{2}-\sum_{\nu=1}^{+\infty} e^{2\pi i\left(n+\frac{1}{2}+iy\right)}\right) dy$$

$$+ \int_{n+\frac{1}{2}}^{-n-\frac{1}{2}} f(x-i\delta)\left(-\frac{1}{2}-\sum_{\nu=1}^{+\infty} e^{2\nu\pi i\,(x+i\delta)}\right) dx.$$

Let first $\delta \to 0$, and then $n \to +\infty$. We obtain (6).

REMARK 1. Theorems 3 and 4 have interesting applications in analytic continuation of potential series. Namely, if a function defined by the potential series $\sum_{n=0}^{+\infty} f(n)z^n$ is to be analytically continued, then start with the integral

$$\oint_\Gamma f(\zeta) z^\zeta \cot g\,\pi\zeta\,d\zeta$$

along a suitably chosen contour. If this integral can be written as a sum of integrals of regular functions in a certain region, this means that the given series can be analytically continued in that region.

EXAMPLE 1. The potential series

$$\sum_{n=1}^{+\infty} (\log^a (n+1))\, z^n \qquad (0 \leq a < 1)$$

can be analytically continued on the whole complex plane, excepting the cut $\{z \mid \operatorname{Re} z > 1, \operatorname{Im} z = 0\}$.

EXAMPLE 2. Suppose that f is a regular function in the region $\{z \mid \operatorname{Re} z \geq \alpha\}$. Furthermore, let

$$|f(a+\varrho\,e^{i\theta})| < e^{(\xi+\epsilon)\varrho} \qquad \left(|\theta| \leq \frac{\pi}{2}\right),$$

where $\xi < \pi$ and ϵ is an arbitrary positive number. Then the series $\sum_{n=0}^{+\infty} f(n)z^n$ can be analytically continued on the whole complex plane, with the exception of the cut $\{z \mid \operatorname{Re} z > 1, \operatorname{Im} z = 0\}$.

In order to prove this statement, start with

$$\sum_{n=q}^p f(n)\, z^n = \oint_\Gamma \frac{f(\zeta)\, z^\zeta}{e^{2\pi i\zeta} - 1}\, d\zeta,$$

where Γ is the contour consisting of the semicircle

$$\left\{ \zeta \,\middle|\, |\zeta - a| = q + \frac{1}{2} - a, \ \mathrm{Re}\,\zeta \geqq a \right\},$$

together with its diameter $(p - 1 < \alpha < p)$.

Using the given hypotheses, it can be proved that

$$\oint_{\Gamma} \frac{f(\zeta)\, z^{\zeta}}{e^{2\pi i \zeta} - 1}\, d\zeta = - \int_{a-i\infty}^{a+i\infty} \frac{f(\zeta)\, z^{\zeta}}{e^{2\pi i \zeta} - 1}\, d\zeta$$

and that the last integral defines a function of z which is regular in the whole z-plane, except the half-line $\{z \mid \mathrm{Re}\,z > 1,\ \mathrm{Im}\,z = 0\}$.

THEOREM 5. *Suppose that f is a regular function in the region*

$$G = \{z \mid \alpha \leqq \mathrm{Re}\,z \leqq \beta,\ |\mathrm{Im}\,z| < \delta\}$$

and let

$$\Gamma = \partial G, \ \Gamma_1 = \Gamma \cap \{z \mid \mathrm{Im}\,z > 0\}, \ \Gamma_2 = \Gamma \cap \{z \mid \mathrm{Im}\,z < 0\}.$$

If $m - 1 < \alpha < m,\ n < \beta < n + 1\ (m,\ n \in \mathbf{Z})$, then

(8) $\quad \displaystyle\sum_{\nu=m}^{n} (-1)^{\nu} f(\nu) = 2 \sum_{\nu=0}^{r-1} \int_{\alpha}^{\beta} f(x) \cos(2\nu+1)\,\pi \cos \pi x \, dx$

$$+ \int_{\Gamma_2 -} \frac{e^{-2r\pi i z}}{e^{\pi i z} - e^{-\pi i z}} f(z)\, dz - \int_{\Gamma_1 +} \frac{e^{2r\pi i z}}{e^{\pi i z} - e^{-\pi i z}} f(z)\, dz,$$

where $r \in \mathbf{N}$.

Proof. Since f is a regular function, from Theorem 1 of 6.2 we obtain

(9) $\quad \displaystyle\sum_{\nu=m}^{n} (-1)^{\nu} f(\nu) = \int_{\Gamma_2 -} \frac{f(z)}{e^{\pi i z} - e^{-\pi i z}}\, dz - \int_{\Gamma_1 +} \frac{f(z)}{e^{\pi i z} - e^{-\pi i z}}\, dz.$

However, we have

$$\frac{1}{e^{\pi iz} - e^{-\pi iz}} = \sum_{v=0}^{r-1} e^{-(2v+1)\pi iz} + \frac{e^{-2r\pi iz}}{e^{\pi iz} - e^{-\pi iz}} \qquad (\text{Im } z < 0),$$

$$\frac{1}{e^{\pi iz} - e^{-\pi iz}} = -\sum_{v=0}^{r-1} e^{(2v+1)\pi iz} + \frac{e^{2r\pi iz}}{e^{\pi iz} - e^{-\pi iz}} \qquad (\text{Im } z > 0),$$

and substituting the above expressions in the corresponding integrals in (9), we get (8).

We give four more theorems which reduce finite or infinite sums to integrals.

THEOREM 6. *Let f be a regular function in the region G defined in Theorem 3, and let F be its primitive function. If* $\Gamma = \partial G$, *then*

$$\sum_{v=m}^{n} f(v) = \frac{1}{2\pi i} \oint_{\Gamma} \left(\frac{\pi}{\sin \pi z} \right)^2 F(z) \, dz.$$

This result follows directly from formula (1) of Theorem 1 in 6.2, after partial integration.

THEOREM 7. *Let F be a primitive function of f, where f satisfies the hypotheses* $1°$, $2°$, *and* $3°$ *of Theorem 3. Then*

$$\sum_{v=m}^{+\infty} f(v) = -\frac{1}{2\pi i} \int_{a-i\infty}^{a+i\infty} \left(\frac{\pi}{\sin \pi z} \right)^2 F(z) \, dz.$$

This theorem can be obtained from the previous one by letting $n \to +\infty$ and $\delta \to +\infty$.

THEOREM 8. *Suppose that f satisfies the conditions* $1°$, $2°$, *and* $3°$ *of Theorem 3, and besides let* $f(x) \in \mathbf{R}$ *when* $x \in \mathbf{R}$. *Put*

$$f(x + iy) = p(x, y) + iq(x, y),$$

$$\frac{1}{e^{-2\pi iz} - 1} = P(x, y) + iQ(x, y) \qquad (z = x + iy),$$

$$R(x, y) = p(x, y) \, Q(x, y) + q(x, y) \, P(x, y).$$

Then

$$\sum_{\nu=m}^{n} f(\nu) = \int_{\alpha}^{\beta} f(x)\,dx - 2 \int_{0}^{+\infty} (R(\alpha, y) - R(\beta, y))\,dy$$

$$\sum_{\nu=m}^{+\infty} f(\nu) = \int_{\alpha}^{+\infty} f(x)\,dx - 2 \int_{0}^{+\infty} R(\alpha, y)\,dy.$$

REMARK 2. In the proof of Theorem 8 we have to use the equalities

$$p(x, y) = \frac{1}{2}(f(x+iy) + f(x-iy)),$$

$$q(x, y) = \frac{1}{2i}(f(x+iy) - f(x-iy)),$$

which hold only if $f(x) \in R$ for $x \in R$.

THEOREM 9. *Let f be a regular function in the region $\{z \mid \alpha \le \mathrm{Re}\, z \le \beta\}$ and suppose that the equality*

$$\lim_{|y|\to+\infty} e^{-\pi|y|} f(x+iy) = 0 \quad (z = x + iy)$$

holds uniformly for $x \in [\alpha, \beta]$. Define the functions P_1, Q_1, and R_1 by

$$\frac{1}{e^{-\pi iz} - e^{\pi iz}} = P_1(x, y) + iQ_1(x, y),$$

$$R_1(x, y) = p(x, y)\, Q_1(x, y) + q(x, y)\, P_1(x, y),$$

where p and q have the same meaning as in Theorem 8. If $m - 1 < \alpha < m$, $n < \beta < n + 1$ ($m, n \in Z$), then

$$\sum_{\nu=m}^{n} (-1)^{\nu} f(\nu) = 2 \int_{0}^{+\infty} (Q_1(\beta, y) - Q_1(\alpha, y))\,dy.$$

If, in addition, we have

$$\lim_{x\to+\infty} \int_{-\infty}^{+\infty} e^{-\pi|y|} |f(x+iy)|\,dy = 0,$$

then

$$\sum_{\nu=m}^{+\infty} (-1)^\nu f(\nu) = -2 \int_0^{+\infty} Q_1(a, y)\, dy.$$

6.4.2. Integration Over More Complicated Contours

THEOREM 1. *Let f be a regular function in the region* $\{z \mid m \le \mathrm{Re}\, z \le n\}$ *and let* $f(x) \in \mathbf{R}$ *for* $x \in \mathbf{R}$. *If the hypothesis* $2°$ *of Theorem 3 of 6.4.1 is fulfilled for* $\alpha = n$, *then we have*

$$(1) \qquad \sum_{\nu=m}^{n} f(\nu) = \frac{1}{2}\left(f(m) + f(n)\right) + \int_m^n f(x)\, dx - 2 \int_0^{+\infty} \frac{q(m, y) - q(n, y)}{e^{2\pi y} - 1}\, dy,$$

where $q(x, y) = \mathrm{Im}\, f(z)$, $z = x + iy$.

Proof. Let

$$\Gamma = \bigcup_{\nu=1}^{10} \Gamma_\nu,$$

where

$$\Gamma_1 = \{z \mid |z - m| = \varepsilon,\ m - \varepsilon \le \mathrm{Re}\, z \le m,\ -\varepsilon \le \mathrm{Im}\, z \le 0\},$$
$$\Gamma_2 = \{z \mid \mathrm{Re}\, z = m,\ -\delta \le \mathrm{Im}\, z \le -\varepsilon\},$$
$$\Gamma_3 = \{z \mid m \le \mathrm{Re}\, z \le n,\ \mathrm{Im}\, z = -\delta\},$$
$$\Gamma_4 = \{z \mid \mathrm{Re}\, z = n,\ -\delta \le \mathrm{Im}\, z \le -\varepsilon\},$$
$$\Gamma_5 = \{z \mid |z - n| = \varepsilon,\ n \le \mathrm{Re}\, z \le n + \varepsilon,\ -\varepsilon \le \mathrm{Im}\, z \le 0\},$$
$$\Gamma_6 = \{z \mid |z - n| = \varepsilon,\ n \le \mathrm{Re}\, z \le n + \varepsilon,\ 0 \le \mathrm{Im}\, z \le \varepsilon\},$$
$$\Gamma_7 = \{z \mid \mathrm{Re}\, z = n,\ \varepsilon \le \mathrm{Im}\, z \le \delta\},$$
$$\Gamma_8 = \{z \mid m \le \mathrm{Re}\, z \le n,\ \mathrm{Im}\, z = \delta\},$$
$$\Gamma_9 = \{z \mid \mathrm{Re}\, z = m,\ \varepsilon \le \mathrm{Im}\, z \le \delta\},$$
$$\Gamma_{10} = \{z \mid |z - m| = \varepsilon,\ m - \varepsilon \le \mathrm{Re}\, z \le m,\ 0 \le \mathrm{Im}\, z \le \varepsilon\},$$

and where $\varepsilon, \delta \in \mathbf{R}^+$ (see Figure 6.4.2).

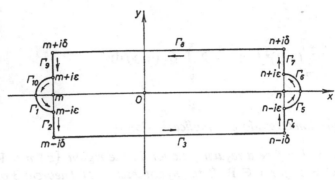

Fig. 6.4.2.

Starting with the integral

$$\oint_{\Gamma} \pi \cot g \, \pi z f(z) \, dz,$$

similarly as in the proof of Theorem 1 from 6.4.1 we get

(2) $$\sum_{\nu=m}^{n} f(\nu) = \int_{m}^{n} f(x) \, dx + \lim_{\delta \to +\infty} \, (J_1(m, \delta) - J_2(n, \delta)),$$

where

$$J_1(m, \delta) = \lim_{\varepsilon \to 0} \left(-\int_{\Gamma_9 \cup \Gamma_{10}} \frac{f(z)}{e^{-2\pi i z} - 1} \, dz + \int_{\Gamma_1 \cup \Gamma_2} \frac{f(z)}{e^{2\pi i z} - 1} \, dz \right),$$

$$J_2(n, \delta) = \lim_{\varepsilon \to 0} \left(\int_{\Gamma_6 \cup \Gamma_7} \frac{f(z)}{e^{-2\pi i z} - 1} \, dz - \int_{\Gamma_4 \cup \Gamma_5} \frac{f(z)}{e^{2\pi i z} - 1} \, dz \right).$$

In order to evaluate $J_1(m, \delta)$, notice the equality

$$\frac{f(z)}{e^{-2\pi i z} - 1} = -\frac{1}{2\pi i} \frac{f(z)}{z - m} + g(z) \qquad (z \in \Gamma_{10} \cup \Gamma_1),$$

where the modulus of g is bounded when $z \to m$. Therefore,

(3) $$-\int_{\Gamma_{10}} \frac{f(z)}{e^{-2\pi i z} - 1} \, dz = \frac{1}{2\pi i} \int_{\Gamma_{10}} \frac{f(z)}{z - m} \, dz - \int_{\Gamma_{10}} g(z) \, dz$$

$$= \frac{1}{2\pi} \int\limits_{\pi/2}^{\pi} f(m + \varepsilon \, e^{i\theta}) \, d\theta + \int\limits_{\pi/2}^{\pi} g(z) \, \varepsilon \, i e^{i\theta} \, d\theta.$$

If $\varepsilon \to 0$ in (3), the second integral on the right-hand side tends to zero, and hence

$$- \int\limits_{\Gamma_{10}} \frac{f(z)}{e^{-2\pi i z} - 1} \, dz = \frac{1}{4} f(m).$$

In a similar manner we prove that

$$\int\limits_{\Gamma_1} \frac{f(z)}{e^{2\pi i z} - 1} \, dz = \frac{1}{4} f(m).$$

Furthermore,

$$- \int\limits_{\Gamma_9} \frac{f(z)}{e^{-2\pi i z} - 1} \, dz + \int\limits_{\Gamma_2} \frac{f(z)}{e^{2\pi i z} - 1} \, dz$$

$$= \int\limits_{\varepsilon}^{\delta} \frac{f(m + iy)}{e^{-2\pi i (m + iy)} - 1} i \, dy + \int\limits_{-\varepsilon}^{-\delta} \frac{f(m + iy)}{e^{2\pi i (m + iy)} - 1} i \, dy$$

$$= \int\limits_{\varepsilon}^{\delta} \frac{f(m + iy)}{e^{2\pi y} - 1} i \, dy - \int\limits_{\varepsilon}^{\delta} \frac{f(m - iy)}{e^{2\pi y} - 1} i \, dy = -2 \int\limits_{\varepsilon}^{\delta} \frac{q(m, y)}{e^{2\pi y} - 1} \, dy$$

and therefore

(4) $\qquad J_1(m, \delta) = \frac{1}{2} f(m) - 2 \int\limits_{0}^{\delta} \frac{q(m, y)}{e^{2\pi y} - 1} \, dy.$

By an analogous procedure we arrive at

(5) $\qquad J_2(n, \delta) = \frac{1}{2} f(n) + 2 \int\limits_{0}^{\delta} \frac{q(n, y)}{e^{2\pi y} - 1} \, dy.$

From (2), (4), and (5) follows (1).

EXAMPLE 1. Put $f(z) = \log z$ and $m = 1$ into (1). We obtain

(6) $$\log n! = \left(n + \frac{1}{2}\right) \log n - n + \frac{1}{2} - 2 \int\limits_0^{+\infty} \operatorname{arctg} y \, \frac{1}{e^{2\pi y} - 1} \, dy$$

$$+ 2 \int\limits_0^{+\infty} \operatorname{arctg} \frac{y}{n} \, \frac{1}{e^{2\pi y} - 1} \, dy.$$

On the other hand, from Stirling's formula follows

(7) $$\log n! = \left(n + \frac{1}{2}\right) \log n - n + \log \sqrt{2\pi} + \varepsilon\,(n),$$

where $\lim_{n \to +\infty} \varepsilon(n) = 0$. From (6) and (7) we get

$$\int\limits_0^{+\infty} \frac{\operatorname{arctg} y}{e^{2\pi y} - 1} \, dy = \frac{1}{2} \, (1 - \log \sqrt{2\pi}).$$

EXAMPLE 2. We shall prove that Euler's constant C can be written in the form

(8) $$C = 1 + \frac{1}{2} + \cdots + \frac{1}{m-1} + \frac{1}{2\,m} - \log m + 2 \int\limits_0^{+\infty} \frac{y}{(m^2 + y^2)(e^{2\pi y} - 1)} \, dy.$$

Indeed, if we put $f(z) = 1/z$ in (1), and if $n \to +\infty$, we get

(9) $$\lim_{n \to +\infty} \left(\sum_{\nu=m}^{n} \frac{1}{\nu} - \log n \right) = \frac{1}{2m} - \log m + 2 \int\limits_0^{+\infty} \frac{y}{(m^2 + y^2)(e^{2\pi y} - 1)} \, dy.$$

The equality (8) is obtained if we add to both sides of (9) the sum

$$1 + \frac{1}{2} + \cdots + \frac{1}{m-1}.$$

In particular, for $m = 1$, we find Poisson's formula

$$C = \frac{1}{2} + 2 \int\limits_0^{+\infty} \frac{y}{(1 + y^2)(e^{2\pi y} - 1)} \, dy,$$

published in 1811.

Using contours similar to the contour used in the proof of Theorem 1, we can prove the following three theorems.

THEOREM 2. *Suppose that f satisfies the conditions* $1°, 2°,$ *and* $3°$ *of Theorem 3 from 6.4.1 for* $\alpha = m$ $(m \in Z)$, *and also that* $f(x) \in R$ *when* $x \in R$. *Then*

$$\sum_{\nu=m}^{+\infty} f(\nu) = \frac{1}{2} f(m) + \int_{m}^{+\infty} f(x)\, dx - 2 \int_{0}^{+\infty} \frac{q(m, y)}{e^{2\pi y} - 1}\, dy,$$

where $q(x, y) = \operatorname{Im} f(z)$ $(z = x + iy)$.

THEOREM 3. *Suppose that f satisfies the conditions* $1°, 2°,$ *and* $3°$ *of Theorem 3 from 6.4.1 for* $\alpha = m - 1$ $(m \in Z)$ *and let* $f(x) \in R$ *when* $x \in R$. *Then*

(10) $$\sum_{\nu=m}^{+\infty} f(\nu) = -\frac{1}{2} f(m-1) + \int_{m-1}^{+\infty} f(x)\, dx - 2 \int_{0}^{+\infty} \frac{q(m-1, y)}{e^{2\pi y} - 1}\, dy,$$

where $q(x, y) = \operatorname{Im} f(z)$ $(z = x + iy)$.

THEOREM 4. *Let f be a regular function in the region* $\{z \mid 0 \leq \operatorname{Re} z \leq n\}$ $(n \in N)$ *and let the equality*

$$\lim_{|y| \to +\infty} e^{-2(r+1)\pi |y|} f(x + iy) = 0 \qquad (z = x + iy)$$

hold uniformly for $x \in [0, n]$, *where r is an integer* > 1. *Then*

$$\sum_{\nu=0}^{n-1} f(\nu) = \frac{1}{2} (f(0) - f(n)) + \int_{0}^{n} f(x)\, dx + 2 \sum_{\nu=1}^{r} \int_{0}^{n} f(x) \cos 2\nu\pi x\, dx$$

$$+ 2 \int_{0}^{+\infty} (q(n, y) - q(0, y)) \frac{e^{-2r\pi y}}{e^{2\pi y} - 1}\, dy,$$

where $q(x, y) = \operatorname{Im} f(z)$.

EXAMPLE 3. If we put $f(z) = e^{-uz}$ $(u > 0)$, $m = 1$ into (10), we find

(11) $$\frac{1}{e^u - 1} = \frac{1}{u} - \frac{1}{2} + 2 \int_{0}^{+\infty} \frac{\sin ut}{e^{2\pi t} - 1}\, dt.$$

Formula (11) is due to Legendre. It can be used for obtaining the integral representation of Bernoulli's numbers. Namely, the Bernoulli's numbers $B_1, B_2, \ldots$ are defined by the expansion

$$(12) \qquad \frac{1}{e^u - 1} = \frac{1}{u} - \frac{1}{2} + \sum_{\nu=1}^{+\infty} (-1)^{\nu-1} \frac{B_\nu}{(2\nu)!} u^{2\nu-1}.$$

Since

$$(13) \qquad \int_0^{+\infty} \frac{\sin ut}{e^{2\pi t} - 1} \, dt = \int_0^{+\infty} \left(\sum_{\nu=1}^{+\infty} (-1)^{\nu-1} \frac{(ut)^{2\nu-1}}{(2\nu-1)!} \right) \frac{1}{e^{2\pi t} - 1} \, dt$$

from (11), (12), and (13) follows

$$(14) \qquad B_\nu = 4 \nu \int_0^{+\infty} \frac{t^{2\nu-1}}{e^{2\pi t} - 1} \, dt.$$

Formula (14) was known to Euler.

EXAMPLE 4. If we apply Theorem 4 to the function f, defined by

$$f(z) = \exp\left(\frac{p\pi i}{n} z^2\right),$$

where positive integers p and n do not have a common factor, for $r = p - 1$ we find

$$\sum_{\nu=0}^{n-1} \exp\left(\frac{p\pi i}{n} \nu^2\right) = e^{\pi i/4} \sqrt{\frac{n}{p}} \sum_{\nu=0}^{p-1} \exp\left(-\frac{n\pi i}{p} \nu^2\right).$$

In particular, for $p = 2$ and odd n we obtain the following formula, due to Gauss

$$\sum_{\nu=0}^{n-1} \exp\left(\frac{2\pi i}{n} \nu^2\right) = \frac{1 - i^n}{1 - i} \sqrt{n}.$$

REFERENCE

E. Lindelöf: *Le calcul des résidus et ses applications à la théorie des fonctions*, Paris 1905, reprinted 1947, pp. 55–75.

6.5. SOME IMPORTANT SUMMATION FORMULAS

6.5.1. *Definitions*

We first give definitions of some special numbers and functions which appear in summation formulas.

(1) Bernoulli's numbers $B_1, B_2, \ldots$ are defined by the expansion

$$\frac{1}{e^z-1} - \frac{1}{z} + \frac{1}{2} = \sum_{\nu=1}^{+\infty} (-1)^{\nu-1} \frac{B_\nu}{(2\nu)!} z^{2\nu-1}.$$

(2) Euler's numbers $E_0, E_1, \ldots$ are defined by the expansion

$$\frac{1}{\cos z} = \sum_{\nu=0}^{+\infty} \frac{E_\nu}{(2\nu)!} z^{2\nu}.$$

(3) Polynomials $\varphi_1, \varphi_2, \ldots$ are defined by the expansion

$$\frac{e^{uz}}{e^z-1} = \frac{1}{z} + \sum_{\nu=1}^{+\infty} \frac{\varphi_\nu(u)}{\nu!} z^{\nu-1} \qquad (0 \le u \le 1).$$

(4) The function $\overline{\varphi}_n$ $(n = 1, 2, \ldots)$ is periodic with period (1) and coincides with the function φ_n when $x \in [0, 1]$.

(5) Bernoulli's polynomials $P_1, P_2, \ldots$ are defined by

$$P_{2k}(x) = \varphi_{2k}(x) + (-1)^k B_k, \qquad P_{2k+1}(x) = \varphi_{2k+1}(x),$$

where $B_1, B_2, \ldots$ are Bernoulli's numbers.

(6) The function $\overline{P}_n$ $(n = 1, 2, \ldots)$ is periodic with period (1) and coincides with Bernoulli's polynomial P_n when $x \in [0, 1]$.

(7) Polynomials χ_n $(n \in N_0)$ are defined by the expansion

$$\frac{e^{uz}}{e^z+1} = \sum_{\nu=0}^{+\infty} \frac{\chi_\nu(u)}{\nu!} z^\nu \qquad (0 \le u \le 1).$$

(8) The function $\overline{\chi}_n$ $(n \in N_0)$ is periodic with period (2). On the segment $[0, 2]$ it is defined by

$$\overline{\chi}_n(x) = \chi_n(x), \qquad x \in [0, 1],$$
$$\overline{\chi}_n(x) = -\chi_n(x-1), \qquad x \in (1, 2].$$

6.5.2. *Summation Formulas*

Let $f(z) = p(x, y) + iq(x, y)$ $(z = x + iy)$ and let $f(x) \in \mathbf{R}$ when $x \in \mathbf{R}$. Then

(1) $\qquad p(x, y) = \dfrac{1}{2} (f(x + iy) + f(x - iy)),$

(2) $\qquad q(x, y) = \dfrac{1}{2i} (f(x + iy) - f(x - iy)).$

From (1) and (2) follows

$$\left.\frac{\partial^{2\nu} p}{\partial y^{2\nu}}\right|_{y=0} = (-1)^\nu f^{(2\nu)}(x), \quad \left.\frac{\partial^{2\nu} q}{\partial y^{2\nu}}\right|_{y=0} = 0 \quad (\nu \in \mathbf{N}_0),$$

$$\left.\frac{\partial^{2\nu+1} p}{\partial y^{2\nu+1}}\right|_{y=0} = 0, \quad \left.\frac{\partial^{2\nu+1} q}{\partial y^{2\nu+1}}\right|_{y=0} = (-1)^\nu f^{(2\nu+1)}(x) \quad (\nu \in \mathbf{N}_0).$$

Hence, if we apply Taylor's formula in the form

$$F(t) = F(0) + F'(0)\, t + \cdots + F^{(n-1)}(0)\, \frac{t^{n-1}}{(n-1)!}$$

$$+ \frac{t^n}{(n-1)!} \int_0^1 (1-u)^{n-1} f^{(n)}(ut)\, du$$

to the functions p and q, we get

(3) $\qquad p(x, y) = f(x) - f''(x)\dfrac{y^2}{2!} + \cdots + (-1)^{k-1} f^{(2k-2)}(x)\, \dfrac{y^{2k-2}}{(2k-2)!}$

$\qquad\qquad + p_{2k}(x, y),$

where

$$p_{2k}(x, y) = \frac{(-1)^k}{2}\, \frac{y^{2k}}{(2k-1)!} \int_0^1 (1-u)^{2k-1} (f^{(2k)}(x+iuy)$$

$$+ f^{(2k)}(x - iuy))\, du$$

and

(4) $\qquad q(x, y) = f'(x)\, y - f'''(x)\dfrac{y^3}{3!} + \cdots + (-1)^{k-1} f^{(2k-1)}(x)\, \dfrac{y^{2k-1}}{(2k-1)!}$

$\qquad\qquad + q_{2k+1},$

where

$$q_{2k+1}(x, y) = \frac{(-1)^k}{2} \frac{y^{2k+1}}{(2k)!} \int_0^1 (1-u)^{2k} \left(f^{(2k+1)}(x+iuy) \right.$$

$$\left. + f^{(2k+1)}(x-iuy) \right) du.$$

Using the expansion (4) and the equality (14) from 6.4.2, we find

$$2 \int_0^{+\infty} \frac{q(x, y)}{e^{2\pi y}-1} \, dy = \frac{1}{2} B_1 f'(x) - \frac{1}{4!} B_2 f'''(x) + \cdots$$

$$+ (-1)^{k-1} \frac{1}{(2k)!} f^{(2k-1)}(x) + 2 \int_0^{+\infty} \frac{q_{2k+1}(x, y)}{e^{2\pi y}-1} \, dy.$$

The following equalities are also valid

$$2 \int_0^{+\infty} \frac{p(x, y)}{e^{\pi y}+e^{-\pi y}} \, dy = \frac{1}{2} \left(E_0 f(x) - \frac{1}{4 \cdot 2!} E_1 f''(x) + \cdots \right.$$

$$\left. + (-1)^{k-1} \frac{1}{4^{k-1}(2k-2)!} E_{k-1} f^{(2k-2)}(x) \right) + 2 \int_0^{+\infty} \frac{p_{2k}(x, y)}{e^{\pi y}+e^{-\pi y}} \, dy,$$

$$2 \int_0^{+\infty} R(x, y) \, dy = -\overline{\varphi}_1(x) f(x) + \frac{\overline{\varphi}_2(x)}{2} \frac{f'(x)}{1!} - \cdots + \frac{\overline{\varphi}_{2k}(x)}{2k} \frac{f^{(2k-1)}(x)}{(2k-1)!}$$

$$+ 2 \int_0^{+\infty} \left(p_{2k}(x, y) Q(x, y) + q_{2k+1}(x, y) P(x, y) \right) dy,$$

$$2 \int_0^{+\infty} R_1(x, y) \, dy = \overline{\chi}_0(x) f(x) - \overline{\chi}_1(x) \frac{f'(x)}{1!} + \cdots - \overline{\chi}_{2k-1}(x) \frac{f^{(2k-1)}(x)}{(2k-1)!}$$

$$+ 2 \int_0^{+\infty} \left(p_{2k}(x, y) Q_1(x, y) + q_{2k+1}(x, y) P_1(x, y) \right) dy,$$

where P, Q, R, P_1, Q_1, R_1 are defined in Theorems 8 and 9 from 6.4.1.

Finally, using those equalities and the Theorem 1 from 6.4.2, and also Theorems 8 and 9 from 6.4.1, we obtain the following summation formulas

(5) $$\sum_{v=m}^{n} f(v) = \frac{1}{2}(f(m)+f(n)) + \int_m^n f(x) \, dx$$

$$+ \sum_{\nu=1}^{k} (-1)^{\nu-1} \frac{B_\nu}{2\nu} \frac{f^{(2\nu-1)}(n)-f^{(2\nu-1)}(m)}{(2\nu-1)!} + R_1{}^*,$$

$$(6) \quad \sum_{\nu=m}^{n} f(\nu) = \int_{a}^{\beta} f(x)\,dx + \sum_{\nu=1}^{2k} \frac{(-1)^\nu}{\nu} \frac{\overline{\varphi_\nu}(\beta)f^{(\nu-1)}(\beta)-\overline{\varphi_\nu}(a)f^{(\nu-1)}(a)}{(\nu-1)!} + R_2{}^*,$$

$$(7) \quad \sum_{\nu=m}^{n} (-1)^\nu f(\nu) = \sum_{\nu=0}^{2k-1} (-1)^\nu \frac{\overline{\chi_\nu}(\beta)f^{(\nu)}(\beta)-\overline{\chi_\nu}(a)f^{(\nu)}(a)}{\nu!} + R_3{}^*,$$

$$(8) \quad \sum_{\nu=m}^{n} (-1)^\nu f(\nu) = \frac{1}{2} \sum_{\nu=0}^{k-1} \frac{E_\nu}{4^\nu} \frac{(-1)^n f^{(2\nu)}\left(n+\frac{1}{2}\right)+(-1)^m f^{(2\nu)}\left(m-\frac{1}{2}\right)}{(2\nu)!} + R_4{}^*.$$

Formula (5) is the Euler–Maclaurin formula, (6) is the Sonin–Hermite formula, and (7) is the Hermite formula.

If some additional conditions are introduced for the function f, it can be proved that the remainders R^* which appear on the right-hand sides of the above formulas are bounded. For instance, if

$$|f^{(n)}(x+iy)| < M_n(x) \qquad (n\in N_0),$$

where $M_0, M_1, \ldots$ are positive functions, then

$$|p_{2k}(x, y)| < M_{2k}(x)\frac{y^{2k}}{(2k)!} \quad i \quad |q_{2k+1}(x, y)| < M_{2k+1}(x)\frac{y^{2k+1}}{(2k+1)!},$$

and hence

$$|R_1{}^*| < \frac{1}{2(2k)!} \frac{E_k}{4^k} \left(M_{2k}\left(m-\frac{1}{2}\right)+M_{2k}\left(n+\frac{1}{2}\right)\right).$$

REFERENCE
E. Lindelöf: *Le calcul des résidus et ses applications à la théorie des fonctions*, Paris 1905, reprinted 1947, pp. 75–84.

6.6. DIVERGENT SERIES

6.6.1. *The Fundamental Formula*

Let F be a complex function of a complex variable and let $C = \{z \mid |z| = R\}$. If $x \in C$ and $|x| < |z|$, then

$$(1) \qquad \frac{1}{2\pi i} \oint_C \frac{F(z)}{z-x} \, dz = \frac{1}{2\pi i} \oint_C F(z) \left(\frac{1}{z} + \frac{x}{z^2} + \cdots \right) dz.$$

Define s_n by

$$(2) \qquad s_n = \frac{1}{2\pi i} \oint_C F(z) \left(\frac{1}{z} + \frac{x}{z^2} + \cdots + \frac{x^n}{z^{n+1}} \right) dz$$

$$= \frac{1}{2\pi i} \oint_C F(z) \frac{z^{n+1} - x^{n+1}}{(z-x) z^{n+1}} \, dz \quad (n \in N_0).$$

Furthermore, let f be a complex function in three complex variables and let Γ be the circle with centre in the origin of w-plane. If $|v| < |w|$, then

$$(3) \qquad \frac{1}{2\pi i} \oint_\Gamma \frac{f(w, v, x)}{w-v} \, dw = \frac{1}{2\pi i} \oint_\Gamma f(w, v, x) \left(\frac{1}{w} + \frac{v}{w^2} + \cdots \right) dw.$$

If we put

$$(4) \qquad c_n = \frac{1}{2\pi i} \oint_\Gamma f(w, v, x) \frac{v^n}{w^{n+1}} \, dw \quad (n \in N_0),$$

then for $n \in N_0$ we have

$$(5) \qquad c_n s_n = \left(\frac{1}{2\pi i} \right)^2 \oint_C \oint_\Gamma F(z) f(w, v, x) \frac{z^{n+1} - x^{n+1}}{z-x} \frac{v^n}{z^{n+1} w^{n+1}} \, dz \, dw.$$

Applying the identity

$$\sum_{n=0}^{+\infty} v^n \frac{z^{n+1}-x^{n+1}}{w^{n+1}\,z^{n+1}} = \sum_{n=0}^{+\infty} \frac{v^n}{w^{n+1}} - x \sum_{n=0}^{+\infty} \frac{(vx)^n}{(wz)^{n+1}} = \frac{v\,(z-x)}{(w-v)\,(wz-vx)},$$

which holds for $|v| < |w|$ and $|vx| < |wz|$, from (5) follows

(6) $\qquad \displaystyle\sum_{n=0}^{+\infty} c_n\, s_n = \frac{1}{(2\pi i)^2} \oint_C \oint_\Gamma \frac{F(z)\,f(w,\,v,\,x)}{(w-v)\left(z-\dfrac{vx}{w}\right)}\, dz\,dw.$

The series which appears on the left-hand side of (6) need not be convergent. Hence the above formula provides a method for summation of divergent series.

6.6.2. *The Case of Regular Functions*

If F is a regular function in the region $C \cup \operatorname{int} C$, formula (6) from 6.6.1 becomes

(1) $\qquad \displaystyle\sum_{n=0}^{+\infty} c_n\, s_n = \frac{1}{2\pi i} \oint_\Gamma \frac{F\left(\dfrac{vx}{w}\right) f(w,\,v,\,x)}{w-v}\, dw.$

If, besides that, the function H defined by

$$H(w,\,v,\,x) = F\left(\frac{vx}{w}\right) f(w,\,v,\,x),$$

is regular with respect to w in the region $\Gamma \cup \operatorname{int} \Gamma$, from (1) follows

(2) $\quad \displaystyle\sum_{n=0}^{+\infty} c_n\, s_n = f(v,\,v,\,x)\,F(x), \quad \text{where} \quad c_n = \frac{1}{2\pi i} \oint_\Gamma \frac{H(w,\,v,\,x)}{F\left(\dfrac{vx}{w}\right)} \frac{v^n}{w^{n+1}}\, dw.$

EXAMPLE 1. Let

$$F(x) = \frac{1}{1-x}, \quad f(w,\,v,\,x) = \left(1 - \frac{w}{vx}\right)^m \quad (m \in \mathbb{N}).$$

Then

$$H(w, v, x) = \frac{1}{1-\frac{vx}{w}} \left(1-\frac{w}{vx}\right)^m = -\frac{w}{vx}\left(1-\frac{w}{vx}\right)^{m-1}.$$

In virtue of (2) we get

$$\left(1-\frac{1}{x}\right)^{-m} \sum_{n=0}^{+\infty} \binom{m}{n}(-1)^n \frac{1}{x^n}(1+x+\cdots+x^n) = \frac{1}{1-x}.$$

Suppose now that the functions F and f are regular in the region $C \cup$ int C, that f is regular also in the region $\Gamma \cup$ int Γ, and that F has only simple poles $a_1, a_2, \ldots$ in int Γ. Then from (6) of 6.6.1 follows

$$(3) \qquad F(x) = \sum_{n=0}^{+\infty} \frac{c_n s_n}{f(v, v, x)} + \sum_k \frac{f\left(\frac{vx}{a_k}, v, x\right)}{f(v, v, x)} \frac{x}{a_k(x-a_k)} \lim_{z \to a_k}(z-a_k)F(z).$$

In particular, if

$$\lim_{|v| \to +\infty} \frac{f\left(\frac{vx}{a}, v, x\right)}{f(v, v, x)} = 0 \qquad (k = 1, 2, \ldots),$$

then from (3) we get

$$(4) \qquad F(x) = \lim_{v \to \infty} \frac{1}{f(v, v, x)} \sum_{n=0}^{+\infty} c_n s_n.$$

In the case when f is a function depending on one variable, (4) reduces to

$$(5) \qquad F(x) = \lim_{v \to \infty} \frac{1}{f(v)} \sum_{n=0}^{+\infty} c_n s_n.$$

Formula (5) is fundamental for Borel's theory of summable series.

EXAMPLE 2. Let $f(w, v, x) = w^p$ ($p \in N$). Applying (3) we obtain

$$F(x) = F(0) + xF'(0) + \cdots + \frac{x^p}{p!} F^{(p)}(0) + \sum_k \left(\frac{x}{a_k}\right)^{p+1} \frac{1}{x-a_k} \lim_{z \to a_k}(z-a_k)F(z).$$

REFERENCES
1. A. Buhl: 'Sur une extension de la méthode de sommation de M. Borel', *Compt. Rend. Acad. Sci. Paris* 144 (1907), 710–712.
2. A. Buhl: 'Sur de nouvelles applications de la théorie des résidus', *Bull. Sci. Math.* (2) 31 (1907), 152–158.

6.7. MISCELLANEOUS SUMMATIONS

6.7.1. Suppose that $a_1, \ldots, a_m$ and $b_1, \ldots, b_n$ are different real numbers. Then

$$(1) \quad \sum_{k=1}^{n} \frac{(b_k-a_1)\cdots(b_k-a_m)}{(b_k-b_1)\cdots(b_k-b_{k-1})(b_k-b_{k+1})\cdots(b_k-b_n)} = \begin{cases} 0 & (m<n-1), \\ 1 & (m=n-1). \end{cases}$$

Proof. For the function f, defined by

$$f(z) = \frac{(z-a_1)\cdots(z-a_m)}{(z-b_1)\cdots(z-b_n)},$$

we have

$$(2) \quad \sum_{k=1}^{n} \operatorname*{Res}_{z=b_k} f(z) = -\operatorname*{Res}_{z=\infty} f(z).$$

Furthermore,

$$(3) \operatorname*{Res}_{z=b_k} f(z) = \lim_{z \to b_k} (z-b_k) f(z) = \frac{(b_k-a_1)\cdots(b_k-a_m)}{(b_k-b_1)\cdots(b_k-b_{k-1})(b_k-b_{k+1})\cdots(b_k-b_n)},$$

$$(4) \quad -\operatorname*{Res}_{z=\infty} f(z) = \lim_{z \to \infty} z f(z) = \begin{cases} 0 & (m<n-1), \\ 1 & (m=n-1). \end{cases}$$

From (2), (3), and (4) follows (1).

REFERENCE
E. Caraman: 'An Application of the Residue Theorem', (Romanian), *Gaz.Mat.Fiz.* 14 (1962), 352–358.

6.7.2. If $|z| < (k-1)^{k-1}/k^k$, then

$$\sum_{n=0}^{+\infty} \binom{kn}{n} z^n = \frac{1+W}{1-(k-1)W},$$

where W is the unique root of the equation $w - z(1 + w)^k = 0$ which is inside the circle

$$\left\{ w \mid |w| = \frac{1}{k-1} \right\}.$$

Proof. Notice first that $\dbinom{kn}{n}$ is the coefficient of w^n in the expansion of $(1 + w)^{kn}$. Hence,

(1) $$\operatorname*{Res}_{w=0} \frac{(1+w)^{kn}}{w^{n+1}} = \binom{kn}{n}.$$

Let Γ be a simple closed contour which encircles the origin. In virtue of (1) and Cauchy's residue theorem, we have

(2) $$\sum_{n=0}^{+\infty} \binom{kn}{n} z^n = \sum_{n=0}^{+\infty} \left(\operatorname*{Res}_{w=0} \frac{(1+w)^{kn}}{w^{n+1}} \right) z^n$$

$$= \sum_{n=0}^{+\infty} \left(\frac{1}{2\pi i} \oint_\Gamma \frac{(1+w)^{kn}}{w^n} \frac{1}{w} \, dw \right) z^n = \frac{1}{2\pi i} \oint_\Gamma \sum_{n=0}^{+\infty} \left(\frac{(1+w)^k z}{w} \right)^n \frac{1}{w} \, dw,$$

where the interchange of summation and integration is justified if the last series is uniformly convergent on Γ.

Since, by hypothesis, $|z| < (k-1)^{k-1}/k^k$, for $n > 1$ and $|w| = 1/(k-1)$ we have

$$\left| \frac{(1+w)^k z}{w} \right| \le \frac{(1+|w|)^k |z|}{|w|} = 1.$$

Therefore,

$$\sum_{n=0}^{+\infty} \left(\frac{(1+w)^k z}{w} \right)^n = \frac{w}{w - z(1+w)^k},$$

and (2) becomes

$$\sum_{n=0}^{+\infty} \binom{kn}{n} z^n = \frac{1}{2\pi i} \oint_\Gamma \frac{1}{w - z(1+w)^k} \, dw.$$

By an application of Rouché's theorem we conclude that the equation $w - z(1 + w)^k = 0$ has only one root in the region int Γ, and so

$$\sum_{n=0}^{+\infty} \binom{kn}{n} z^n = \operatorname*{Res}_{w=W} \frac{1}{w - z(1 + w)^k} = \frac{1}{1 - kz(1 + W)^{k-1}}$$

$$= \frac{1 + W}{1 + W - kz(1 + W)^k} = \frac{1 + W}{1 - (k-1)W}.$$

EXAMPLES

1. $\displaystyle\sum_{n=0}^{+\infty} \binom{2n}{n} 5^{-n} = \sqrt{5}$; 2. $\displaystyle\sum_{n=0}^{+\infty} \binom{2n}{n} 10^{-n} = \frac{\sqrt{15}}{3}$; 3. $\displaystyle\sum_{n=0}^{+\infty} \binom{3n}{n} \frac{2^n}{27^n} = \frac{1 + \sqrt{3}}{2}$.

REFERENCE
H. J. Ricardo: 'Summation of Series by the Residue Theorem', *Math. Mag.* 44 (1971), 24–26.

6.7.3. If $a > 0$ and $m \in \mathbf{N}$, then

$$\sum_{n=1}^{+\infty} \frac{(-1)^n}{n^{2m+1}} \left(\frac{1}{\sin a\pi n} + \frac{a^{2m}}{\sin \dfrac{\pi n}{a}} \right)$$

$$= \frac{(-1)^m \pi^{2m+1}}{2a} \sum_{k=0}^{m+2} \frac{(2k-2)(2^{2m+2-2k}-2)a^{2k}}{(2k)!(2m+2-2k)!} B_{2k} B_{2m+2-2k}.$$

REFERENCE
G. Bach: 'Über die Summierung von Kosekantenreihen mit Hilfe des Residuensatzes', *Elemente Math.* 29 (1974), 139–141.

6.7.4. Suppose that the function a has a finite number of simple poles $z_1, \ldots,$ z_K in the z-plane, where $z_\nu \notin \mathbf{Z}$ and $z_\nu \neq \infty$. Then

$$(1) \qquad \sum_{n=0}^{N} a(n) = (N+1)a(\infty) - \sum_{k=1}^{K} \zeta_N^{(1)}(z_k) \operatorname*{Res}_{z=z_k} a(z),$$

$$(2) \qquad \sum_{n=0}^{N} (-1)^n a(n) = \frac{1}{2}(1 + (-1)^N)a(\infty) - \sum_{k=1}^{K} \zeta_N^{(2)}(z_k) \operatorname*{Res}_{z=z_k} a(z),$$

where

$$\zeta_N^{(1)}(z) = \sum_{n=0}^{N} \frac{1}{z-n}, \quad \zeta_N^{(2)}(z) = \sum_{n=0}^{N} \frac{(-1)^n}{z-n}.$$

Proof. The function $z \mapsto a(z)\zeta_N^{(1)}(z)$ has only simple poles at $z = z_1, \ldots, z_K, 0, 1, \ldots, N$. Therefore

$$\sum_{n=0}^{N} \mathop{\mathrm{Res}}_{z=n} a(z)\zeta_N^{(1)}(z) + \sum_{k=1}^{K} \mathop{\mathrm{Res}}_{z=z_k} a(z)\zeta_N^{(1)}(z) = -\mathop{\mathrm{Res}}_{z=\infty} a(z)\zeta_N^{(1)}(z).$$

However, we have

$$\mathop{\mathrm{Res}}_{z=n} a(z)\zeta_N^{(1)}(z) = a(n),$$

$$\mathop{\mathrm{Res}}_{z=\infty} a(z)\zeta_N^{(1)}(z) = \sum_{n=0}^{N} \mathop{\mathrm{Res}}_{z=\infty} \frac{a(z)}{z-n} = -(N+1)a(\infty),$$

and so

$$\sum_{n=0}^{N} a(n) + \sum_{k=1}^{K} \zeta_N^{(1)}(z_k) \mathop{\mathrm{Res}}_{z=z_k} a(z) = (N+1)a(\infty),$$

implying (1).

The equality (2) can be proved analogously.

REMARK. The given formulas are of practical use when K is much smaller than N.

REFERENCE
G. G. Comisar: 'On the Evaluation of Finite Sums by Residues', *Am. Math. Monthly* **67** (1960), 775–776.

6.7.5. Let (a_n) be a sequence of complex numbers with period k, i.e. $a_n = a_{n+k}$ for every $n \in \mathbb{N}$. It is known that a_n may be expanded in a finite Fourier series

(1) $\qquad a_n = \sum_{r=0}^{k-1} b_r \exp \frac{2\pi i r n}{k},$

with

$$b_n = \frac{1}{k} \sum_{r=0}^{k-1} a_r \exp\left(-\frac{2\pi i r n}{k}\right).$$

Put

(2) $$G(z, B) = \sum_{r=0}^{k-1} b_r \exp \frac{2\pi i r z}{k},$$

(3) $$F(z, B) = \sum_{|r| \le \frac{1}{2}(k-1)} b_r \exp \frac{2\pi i r z}{k},$$

where in (3) k is odd.

Let f be a rational function such that $|f(z)| \le C|z|^{-c}$, where $C > 0$, $c > 1$, uniformly as $|z| \to +\infty$. Furthermore, let $S = S(f) = \{z_1, \ldots, z_m\}$ denote the set of all poles of f. Then

(4) $$\sum_{\substack{n=-\infty \\ n \notin S}}^{+\infty} a_n f(n) = - \sum_{r=1}^{m} \operatorname*{Res}_{z=z_r} \frac{\pi e^{-\pi i z} G(z, B) f(z)}{\sin \pi z};$$

(5) $$\sum_{n=-\infty}^{+\infty} (-1)^n a_n f(n) = - \sum_{r=1}^{m} \operatorname*{Res}_{z=z_r} \frac{\pi F(z, B) f(z)}{\sin \pi z},$$

where in (5) k is an odd number.

Proof. Let C_N denote the square whose centre is the origin and whose sides of length $2N + 1$ are parallel to the real and imaginary axes. The positve integer N is chosen large enough so that $S \subset \operatorname{int} C_N$. Then

(6) $$\oint_{C_N} \frac{\pi e^{-\pi i z} G(z, B) f(z)}{\sin \pi z} dz = \sum_{r=1}^{m} \operatorname*{Res}_{z=z_r} \frac{\pi e^{-\pi i z} G(z, B) f(z)}{\sin \pi z}$$
$$+ \sum_{\substack{n=-N \\ n \notin S}}^{N} \operatorname*{Res}_{z=n} \frac{\pi e^{-\pi i z} G(z, B) f(z)}{\sin \pi z}.$$

If $n \notin S$,

$$\operatorname*{Res}_{z=n} \frac{\pi e^{-\pi i z} G(z, B) f(z)}{\sin \pi z} = G(n, B) f(n) = a_n f(n).$$

On the other hand, if $z = x + iy$ $(x, y \in \mathbf{R})$, then

$$\left| \frac{e^{-\pi i z} G(z, B) f(z)}{\sin \pi z} \right| \leq \frac{2 e^{\pi y} \sum\limits_{r=0}^{k-1} |b_r| e^{\frac{-2\pi r y}{k}}}{|e^{\pi y} - e^{-\pi y}|},$$

which is bounded as $|y| \to +\infty$. Furthermore, the function $z \mapsto [e^{-\pi i z} G(z, B)] / (\sin \pi z)$ has period $2k$, and hence there exists a positive constant M, independent of N, such that

$$\left| \frac{e^{-\pi i z} G(z, B)}{\sin \pi z} \right| \leq M \qquad (z \in C_N).$$

Therefore

$$\left| \oint\limits_{C_N} \frac{\pi e^{-\pi i z} G(z, B) f(z)}{\sin \pi z} \, dz \right| \leq \frac{4 \pi M C (2N+1)}{\left(N + \dfrac{1}{2}\right)^c}.$$

Thus, if $N \to +\infty$ in (6), we obtain (4).

If k is an even number, then $(-1)^n a_n$ also has period k, and we can apply formula (4). If k is odd, then we integrate the function $z \mapsto [\pi F(z, B) f(z)] / (\sin \pi z)$ along the square C_N, and we arrive at (5) in a similar manner.

EXAMPLE. Let $f(z) = 1/z^2$ and let

$$M_n(B) = \sum_{r=0}^{k-1} b_r r^n, \quad N_n(B) = \sum_{|r| \leq \frac{1}{2}(k-1)} b_r r^n,$$

where in the last case k is odd. Then

$$\sum_{|n| \geq 1} \frac{a_n}{n^2} = \frac{\pi^2}{3 k^2} \left(6 M_2(B) - 6 k M_1(B) + k^2 M_0(B) \right)$$

$$\sum_{|n| \geq 1} \frac{(-1)^n a_n}{n^2} = \frac{\pi^2}{6 k^2} \left(12 N_2(B) - k^2 N_0(B) \right).$$

REFERENCE

B. C. Berndt: 'The Evaluation of Infinite Series by Contour Integration', *Univ. Beograd. Publ. Elektrotehn. Fak. Ser. Mat. Fiz.* 412–460 (1973), 119–122.

6.7.6. If $x \neq 0$, then

(1) $$\sum_{k=1}^{+\infty} \frac{1}{k^3}\left(\coth k\pi x + x^2 \coth \frac{k\pi}{x}\right) = \frac{\pi^3}{90x}(x^4 + 5x^2 + 1).$$

Proof. The function $z \mapsto \pi \cotg \pi z \coth \pi x z$ is bounded when $|z| = N + \frac{1}{2}$, and $N \to +\infty$ through integral values; consequently the sum of the residues of the function $z \mapsto f(z) = \pi z^{-3} \cotg \pi z \coth \pi x z$ is zero.

On the other hand, we have

$$\operatorname*{Res}_{z=m} f(z) = \frac{\coth m\pi x}{m^3}, \qquad \operatorname*{Res}_{z=mi/x} f(z) = \frac{x^2 \coth (m\pi)/x}{m^3} \qquad (m \in \mathbf{Z}),$$

$$\operatorname*{Res}_{z=0} f(z) = -\frac{\pi^3}{x}\left(\frac{1+x^4}{45} + \frac{x^2}{9}\right),$$

which implies (1).

Formula (1) was stated by Ramanujan, and proved by Preece [1].

REFERENCE
1. C. T. Preece: 'Theorems Stated by Ramanujan, XIII', *J. London Math. Soc.* 6 (1931), 95–99.

6.7.7. If k is an odd number, then

$$\sum_{k=1}^{+\infty} \frac{(-1)^{k-1}}{(2k-1)\left(\ch\dfrac{(2k-1)\pi}{2n} + \cos\dfrac{(2k-1)\pi}{2n}\right)} = \frac{\pi}{8}.$$

This result was stated by Ramanujan. The proof by residues was given by Preece [1].

REFERENCE
1. C. T. Preece: 'Theorems Stated by Ramanujan, XIII', *J. London Math. Soc.* 6 (1931), 95–99.

6.7.8. The following formula is valid

$$\sum_{r=1}^{+\infty} \sum_{s=1}^{+\infty} \frac{1}{r+s-1}\binom{r+s-1}{r}\binom{r+s-1}{s} u^r v^s$$

$$= \frac{1}{2}\left(1 - u - v - \sqrt{1 - 2(u+v) + (u-v)^2}\right).$$

Hint. Apply the equality

$$\binom{r+s-1}{s} = \frac{1}{2\pi i} \oint_C \frac{(1+z)^{r+s-1}}{z^{s+1}} \, dz,$$

where C is a circle with centre at origin.

REFERENCE
G. H. Weiss and M. Dishan: 'A Method for the Evaluation of Certain Sums Involving Binomial Coefficients', *Fibonacci Quart.* 14 (1976), 75–77.

6.7.9. Let C_1 and C_2 be concentric circles with centre at the origin and with radii r_1 and r_2 ($r_1 > r_2$), and suppose that F is a regular function in the annulus $\{z \mid r_2 < |z| < r_1 \}$. Put

$$s'_n = \frac{1}{2\pi i} \oint_{C_1} F(z) \frac{z^{n+1} - x^{n+1}}{z^{n+1}(z-x)} \, dz, \qquad s''_n = \frac{1}{2\pi i} \oint_{C_2} F(z) \frac{x^{n+1} - z^{n+1}}{x^{n+1}(x-z)} \, dz.$$

Let Γ_1 and Γ_2 be concentric circles with centre at the origin and with radii ρ_1 and ρ_2 ($\rho_1 > \rho_2$), and suppose that f is a regular function in the annulus $\{z \mid \rho_2 < |z| < \rho_1 \}$. Put

$$c'_n = \frac{1}{2\pi i} \oint_{\Gamma_1} f(\zeta) \frac{\xi^n}{\zeta^{n+1}} \, d\zeta, \qquad c''_n = \frac{1}{2\pi i} \oint_{\Gamma_2} f(\zeta) \frac{\zeta^{n+1}}{\xi^{n+2}} \, d\zeta.$$

The following formulas are valid

$$\sum_{n=0}^{+\infty} c'_n s'_n = \left(\frac{1}{2\pi i}\right)^2 \oint_{C_1} \oint_{\Gamma_1} \frac{F(z)f(\zeta)}{(\zeta-\xi)\left(z - \dfrac{\xi x}{\zeta}\right)} \, dz \, d\zeta \qquad (|\xi| < \varrho_1; \ |\xi x| < r_1 \varrho_1),$$

$$\sum_{n=0}^{+\infty} c''_n s''_n = \left(\frac{1}{2\pi i}\right)^2 \oint_{C_2} \oint_{\Gamma_2} \frac{F(z)f(\zeta)}{(\zeta-\xi)\left(z - \dfrac{\xi x}{\zeta}\right)} \, dz \, d\zeta \qquad (|\xi| > \varrho_2; \ |\xi x| > r_2 \varrho_2).$$

REFERENCES
1. A. Buhl: 'Sur la sommabilité des séries de Laurent', *Compt. Rend. Acad. Sci. Paris* 145 (1907), 614–616.
2. A. Buhl: 'Sur de nouvelles formules de sommabilité', *Bull. Sci. Math.* 31 (1907), 340–346.

6.7.10. Let f and F be functions in two real variables x and y, such that the function f has first order partial derivatives. Put $u = f(x, y)$, $v = f(y, x)$, $w = F(x, y)$. Using calculus of residues, Cauchy [1] proved the following formula

$$\sum_{k=0}^{n} \frac{n!}{k!(n-k)!} \left(\frac{d^n (u^k v^{n-k} w)}{dx^k \, dy^{n-k}} \right) \Big|_{y=x} = \left(\frac{d^n}{dx^n} \frac{f(x, x)^{n+1} F(x, x)}{f(x, x) - (x-z)(u_x(x, x) - u_y(x, x))} \right) \Big|_{z=x}$$

This formula was generalized by Amanzio [2]. He proved the following result:

Suppose that f and F are functions depending on $x_1, \ldots, x_m$, such that the function f has first order partial derivatives. If we put

$$u_1 = f(x_1, x_2, \ldots, x_m), \quad u_2 = f(x_2, x_1, \ldots, x_m), \ldots,$$

$$u_m = f(x_m, x_{m-1}, \ldots, x_1), \quad w = F(x_1, x_2, \ldots, x_m),$$

$$U = f(x, x, \ldots, x), \quad V = F(x, x, \ldots, x).$$

then

$$\sum_{k_1, \ldots, k_m} \frac{n!}{k_1! \cdots k_m!} \left(\frac{d^n (u_1^{k_1} \cdots u_m^{k_m} w)}{dx_1^{k_1} \cdots dx_m^{k_m}} \right) \Big|_{x_1 = \cdots = x_m = x}$$

$$= \left(\frac{d^n}{dx^n} \frac{u^{m+n-1} V}{(u - (x-z)(v_1 - v_2)) \cdots (u - (x-z)(v_1 - v_m))} \right) \Big|_{z=x},$$

where

$$v_r = \frac{\partial u_1}{\partial x_r} (x, x, \ldots, x).$$

REFERENCES
1. A.-L. Cauchy: 'Application du calcul des résidus à la sommation de plusieurs suites', *Exercices de mathématiques*, Paris 1826. $\equiv$ *Oeuvres* (2) 6, 62–73.
2. D. Amanzio: 'Sopra Alcune Formule', *Giornale Mat. Battaglini* (1) 15 (1877), 257–267.

6.7.11. Let P and Q be polynomials such that $\mathrm{dg}\, Q \geq 2 + \mathrm{dg}\, P$, and the zeros of Q do not belong to the set of nonnegative integers. If

$$\frac{P(z)}{Q(z)} = \sum_{j=1}^{n} \sum_{i=1}^{m} \frac{C_{ij}}{(z + a_i)^j},$$

where $n = \{\max s \mid (z + a_i)^s$ divides $Q(z)\} \leq \mathrm{dg}\, Q$, then

$$(1) \qquad \sum_{\nu=0}^{+\infty} \frac{P(\nu)}{Q(\nu)} = \sum_{j=1}^{n} \frac{(-1)^j}{(j-1)!} \sum_{i=1}^{m} C_{ij}\, \psi^{(j-1)}(a_i),$$

where ψ is the psi function, i.e. $\psi(z) = \Gamma'(z)/\Gamma(z)$.

The above result was obtained by State [1], who integrated the function $z \mapsto P(z)\psi(-z)/Q(z)$ along the contour $C_\nu = \{z \mid |z| = \nu + \frac{1}{2}\}$ ($\nu \in \mathbb{N}$). We quote two special cases of (1):

1° If g is periodic with period m, and if $\sum_{\nu=0}^{m-1} g(\nu) = 0$, then

$$\sum_{\nu=0}^{+\infty} \frac{g(\nu)}{\nu+a} = -\sum_{\nu=0}^{m-1} \frac{g(\nu)}{m} \psi\left(\frac{a+\nu}{m}\right),$$

where $a \in C$ and $a \neq 0, -1, -2, \dots$.

2° If $k\, (\geq 2)$ is a positive integer, then

$$2 \sum_{\nu=1}^{+\infty} \nu^{-k} \psi(\nu) = k\zeta(k+1) - 2\gamma\zeta(k) - \sum_{\nu=2}^{k-1} \zeta(\nu)\, \zeta(k-\nu+1),$$

where ζ is the Riemann zeta function, and γ is Euler's constant.

REFERENCE

1. E. Y. State: 'A One-Sided Summatory Function', *Proc. Am. Math. Soc.* 60 (1976), 134–138.

6.7.12. In the first part of his paper [1] Košljakov generalizes various summation formulas for the series of the form $\sum_{n=1}^{+\infty} f(n)$ and $\sum_{n=1}^{+\infty} (-1)^n f(n)$ from the book [2], to the series of the form $\sum_{n=1}^{+\infty} \sigma(n)f(n)$, where $\sigma(n)$ denotes the Legendre symbol (n/k), or character mod k, where $k\ (>1)$ is a positive integer. In particular, the series $\sum_{n=1}^{+\infty} n\sigma(n)\exp(-(\pi x/k)n^2)$ and $\sum_{n=1}^{+\infty} \sigma(n)n^{-s}$ are summed; the special cases $\sigma(n) = 1$ or $\sigma(n) = (-1)^n$ were known. A generalization of the gamma function, namely the expression $\prod_{n=1}^{+\infty} (1 + (x/n))\sigma(n)$ is also considered, and some formulas analogous to the Euler–Maclaurin formula are obtained.

REMARK. Quoted according to the referative journal *Jahrbuch über die Fortschritte der Mathematik* 48 (1921/22), 156.

REFERENCES
1. N. S. Kochliakoff: 'Sur quelques applications du calcul des résidus à la théorie des nombres', *Proc. Math. Labor. Crimean (Tauric) Univ.* 3 (1921), 101–128.
2. E. Lindelöf: Calcul des résidus, Paris 1905.

6.7.13. By means of residues Košljakov [1] deduced some summation formulas which generalize the Abel and the Euler–Maclaurin formulas. He then applied the obtained formulas to:

$1°$ the summation of Dirichlet's series of the form $\Sigma_{n=1}^{+\infty} e^{-\lambda_n x}$;

$2°$ the evaluation of the limit

$$\lim_{n \to +\infty} \left(\frac{1}{w} + \frac{1}{w+\lambda_1} + \cdots + \frac{1}{w+\lambda_{n-1}} - \log (w+\lambda_n) \right);$$

$3°$ generalization of the gamma function, i.e. defining of the function Γ_1, given by

$$\Gamma_1(x) = \frac{1}{x} e^{-c_1 x} \prod_{n=1}^{+\infty} \frac{e^{x/\lambda_n}}{1 + x/\lambda_n};$$

$4°$ the proof of

$$\lim_{s \to 1} \left(\frac{1}{w^s} + \frac{1}{(w+\lambda_1)^s} + \cdots - \frac{1}{s-1} \right) = -\frac{\Gamma_1'(w)}{\Gamma_1(w)}.$$

REMARK. Quoted according to the referative journal *Jahrbuch über die Fortschritte der Mathematik* 48 (1921/22), 350–351.

REFERENCE
1. N. S. Košljakov: 'On Certain Applications of the Theory of Integral Residues', (Russian), *Simferopol', Zap.Matem.Kab.Krym.Univ.* 1 (1919), 189–254; and 2 (1921), 1–79.

6.7.14. If $m \in \mathbf{N}_0$, then

$$\sum_{n=0}^{+\infty} \frac{(-1)^n}{(2n+1)^{m+2}} = \frac{1}{2} \left(\frac{\pi}{2} \right)^{m+2} C_{m+1}; \qquad \sum_{n=0}^{+\infty} \frac{1}{(2n+1)^{m+2}} = \frac{1}{2} \left(\frac{\pi}{2} \right)^{m+2} T_{m+1},$$

where the coefficients C_k, T_k are defined by:

$$\left(\cos \frac{1}{x} \right)^{-1} = \sum_{k=0}^{+\infty} C_k x^{-k}; \qquad \left(\text{tg} \frac{1}{x} \right)^{-1} = \sum_{k=0}^{+\infty} T_k x^{-k}.$$

REFERENCE

G. Oltramare: 'Mémoire sur quelques propositions du calcul des résidus', *Mémoires de l'Institut National Genevois* **3** (1855), 15 pp.

6.7.15. 1° Suppose that $m_1, \ldots, m_n$ are natural, and $z_1, \ldots, z_n$ complex numbers. Then

$$(1) \qquad \sum_{k_1=0}^{m_1} \cdots \sum_{k_n=0}^{m_n} (-1)^K (k_1 z_1 + \cdots + k_n z_n)^M \binom{m_1}{k_1} \cdots \binom{m_n}{k_n}$$

$$= (-1)^M M! \; z_1^{m_1} \cdots z_n^{m_n},$$

where

$$K = \sum_{\nu=1}^{n} k_\nu, \quad M = \sum_{i=1}^{n} m_{..}.$$

2° If $m \in N_0$ and if $a_0, a_1, \ldots, a_{2m+1}, b_0, b_1, \ldots, b_{2m+1}$ are complex numbers, then

$$(2) \qquad \sum_{k=0}^{m} \binom{m+k}{k} \sum_{\nu=0}^{m-k} \binom{m-k}{\nu} (a_{k+\nu} b_{2m+1-k-\nu} + a_{m+1+\nu} b_{m-\nu})$$

$$= \sum_{k=0}^{2m+1} \binom{2m+1}{k} a_k b_{2m+1-k}.$$

Equalities (1) and (2) were proved by Egoryčev and Južakov who applied residues of functions in several variables.

REFERENCE

G. P. Egoryčev and A. P. Južakov: 'On the Computation of Generating Functions and Combinatorial Sums by Multi-dimensional Residues', *Sibir. Mat. Ž.* **15** (1974), 1049–1060.

6.7.16. Let f be a rational function and let $a: N \to C$ be an arithmetical function. Denote by P_f the set consisting of the origin and the poles of f, and denote by a^* the Dirichlet convolution of the Möbius function with a. Suppose that the series

$$\sum_{n \notin P_f} f(n) \quad \text{and} \quad \sum_{n \notin P_f} a(n) f(n)$$

are absolutely convergent. Then, by an application of the residue theorem, it can be shown that

$$\sum_{\substack{n=-\infty \\ n \notin P_f}}^{+\infty} a(|n|)f(n) = -\sum_{n=1}^{+\infty} \frac{a^*(n)s(n)}{n} \; ,$$

where

$$s(n) = 2\pi i \sum_{w \in P_f} \operatorname*{Res}_{z=w} \frac{f(z)}{e^{2\pi i z/n} - 1} \; ,$$

and a similar formula holds for

$$\sum_{\substack{n=-\infty \\ n \notin P_f}}^{+\infty} (-1)^n a(|n|)f(n).$$

From those results it is possible to derive several identities by specializing the functions f and a.

REMARK. This is an extension of Berndt's method (see 6.7.5).

REFERENCE
P. V. Krishnaiah and R. Sita Rama Chandra Rao: 'On Berndt's Method in Arithmetical Functions and Contour Integration', *Canad.Math.Bull.* **22** (1979), 177–185.

Chapter 7

Differential and Integral Equations

7.1. ORDINARY DIFFERENTIAL EQUATIONS

7.1.1. *The General Solution of Linear Differential Equations with Constant Coefficients*

THEOREM 1. *Consider the differential equation with constant coefficients*

$$(1) \qquad y^{(n)} + a_1 y^{(n-1)} + \cdots + a_n y = 0,$$

Let f be an arbitrary regular function of the complex variable z, whose zeros do not coincide with the zeros of the polynomial $z \mapsto g(z) = z^n + a_1 z^{n-1} + \cdots + a_n$.
The general solution of the Equation (1) is given by

$$(2) \qquad \sum \text{Res} \frac{f(z) e^{zx}}{g(z)},$$

where the summation is taken over all the singularities of the function

$$z \mapsto \frac{f(z) e^{zx}}{g(z)},$$

i.e. over all the zeros of the polynomial g.
   Proof. If

$$y = \sum \text{Res} \frac{f(z) e^{zx}}{g(z)},$$

then

$$y^{(k)} = \sum \text{Res} \frac{f(z) e^{zx}}{g(z)} z^k \qquad (k = 1, \ldots, n).$$

Hence,

$$y^{(n)} + a_1 y^{(n-1)} + \cdots + a_n y$$

$$= \sum \operatorname{Res} \frac{f(z) e^{zx}}{g(z)} (z^n + a_1 z^{n-1} + \cdots + a_n) = \sum \operatorname{Res} f(z) e^{zx} = 0,$$

since the function $z \mapsto f(z)e^{zx}$ is, by hypothesis, regular.

Thus, (2) is a solution of (1). Let us prove that this solution is general.

If r is a simple root of the equation $g(z) = 0$, i.e. a simple pole of the function

$$z \mapsto \frac{f(z) e^{zx}}{g(z)} \ ,$$

then

$$\operatorname*{Res}_{z=r} \frac{f(z) e^{zx}}{g(z)} = \lim_{z \to r} (z - r) \frac{f(z) e^{zx}}{g(z)} = \frac{f(r)}{g'(r)} e^{rx}.$$

Since f is an arbitrary function, $f(r)/g'(r)$ is an arbitrary constant, which means that to a simple root of the characteristic equation $g(z) = 0$ corresponds the term Ce^{rx} (C arbitrary constant).

If r is a multiple root, of order s, of the equation $g(z) = 0$, then

$$(3) \qquad \operatorname*{Res}_{z=r} \frac{f(z) e^{zx}}{g(z)} = \frac{1}{(s-1)!} \lim_{z \to r} \frac{\partial^{s-1}}{\partial z^{s-1}} \frac{f(z) e^{zx}}{g_1(z)},$$

where $g_1(z)(z-r)^s = g(z)$.

Since f is an arbitrary function, $f(r), f'(r), \ldots, f^{(s-1)}(r)$ are arbitrary constants, so that (3) becomes

$$(4) \qquad (C_1 + C_2 x + \cdots + C_s x^{s-1}) e^{rx},$$

where $C_1, \ldots, C_s$ are arbitrary constants.

In other words to the root r of order s of the characteristic equation $g(z) = 0$ corresponds the term (4).

Therefore, (2) is a solution of (1) containing n arbitrary constants, which implies that (2) is the general solution of the Equation (1).

THEOREM 2. *The general solution of the equation*

$$(5) \qquad y^{(n)} + a_1 y^{(n-1)} + \cdots + a_n y = F(x)$$

is given by

$$(6) \qquad y = \sum \mathrm{Res} \left(\frac{f(z)}{g(z)} e^{zx} \right) + \sum \mathrm{Res} \left(e^{zx} \int_{x_0}^{x} e^{-zt} \frac{F(t)}{g(z)} \, dt \right),$$

where f and g have the same meaning as in Theorem 1.

Proof. We look for the solution of (5) in the form

$$(7) \qquad y = \sum \mathrm{Res} \left(\frac{h(z, x)}{g(z)} e^{zx} \right),$$

where the sum is, as before, taken over all the zeros of the polynomial g. We now require that the first $n-1$ derivatives of y, defined by (7), be the same as in the case when h is only a function of z. This implies

$$(8) \qquad \sum \mathrm{Res} \left(\frac{\partial h}{\partial x} \frac{z^k}{g(z)} e^{zx} \right) = 0 \qquad (k = 0, 1, \dots, n-2).$$

If the above conditions are satisfied, then

$$(9) \qquad y^{(k)} = \sum \mathrm{Res} \left(\frac{h(z, x)}{g(z)} z^k e^{zx} \right) \qquad (k = 0, 1, \dots, n-1),$$

$$(10) \qquad y^{(n)} = \sum \mathrm{Res} \left(\frac{h(z, x)}{g(z)} z^n e^{zx} \right) + \sum \mathrm{Res} \left(\frac{\partial h}{\partial x} \frac{z^{n-1}}{g(z)} e^{zx} \right).$$

Substituting (9) and (10) into (5), we obtain

$$(11) \qquad \sum \mathrm{Res} \left(\frac{\partial h}{\partial x} \frac{z^{n-1}}{g(z)} e^{zx} \right) = F(x).$$

Therefore, we have to find a function h which satisfies (8) and (11). Since

$$\sum \mathrm{Res} \frac{z^k}{g(z)} = \begin{cases} 0 & (0 < k < n-1), \\ 1 & (k = n-1), \end{cases}$$

we conclude that h can be determined from $(\partial h / \partial x) e^{zx} = F(x)$, which implies

$$(12) \qquad h(z, x) = f(z) + \int_{x_0}^{x} e^{-zt} F(t) \, dt,$$

where f is an arbitrary regular function.

Finally, if the function h, given by (12), is substituted into (7), we obtain (6).

EXAMPLE 1. For the equation

$$y'' - 2y' + y = F(x)$$

we have $g(z) = (z-1)^2$, and formula (6) becomes

$$y = \operatorname*{Res}_{z=1}\left(\frac{f(z)}{(z-1)^2}\,e^{zx}\right) + \operatorname*{Res}_{z=1}\left(\frac{e^{zx}}{(z-1)^2}\int_{x_0}^{x} e^{-zt}F(t)\,dt\right)$$

$$= \lim_{z\to 1}\frac{d}{dz}(f(z)\,e^{zx}) + \lim_{z\to 1}\frac{d}{dz}\int_{x_0}^{x} e^{z(x-t)}F(t)\,dt$$

$$= (f'(1) + x f(1))\,e^x + \int_{x_0}^{x}(x-t)\,e^{x-t}F(t)\,dt,$$

or

$$y = (C_1 + C_2 x)\,e^x + \int_{x_0}^{x}(x-t)\,e^{x-t}F(t)\,dt.$$

on putting $f'(1) = C_1$, $f(1) = C_2$, where C_1 and C_2 are arbitrary constants.

REMARK 1. Theorems 1 and 2 were proved by Cauchy [1]. Detailed surveys of Cauchy's method are given in papers [2] – [5].

REFERENCES
1. A.-L. Cauchy: 'Application du calcul des résidus a l'intégration des équations différentialles linéaires et a coefficients constants', *Exercices de mathématiques*, Paris 1826. ≡ *Oeuvres* (2) 6 (1887), pp. 252–255.
2. B. Tortolini: 'Sur calcolo dei residui', (Memoria 2ª), *Giornale Arcad.* 67 (1836), 179–198;
3. J. Collet: 'Sur l'intégration des équations différentielles linéaires à coefficients constants', *Ann. École Norm. Sup.* (3) 4 (1887), 129–144.
4. Ch. Hermite: 'Équations différentielles linéaires', *Bull. Sci. Math.* (2) 3 (1879), 311–325.
5. G. Darboux: 'Application de la méthode précédente à l'équation linéaire à coefficients constants avec second membre', *Bull. Sci. Math.* (2) 3 (1879), 325–328.

7.1.2. The General Solution of Euler's Equation

Consider the equations

$$(1) \qquad y^{(n)} + \frac{a_1}{Ax+B}y^{(n-1)} + \cdots + \frac{a_n}{(Ax+B)^n}y = 0,$$

$$(2) \qquad y^{(n)} + \frac{a_1}{Ax+B} y^{(n-1)} + \cdots + \frac{a_n}{(Ax+B)^n} y = F(x),$$

where $a_1, \ldots, a_n, A, B$ are given constants and F is a given function.

THEOREM 1. *The general solution of the Equation* (1) *is*

$$y = \sum \mathrm{Res}\, \frac{f(z)\,(Ax+B)^z}{g(z)},$$

where f is an arbitrary regular function, and where the sum is taken over all the zeros of the polynomial g, where

$$g(z) = A^n z (z-1) \cdots (z-n+1) + a_1 A^{n-1} z (z-1) \cdots (z-n+2)$$
$$+ \cdots + a_{n-1} Az + a_n.$$

THEOREM 2. *The general solution of the Equation* (2) *is*

$$(3) \qquad y = \sum \mathrm{Res}\, \frac{f(z)\,(Ax+B)^z}{g(z)}$$

$$+ A \sum \mathrm{Res}\left(\frac{1}{g(z)} \int_{x_0}^{x} \left(\frac{Ax+B}{At+B}\right)^z (At+B)^{n-1} F(t)\, dt \right),$$

where f is an arbitrary regular function, and where both sums are taken over all the zeros of the polynomial g.

EXAMPLE 1. For the equation

$$(4) \qquad y'' - \frac{1}{x+1} y' + \frac{1}{(x+1)^2} y = F(x)$$

we have $g(z) = z(z-1) - z + 1 = (z-1)^2$. Therefore formula (3) becomes

$$y = \mathrm{Res}_{z=1} \frac{f(z)\,(x+1)^z}{(z-1)^2} + \mathrm{Res}_{z=1}\left(\frac{1}{(z-1)^2} \int_{x_0}^{x} \left(\frac{x+1}{t+1}\right)^z (t+1)\, F(t)\, dt \right)$$

$$= \lim_{z \to 1} \frac{d}{dz}\left(f(z)\,(x+1)^z\right) + \lim_{z \to 1} \frac{d}{dz} \int_{x_0}^{x} \left(\frac{x+1}{t+1}\right)^z (t+1)\, F(t)\, dt$$

$$= -f'(1) \; (x+1) + f(1) \; (x+1) \; \log{(x+1)} + \int\limits_{x_0}^{x} (x+1) \; \log{\frac{x+1}{t+1}} \, F(t) \, dt,$$

or, putting $f'(1) = C_1$, $f(1) = C_2$, where C_1 and C_2 are arbitrary constants

(5) $$y = (\dot{x}+1) \, (C_1 + C_2 \log{(x+1)}) + (x+1) \int\limits_{x_0}^{x} F(t) \, \log{\frac{x+1}{t+1}} \, dt.$$

Formula (5) presents the general solution of (4).

REFERENCE
A.-L. Cauchy: 'Application du calcul des résidus à l'intégration de quelques équations différentielles linéaires et à coefficients variables', *Exercices de mathématiques*, Paris 1826. ≡ *Oeuvres* (2) 6, 316–319.

7.1.3. *Determination of Particular Solutions*

THEOREM 1. *The solution of the equation*

(1) $$y^{(n)} + a_1 y^{(n-1)} + \cdots + a_n y = 0,$$

which satisfies Cauchy's initial conditions

(2) $$y^{(k)} (a) = b_k \qquad (k = 0, \, 1, \, \ldots, \, n-1),$$

is obtained by writing the expression

$$y = \sum \text{Res} \left(\frac{g(z) - g(b)}{(z-b) \, g(z)} \, e^{z(x-a)} \right)$$

in the form

$$y = P_0 b^0 + P_1 b^1 + \cdots + P_{n-1} b^{n-1}$$

and by replacing the exponents of b by corresponding indices.
    Proof. We first treat the special case when $y^{(k)} (a) = b^k$. The general solution of (1) is given by

(3) $$y = \sum \text{Res} \frac{f(z) \, e^{zx}}{g(z)},$$

where f is an arbitrary regular function, whose zeros do not coincide with the zeros of the polynomial $z \mapsto g(z) = z^n + a_1 z^{n-1} + \cdots + a_n$.

The function f must be chosen so that

$$(4) \qquad \sum \text{Res} \, \frac{f(z)}{g(z)} \, z^k \, e^{az} = b^k \qquad (k = 0, \, 1, \, \ldots, \, n-1).$$

However, since

$$b^k = \sum \text{Res} \, \frac{z^k}{z-b},$$

the condition (4) can be replaced by

$$\sum \text{Res} \left(\frac{f(z)}{g(z)} \, z^k \, e^{az} - \frac{z^k}{z-b} \right) = 0 \qquad (k = 0, \, 1, \, \ldots, \, n-1)$$

or

$$(5) \qquad \sum \text{Res} \, \frac{f(z) \, e^{az} \, (z-b) - g(z)}{(z-b) \, g(z)} \, z^k = 0 \qquad (k = 0, \, 1, \, \ldots, \, n-1).$$

Since $z \mapsto (z-b)g(z)$ is a polynomial of degree $n+1$, the conditions (5) will be satisfied if the function f is chosen so that

$$z \mapsto \left(f(z) \, e^{az} \, (z-b) - g(z) \right) z^k$$

is a polynomial of degree at most $n-1$ for every $k = 0, 1, \ldots, n-1$. We therefore take $f(z) e^{az} (z-b) - g(z) = C$, where C is a constant, i.e.

$$(6) \qquad f(z) = \frac{g(z) + C}{z-b} \, e^{-az}.$$

However, the function f must be regular at $z = b$, and so we take $C = -g(b)$, and finally

$$f(z) = \frac{g(z) - g(b)}{z-b} \, e^{-az},$$

The formula (3) becomes

$$(7) \qquad y = \sum \text{Res} \left(\frac{g(z) - g(b)}{(z-b) \, g(z)} \, e^{z(x-a)} \right).$$

Since $[g(z)-g(b)]/(z-b)$ is a polynomial of degree $n-1$ in z and b, we can write it in the form

$$\frac{g(z)-g(b)}{z-b} = A_0 + A_1 b + \cdots + A_{n-1} b^{n-1},$$

where $A_0, \ldots, A_{n-1}$ depend on z. Therefore, in view of (7), we conclude that

(8) $\qquad y = P_0 + P_1 b + \cdots + P_{n-1} b^{n-1},$

where $P_0, \ldots, P_{n-1}$ are functions of x.

Since (7) is a solution of (1) for arbitrary b, we see that $P_0, \ldots, P_{n-1}$ are particular solutions of (1). Besides, (7) satisfies the conditions (2), and hence

$$P_0(a) = 1, \qquad P_1(a) = 0, \ldots, \qquad P_{n-1}(a) = 0$$

(9) $\qquad P_0'(a) = 0, \qquad P_1'(a) = 1, \qquad P_{n-1}'(a) = 0$

$$\vdots$$

$$P_0^{(n-1)}(a) = 0, \qquad P_1^{(n-1)}(a) = 0, \qquad P_{n-1}^{(n-1)}(a) = 1.$$

We therefore conclude that when (8) is written as

$$y = P_0 b^0 + P_1 b^1 + \cdots + P_{n-1} b^{n-1},$$

and when the exponents of b are replaced by corresponding indices:

$$y = P_0 b_0 + P_1 b_1 + \cdots + P_{n-1} b_{n-1},$$

then the above expression is the solution of (1) which also satisfies the conditions (2).

The theorem is proved.

In the case of the nonhomogeneous equation

(10) $\qquad y^{(n)} + a_1 y^{(n-1)} + \cdots + a_n y = F(x),$

with the same conditions (2), we start with its general solution

(11) $\qquad y = u + v,$

where

$$u = \sum \operatorname{Res} \frac{f(z)}{g(z)} e^{zx}, \qquad v = \sum \operatorname{Res} \frac{1}{g(z)} \int_a^x e^{z(x-t)} F(t)\, dt.$$

Since

$$v^{(k)} = \sum \operatorname{Res} \frac{1}{g(z)} \left(z^k \int_a^x e^{z(x-t)} F(t)\, dt + z^{k-1} F(x) + \cdots + F^{(k-1)}(x) \right),$$

we have $v^{(k)}(a) = 0$, which means that the term v in (11) has no influence on the conditions (2).

Hence, the solution of the problem (10) − (2) is obtained in the following way:
   If

$$\sum \operatorname{Res} \frac{g(z) - g(b)}{(z-b) g(z)} e^{z(x-a)} = P_0 b^0 + P_1 b^1 + \cdots + P_{n-1} b^{n-1},$$

then the solution of the proposed problem is

$$y = P_0(x) b_0 + P_1(x) b_1 + \cdots + P_{n-1}(x) b_{n-1} + \sum \operatorname{Res} \frac{1}{g(z)} \int_a^x e^{z(x-t)} F(t)\, dt.$$

7.1.4. First Order Systems of Linear Differential Equations with Constant Coefficients

The method given in 7.1.1 can be extended to handle systems of first order linear differential equations with constant coefficients. We only quote the result for homogeneous systems.

THEOREM 1. *The general solution of the system*

$$\frac{dx_1}{dt} = a_{11} x_1 + \cdots + a_{1n} x_n,$$

$$\vdots$$

$$\frac{dx_n}{dt} = a_{n1} x_1 + \cdots + a_{nn} x_n,$$

where a_{ij} are constants, is

$$x_k = \sum_\lambda \operatorname{Res}_\lambda \frac{f_k(\lambda)}{F(\lambda)} e^{\lambda t} \qquad (k = 1, \ldots, n),$$

where

$$F(\lambda) = \begin{vmatrix} a_{11} - \lambda & a_{12} & \cdots & a_{1n} \\ a_{21} & a_{22} - \lambda & & a_{2n} \\ \vdots & & & \\ a_{n1} & a_{n2} & & a_{nn} - \lambda \end{vmatrix}.$$

The functions $f_1, \ldots, f_n$ are solutions of the following algebraic system of equations

$$(a_{11} - \lambda) f_1(\lambda) + \qquad a_{12} f_2(\lambda) + \cdots + \qquad a_{1n} f_n(\lambda) = -C_1 F(\lambda),$$

$$a_{21} f_1(\lambda) + (a_{22} - \lambda) f_2(\lambda) + \cdots + \qquad a_{2n} f_n(\lambda) = -C_2 F(\lambda),$$

$$\vdots$$

$$a_{n1} f_1(\lambda) + \qquad a_{n2} f_2(\lambda) + \cdots + (a_{nn} - \lambda) f_n(\lambda) = -C_n F(\lambda),$$

where $C_1, \ldots, C_n$ are arbitrary constants.

EXAMPLE 1. Consider the system

$$(1) \qquad \frac{dx}{dt} = 2x - y - z; \quad \frac{dy}{dt} = -y; \quad \frac{dz}{dt} = 2y + z.$$

Then $F(\lambda) = -(\lambda - 1)(\lambda - 2)(\lambda + 1)$, and f, g, h are determined from the system of algebraic equations

$$(2 - \lambda) f + g + h = C_1 (\lambda - 1)(\lambda - 2)(\lambda + 1),$$

$$(-1 - \lambda) g = C_2 (\lambda - 1)(\lambda - 2)(\lambda + 1),$$

$$2g + (1 - \lambda) h = C_3 (\lambda - 1)(\lambda - 2)(\lambda + 1),$$

where C_1, C_2, C_3 are arbitrary constants.

We easily find

$$f(\lambda) = -C_1 \lambda^2 + (C_2 + C_3) \lambda + (C_1 + C_2 + C_3),$$

$$g(\lambda) = -C_2 \lambda^2 + 3 C_2 \lambda - 2 C_2,$$

$$h(\lambda) = -C_3 \lambda^2 + (C_3 - 2 C_2) \lambda + (2 C_3 + 4 C_2)$$

and hence

$$x(t) = \sum_\lambda \operatorname*{Res} \frac{C_1 \lambda^2 - (C_2 + C_3) \lambda - (C_1 + C_2 + C_3)}{(\lambda - 1)(\lambda - 2)(\lambda + 1)} e^{\lambda t};$$

$$y(t) = \sum_\lambda \operatorname*{Res} \frac{C_2 \lambda^2 - 3 C_2 \lambda + 2 C_2}{(\lambda - 1)(\lambda - 2)(\lambda + 1)} e^{\lambda t};$$

$$z(t) = \sum_\lambda \operatorname*{Res} \frac{C_3 \lambda^2 - (C_3 - 2 C_2) \lambda - (2 C_3 + 4 C_2)}{(\lambda - 1)(\lambda - 2)(\lambda + 1)} e^{\lambda t},$$

implying the general solution of the system (1):

$$x(t) = (C_2 + C_3)e^t + (C_1 - C_2 - C_3) e^{2t},$$

$$y(t) = C_2 e^{-t},$$

$$z(t) = (C_2 + C_3) e^t - C_2 e^{-t},$$

where C_1, C_2, C_3 are arbitrary constants.

REMARK. This method is due to Cauchy. A detailed survey of the method is given in [1].

REFERENCE
1. B. Tortolini: 'Memoria sull'applicazione del calcolo dei residui all'integrazione dell'equazioni differenziali lineari', *Giornale Arcad.* 92 (1842), 129–152.

7.1.5. An Application in the Theory of Linear Differential Equations

Consider the complex differential equation

(1) $$w'' + p(z) w' + q(z) w = 0,$$

where p and q are analytic functions whose only singularities in the finite plane are poles.

The following theorem on the existence of solutions of (1) was proved in 1866 by Fuchs.

THEOREM 1. *If p and q are regular functions in a neighbourhood of a point* z_0 *and if* a_0 *and* a_1 *are arbitrary constants, there exists a unique function w which is regular and satisfies* (1) *in a neighbourhood of* z_0, *and which also satisfies the initial conditions* $w(z_0) = a_0$, $w'(z_0) = a_1$.

This solution of (1) can be obtained as a potential series

$$w(z) = \sum_{n=0}^{+\infty} a_n (z - z_0)^n,$$

and it can be expressed in the form

$$w(z) = a_0 w_0(z) + a_1 w_1(z),$$

where w_0 and w_1 are solutions of (1) which satisfy $w_0(z_0) = 1$, $w_0'(z_0) = 0$ and $w_1(z_0) = 0$, $w_1'(z_0) = 1$.

The functions w_0 and w_1 are linearly independent, since their Wronskian $\Delta(w_0, w_1)$ defined by

$$\Delta(w_0, w_1) = \begin{vmatrix} w_0(z) & w_1(z) \\ w_0'(z) & w_1'(z) \end{vmatrix},$$

does not vanish.

So far the functions w_0 and w_1 are defined only in a neighbourhood of z_0. However, they can be analytically continued along any path which does not pass through a singular point of the equation, i.e. through a singularity of p or q. The question is whether the obtained functions are also linearly independent.

In connection with that the following theorem is useful.

THEOREM 2. *If the functions* w_0 *and* w_1 *becomes* W_0 *and* W_1 *after analytic continuation along a closed curve* Γ, *then*

$$\Delta(W_0, W_1) = e^{-2\pi i R} \Delta(w_0, w_1),$$

where R is the sum of the residues of p at its poles within Γ.

Hence, the linear independence of w_0 and w_1 implies the linear independence of W_0 and W_1.

7.1.6. Residue of a Function Defined by a Differential Equation

Consider the differential equation

(1) $$\sum_{k=1}^{s} f_k(x) y^{m_k} (y')^{n_k} = 0.$$

Let $\xi_k = m_k + n_k, \eta_k = n_k$ $(k = 1, \ldots, s)$ and let M_k be points in the $\xi\eta$-plane defined by $M_k = (\xi_k, \eta_k)$. Join the points which are nearest to and farthest from the η-axis by a polygonal line P with vertices M_k such that none of those points is above the line P.

In his thesis [1] Petrović proved the following theorem:

The integral of (1) *will have simple mobile poles if and only if the line P has a side with slope* -1.

In [2] Petrović gave a method for evaluating residues of the integrals of (1) with respect to simple mobile poles. His result reads:

Suppose that y is a solution of (1) *which at $x = a$ has a simple pole, and let $A = \operatorname{Res}_{x=a} y(x)$. Then A is the nontrivial solution of the algebraic equation*

$$\Sigma (-1)^{n_k} f_k(a) A^{m_k + n_k} = 0,$$

where the sum is taken over all indices k, such that the point (ξ_k, η_k) belongs to the side of P whose slope is -1.

EXAMPLE 1. The equation

$$(2) \qquad (y')^2 - 4x^2(1-y^2)(1-k^2y^2) = 0 \qquad (k = \text{const})$$

can be written in the form

$$(3) \qquad (y')^2 - 4x^2 + 4(k^2+1)x^2y^2 - 4k^2x^2y^4 = 0,$$

and comparing (1) and (3) we find:

$f_1(x) = 1, m_1 = 0, n_1 = 2,$	$M_1 = (2, 2);$
$f_2(x) = -4x^2, m_2 = 0, n_2 = 0,$	$M_2 = (0, 0);$
$f_3(x) = 4(k^2+1)x^2, m_3 = 2, n_3 = 0,$	$M_3 = (2, 0);$
$f_4(x) = -4k^2x^2, m_4 = 4, n_4 = 0,$	$M_4 = (4, 0).$

The polygonal line P (joining M_2, M_1, and M_4) has one side with slope -1. Hence, if $x = a$ is a simple pole of the solution y of (2) and if $A = \operatorname{Res}_{x=a} y(x)$, then

$$(-1)^{n_1} f_1(a) A^{m_1+n_1} + (-1)^{n_4} f_4(a) A^{m_4+n_4} = 0,$$

i.e.

$$4k^2 a^2 A^2 - 1 = 0, \qquad \text{or} \qquad A = \pm \frac{1}{2ak}.$$

This example shows that the residue A depends on the pole a. However, if the coefficients f_k of (1) are constant, then the residues can be evaluated without knowing the poles.

EXAMPLE 2. For the equation

$$(y')^3 + 3(y')^2 + y^6 - 4 = 0$$

we have

$$f_1(x) = 1, \qquad m_1 = 0, n_1 = 3, \qquad M_1 = (3, 3);$$
$$f_2(x) = 3, \qquad m_2 = 0, n_2 = 2, \qquad M_2 = (2, 2);$$
$$f_3(x) = 1, \qquad m_3 = 6, n_3 = 0, \qquad M_3 = (6, 0);$$
$$f_4(x) = -4, \qquad m_4 = 0, n_4 = 0, \qquad M_4 = (0, 0).$$

The polygonal line P (joining M_4, M_1, and M_3) has one side, $M_1 M_3$, with slope -1. Hence, the residue of y at a simple pole must satisfy the equation $A^3 = 1$.

This result was later (see [3]) extended to mobile poles of higher order equations.

REFERENCES

1. M. Petrovitch: 'Sur les zéros et les infinis des intégrales des équations différentielles algébriques', Thèse, Paris 1894, 109 pp.
2. M. Petrovitch: 'Sur les résidus des fonctions définies par les équations différentielles', *Math. Ann.* 48 (1896), 75–80.
3. M. Petrovitch: 'Sur les résidus des fonctions définies par les équations différentielles d'ordre supérieur', *Věstnik Král. České Společnosti Náuk. Třída Math. Přir.* 1898, (VI), 24 pp.

7.2. THE LAPLACE TRANSFORM

7.2.1. *Introduction*

We begin by quoting, without proof, the following theorem, usually called the Fourier integral theorem.

THEOREM 1. *If f is an absolutely integrable function on R, i.e. if*

$$\int\limits_{-\infty}^{+\infty} |f(t)| \, dt < +\infty \, ,$$

and if f is piecewise smooth on every finite segment of R, then

$$f(t) = \frac{1}{2\pi} \int\limits_{-\infty}^{+\infty} e^{ist} F(s) \, ds,$$

where

$$(1) \qquad F(s) = \int\limits_{-\infty}^{+\infty} e^{-ist} f(t) \, dt.$$

REMARK 1. The function F defined by (1) is called the Fourier transform of f. The condition that f is absolutely integrable is rather strong, and so in applications we use an other transform which does not require absolute integrability.

DEFINITION 1. Let f be a complex function of a real variable which satisfies the conditions:

1° The function f, together with its derivatives up to order n is piecewise continuous;

2° $f(t) = 0$ for $t < 0$;

3° There exist positive numbers M and s such that $|f(t)| < M e^{st}$.

The Laplace transform of the original f is the function $F : C \to C$ defined by

$$(2) \qquad F(p) = \int\limits_{0}^{+\infty} f(t) e^{-pt} \, dt.$$

The integral in (2) is called the Laplace integral.
The following notation is often used:

$$F(p) = Lf(t).$$

The Laplace transform (as well as other integral transforms) plays an important role in various branches of mathematics, particularly in solving ordinary and partial differential equations. An extensive literature exists on that subject (see, for example [1], [2], [3], [4]).

We shall only be concerned with one problem in connection with the Laplace transform, namely if the transform $p \mapsto F(p)$ is given, how to determine the original $t \mapsto f(t) = L^{-1}F(p)$, since that problem is successfully solved by an application of the calculus of residues.

REFERENCES
1. G. Doetsch: *Einführung in Theories und Anwendung der Laplace-Transformation*, Basel-Stuttgart 1958.
2. G. Doetsch: *Handbuch der Laplace-Transformation I, II, III*, Berlin 1950, 1955, 1956.
3. J. C. Jaeger: *An Introduction to the Laplace Transformation*, London-New York 1951.
4. D. V. Widder: *The Laplace Transform*, Princeton 1941.

7.2.2. *Formula for the Inverse Laplace Transform*

Suppose that f satisfies the conditions of Definition 1 and let $s_0 = \inf \{s \mid |f(t)| < Me^{st}\}$. The function g, defined by

$$(1) \qquad g(t) = e^{-ut}f(t),$$

where $u > s_0$, satisfies all the conditions of Fourier's theorem (Theorem 1 from 7.2.1), and so we have

$$(2) \qquad g(t) = \frac{1}{2\pi} \int_{-\infty}^{+\infty} e^{ivt} G(v)\, dv,$$

where

$$G(v) = \int_{-\infty}^{+\infty} e^{-ivt} g(t)\, dt, \qquad \text{i.e.} \qquad G(v) = \int_{0}^{+\infty} e^{-ivt} g(t)\, dt,$$

since $g(t) = 0$ for $t < 0$.

Having in mind (1), we see that

$$\int_{0}^{+\infty} e^{-ivt} g(t)\, dt = \int_{0}^{+\infty} e^{-(u+iv)t} f(t)\, dt = \int_{0}^{+\infty} e^{-pt} f(t)\, dt,$$

where $p = u + iv$. Therefore, $F(p) = G(v)$, which together with (2) implies

$$(3) \qquad g(t) = \frac{1}{2\pi} \int_{-\infty}^{+\infty} e^{ivt} F(u+iv)\, dv.$$

Multiplying (3) by e^{ut}, and taking into account (1), we obtain

$$f(t) = \frac{1}{2\pi} \int_{-\infty}^{+\infty} e^{(u+iv)t} F(u+iv)\, dv,$$

i.e.

$$(4) \qquad f(t) = \frac{1}{2\pi i} \int_{L} e^{pt} F(p)\, dp,$$

where L is the straight line $\{p \mid \operatorname{Re} p = u\}$, oriented upwards.

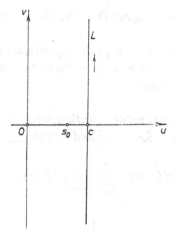

Fig. 7.2.2.

Since F has no singularities on the right-hand side of L, we can move that line towards right, and the value of (4) will remain unaltered. Hence, the contour L may be taken to be any line $\{p \mid \operatorname{Re} p = c\}$, such that all the singularities of F are on the left of L.

Formulas (4) is therefore often written in the form

$$(5) \qquad f(t) = \frac{1}{2\pi i} \int\limits_{c-i\infty}^{c+i\infty} e^{pt} F(p)\, \mathrm{d}p.$$

REMARK 1. The integral of a function Φ along the line $L = \{z \mid \operatorname{Re} z = c\}$:

$$(6) \qquad \lim_{h \to +\infty} \int\limits_{c-ib}^{c+ib} \Phi(z)\, \mathrm{d}z = \int\limits_{c-i\infty}^{c+i\infty} \Phi(z)\, \mathrm{d}z$$

is often called the Bromwich integral or the Bromwich-Wagner integral, since it is considered that Bromwich [1] and Wagner [2] introduced such integrals independently from each other. Integrals of the form (6) can be found, however, in an earlier paper [3] by Mellin.

REFERENCES
1. T. J. I'a Bromwich: 'Normal Coordinates in Dynamical Systems', *Proc. London Math. Soc.* (2) 15 (1916), 401–448.
2. K. W. Wagner: 'Über eine Formel von Heaviside zur Berechnung von Einschaltvergängen (Mit Anwendungsbeispielen)', *Archiv für Elektrotechnik* 4 (1916), 159–193.
3. Hj. Mellin: 'Über die fundamentale Wichtigkeit des Satzes von Cauchy für die Theorien der Gamma-und der hypergeometrischen Functionen', *Acta. Soc. Sci. Fennicae* 21 (1896), 1–115.

7.2.3. Determination of the Original from the Given Transform

Formula (5) from 7.2.2 gives an explicit expression for the original f, when the transform F is known. We now show how the integral which appears in that expression can be evaluated.

THEOREM 1. *Let F be an analytic funtion whose singularities $z_1, \ldots, z_n$ belong to the half-plane $\{z \mid \operatorname{Re} z < c\}$ and let $\lim_{z \to \infty} F(z) = 0$. Then*

(1) $$\frac{1}{2\pi i} \int_{c-i\infty}^{c+i\infty} e^{zt} F(z)\,dz = \sum_{k=1}^{n} \operatorname*{Res}_{z=z_k} e^{zt} F(z).$$

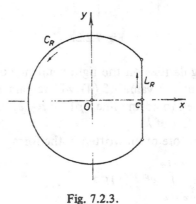

Fig. 7.2.3.

Proof. Let

$$C_R = \{z \mid |z| = R, \ \operatorname{Re} z \le c\}, \quad L_R = \{z \mid \operatorname{Re} z = c, \ |\operatorname{Im} z| \le \sqrt{R^2 - c^2}\},$$

and let $\Gamma_R = C_R \cup L_R$, where R is chosen so large that all the singularities of F are contained inside the contour Γ_R. Then

(2) $$\oint_{\Gamma_R} e^{zt} F(z)\,dz = \int_{C_R} e^{zt} F(z)\,dz + \int_{L_R} e^{zt} F(z)\,dz.$$

On the other hand, using Cauchy's residue theorem, we get

(3) $$\oint_{\Gamma_R} e^{zt} F(z)\,dz = 2\pi i \sum_{k=1}^{n} \operatorname*{Res}_{z=z_k} e^{zt} F(z).$$

From (2) and (3) follows

$$(4) \qquad \frac{1}{2\pi i} \int_{LR} e^{zt} F(z)\,dz = \sum_{k=1}^{n} \operatorname*{Res}_{z=z_k} e^{zt} F(z) - \frac{1}{2\pi i} \int_{CR} e^{zt} F(z)\,dz.$$

Consider the integral $\int_{CR} e^{zt} F(z)\,dz$. Since $\lim_{z\to\infty} F(z) = 0$, on basis of Jordan's lemma (Theorem 2 from 3.1.4) we conclude that

$$\lim_{R\to+\infty} \int_{CR} e^{zt} F(z)\,dz = 0.$$

Therefore, if $R \to +\infty$ from (4) follows

$$\frac{1}{2\pi i} \int_{L} e^{zt} F(z)\,dz = \sum_{k=1}^{n} \operatorname*{Res}_{z=z_k} (e^{zt} F(z)),$$

where $L = \{z|\ \operatorname{Re} z = c\}$, and that is precisely formula (1).

In particular, suppose that $F(z) = P(z)/Q(z)$, where P and Q are polynomials such that dg $P <$ dg Q. Suppose, further, that P and Q do not have common zeros and that all the zeros of Q are mutually distinct. Then

$$\operatorname*{Res}_{z=z_k} e^{zt} \frac{P(z)}{Q(z)} = \lim_{z\to z_k} \frac{e^{zt}(z-z_k)P(z)}{Q(z)} = \frac{e^{z_k t} P(z_k)}{Q'(z_k)},$$

which, in view of (1), implies

$$(5) \qquad f(t) = \mathbf{L}^{-1} \left(\frac{P(z)}{Q(z)} \right) = \sum_{k=1}^{n} \frac{P(z_k)}{Q'(z_k)} e^{z_k t}.$$

Formula (5) is called Heaviside's formula.

7.2.4. A More General Bromwich-Wagner Integral

THEOREM 1. *Suppose that F is a function satisfying the conditions:*
   1° F is analytic in the region $G = \{z|\ \operatorname{Re} z < c\}$ where it can have a finite number of singularities $z_1, \ldots, z_n$;
   2° On the line $\{z|\ \operatorname{Re} z = c\}$, F can have only simple poles $a_1, \ldots, a_m$;
   3° $\lim_{|z|\to+\infty} F(z) = 0 \qquad (z \in G).$

If $a > 1$, then

$$(1) \qquad \text{v.p.} \int_{c-i\infty}^{c+i\infty} a^z F(z)\,dz = 2\pi i \sum_{k=1}^{n} \operatorname*{Res}_{z=z_k} a^z F(z) + \pi i \sum_{k=1}^{m} \operatorname*{Res}_{z=a_k} a^z F(z).$$

Proof. Put $z = c + ix$. Then

(2) $$\text{v.p.} \int_{c-i\infty}^{c+i\infty} a^z F(z)\,dz = \text{v.p.} \int_{-\infty}^{+\infty} e^{(c+ix)\log a} F(c+ix) i\,dx$$

$$= ia^c \,\text{v.p.} \int_{-\infty}^{+\infty} e^{ix\log a} F(c+ix)\,dx.$$

Further,

(3) $$\text{v.p.} \int_{-\infty}^{+\infty} e^{ix\log a} F(c+ix)\,dx$$

$$= 2\pi i \sum \text{Res}\left(e^{ix\log a} F(c+ix)\right)$$
$$+ \pi i \sum \text{Res}\left(e^{ix\log a} F(c+ix)\right),$$

where the first sum is taken over all the singularities of the function $x \mapsto e^{ix\log a} F(c+ix)$ in the upper open half-plane, and the second over the simple poles of the same function which belong to the real axis.

From (2) and (3) follows

(4) $$\text{v.p.} \int_{c-i\infty}^{c+i\infty} a^z F(z)\,dz = -2\pi a^c \sum \text{Res}\left(e^{ix\log a} F(c+ix)\right)$$

$$- \pi a^c \sum \text{Res}\left(e^{ix\log a} F(c+ix)\right).$$

However, we have

(5) $$-2\pi a^c \sum \text{Res}\left(e^{ix\log a} F(c+ix)\right)$$

$$= 2\pi i \sum \text{Res}\left(ie^{(c+ix)\log a} F(c+ix)\right) = 2\pi i \sum \text{Res}\, a^z F(z),$$

where the second sum is taken over all the singularities of the function $x \mapsto ie^{(c+ix)\log a} F(c+ix)$ in the open upper half-plane. We also have

(6) $$-\pi a^c \sum \text{Res}\left(e^{ix\log a} F(c+ix)\right)$$

$$= \pi i \sum \text{Res}\left(ie^{(c+ix)\log a} F(c+ix)\right) = \pi i \sum \text{Res}\, a^z F(z),$$

where the second sum is taken over all the simple poles of the function $x \mapsto ie^{(c+ix)\log a} F(c+ix)$ on the x-axis.

Form (4), (5), and (6) follows (1).

REMARK 1. If $0 < a < 1$, then the region G should be replaced by the region $G_1 = \{z \mid \text{Re } z > c\}$ and we get

$$\text{v.p.} \int_{c-i\infty}^{c+i\infty} a^z F(z) \, dz = -2\pi i \sum_{k=1}^{n} \operatorname*{Res}_{z=z_k} a^z F(z) - \pi i \sum_{k=1}^{m} \operatorname*{Res}_{z=a_k} a^z F(z).$$

If we suppose that F has no singularities on the line $\{z \mid \text{Re } z = c\}$, and if we replace a by e^t, then from (1) we obtain

$$\int_{c-i\infty}^{c+i\infty} e^{zt} F(z) \, dz = 2\pi i \sum_{k=1}^{n} \operatorname*{Res}_{z=z_k} e^{zt} F(z),$$

and that is formula (1) from 7.2.3.

7.2.5. The Discrete Laplace Transform

In solving linear difference equations with constant coefficients or systems of such equations, the so-called discrete Laplace transform proved to be useful.

DEFINITION 1. The discrete Laplace transform of a function $f: \mathbf{N}_0 \to \mathbf{C}$ is defined by

$$(1) \qquad \mathbf{L}f(n) = F(p) = \sum_{n=0}^{+\infty} e^{-pn} f(n) \qquad (p = s + it; \ s, \ t \in \mathbf{R}).$$

If the original f satisfies the condition $\mid f(n) \mid < M e^{s_0 n}$, where $M > 0$ and s_0 are real numbers, then the series in (1) is convergent for every $s < s_0$, and hence it defines the function F.

If the transform F is known, then the original f can be determined from the formula

$$f(n) = \frac{1}{2\pi i} \int_{c-i\pi}^{c+i\pi} F(p) \, e^{pn} \, dp = \sum \operatorname{Res} F(p) \, e^{(n-1)p},$$

where the sum is taken over all the singularities of F from the region $\{p \mid \text{Re } p < c, \mid \text{Im } p \mid < \pi\}$.

REFERENCES
1. H. Dobesch: *Laplace-Transformation von Abtastfunktionen*, Berlin 1970, pp. 36–37.
2. Ya. Z. Cypkin: *Theory of Impulse Systems*, (Russian), Moscow 1958, pp. 157–175.

7.3. APPLICATION TO INTEGRAL EQUATIONS

Calculus of residues can also be applied to integral equations. We give a procedure due to Morduhaj-Boltovskoj [1].

Consider the Fredholm integral equation of the first kind

$$(1) \qquad \int_0^1 t^{x-1}\left(x \log t + \frac{1}{2}\right) f(t)\, dt = \frac{1}{\sqrt[3]{x}}$$

where f is the unknown function.

The Equation (1) can be replaced by the equivalent equation

$$(2) \qquad \int_0^1 e^{(x-1)\log t}\left(x \log t + \frac{1}{2}\right) f(t)\, dt = \frac{1}{\sqrt[3]{x}}.$$

Start with the obvious eqaulity

$$(3) \qquad F(x, u) = \int_0^1 e^{(x-1-u)\log t}\, dt = \frac{1}{x-u},$$

which, after differentiation with respect to x, becomes

$$(4) \qquad F_x(x, u) = \int_0^1 e^{(x-1-u)\log t} \log t\, dt = -\frac{1}{(x-u)^2}.$$

From (3) and (4) follows

$$\int_0^1 e^{(x-1-u)\log t}\left(x \log t + \frac{1}{2}\right) dt$$

$$= xF_x(x, u) + \frac{1}{2} F(x, u) = -\frac{x}{(x-u)^2} + \frac{1}{2(x-u)}.$$

Let Γ be a contour, which will later be defined, such that the function G is continuous along Γ. We then have

$$(5) \qquad \int_\Gamma G(u)\, du \int_0^1 e^{(x-1-u)\log t}\left(x \log t + \frac{1}{2}\right) dt$$

$$= x \int_{\Gamma} F_x(x, u) G(u) \, du + \frac{1}{2} \int_{\Gamma} F(x, u) G(u) \, du$$

$$= x \frac{\partial}{\partial x} \int_{\Gamma} F(x, u) G(u) \, du + \frac{1}{2} \int_{\Gamma} F(x, u) G(u) \, du.$$

If we define the function Y by

$$Y(x) = \int_{\Gamma} F(x, u) G(u) \, du,$$

then (5) becomes

(6) $$\int_{0}^{1} e^{(x-1) \log t} \left(x \log t + \frac{1}{2} \right) \left(\int_{\Gamma} e^{-u \log t} G(u) \, du \right) dt$$

$$= x Y'(x) + \frac{1}{2} Y(x).$$

This last equation coincides with the Equation (1) provided that

(7) $$\int_{\Gamma} e^{-u \log t} G(u) \, du = f(t),$$

(8) $$x Y'(x) + \frac{1}{2} Y(x) = x^{-1/3}.$$

The general solution of the Equation (8) is

$$Y(x) = \frac{C}{\sqrt{x}} + \frac{6}{\sqrt[3]{x}},$$

where C is an arbitrary constant. We now let Γ be a closed contour such that $u = x$ belongs to int Γ, and the singularities of G do not lie in int Γ. Then

(9) $$\oint_{\Gamma} F(x, u) G(u) \, du = \oint_{\Gamma} \frac{G(u)}{x - u} \, du = 2 \pi i \operatorname*{Res}_{u = x} \frac{G(u)}{x - u} = - 2 \pi i G(x)$$

and hence

$$Y(x) = - 2 \pi i G(x), \qquad \text{i.e.} \qquad G(x) = - \frac{1}{2 \pi i} \left(\frac{C}{\sqrt{x}} + \frac{6}{\sqrt[3]{x}} \right).$$

In view of (7) we find the solution of the given equation (1) in the form of the contour integral

$$(10) \qquad f(t) = -\frac{1}{2\pi i} \oint_{\Gamma} e^{-u \log t} \left(\frac{C}{\sqrt{u}} + \frac{6}{\sqrt[3]{u}} \right) du,$$

where C is an arbitrary constant.

It can be proved that the integral of the right-hand side of (10) is equal to

$$(11) \qquad f(t) = A \log^{-\frac{1}{2}} t + \frac{3}{\pi} \Gamma \left(\frac{2}{3} \right) \log^{-\frac{2}{3}} t,$$

and this is the general solution of (1).

REMARK 1. The exposed method is more or less formal. Various justifications are needed, e.g. an explanation why it is possible to differentiate under the integral. On the other hand, the correct result is, in a way, a justification of the method. Namely, it is easily verified that (11) is indeed a solution of (1).

REMARK 2. This method can be extended to more general integral equations of the form

$$\int_{a}^{b} \sum_{k=1}^{n} e^{a(t) b_k(x)} t^{\alpha_k} F_k(a(t), x) f(t) \, dt = F(x),$$

where a, b_k, F are given functions, α_k are constants, F_k are polynomials in both variables and f is the unknown function.

REMARK 3. The author of [1] states that Murphy as far back as 1833 (see [2]) solved the integral equation

$$(12) \qquad \int_{0}^{1} t^x f(t) \, dt = F(x),$$

i.e. determined the function f, considering F to be known. Murphy's result reads:

"When the known function $F(x)$ is rational, seek the coefficient of $1/x$ in $F(x) t^{-x}$; dividing it by t, the quotient will be the required function $f(t)$."

Later on Murphy modified this procedure so that it could be applied to the cases when F is not a rational function. We quote a few examples from Murphy's paper [2]:

$1°$ $F(x) = (x+m)^{-n}$ $(m>0, n \in N)$. Then $f(t) = \frac{1}{(n-1)!} t^{m-1} \log^{n-1} \frac{1}{t}$.

$2°$ $F(x) = \log\left(1 + \dfrac{a}{x+m}\right)$ $(m > 0,\ a > 0\ \text{ili}\ -m < a < 0).$

Then $f(t) = t^{m-1}\dfrac{t^a - 1}{\log t}.$

$3°$ $F(x) = \dfrac{\pi}{2}\,\operatorname{tg}\dfrac{\pi}{2}\,x.$ Then $f(t) = (t^2 - 1)^{-1}.$

REMARK 4. Since Murphy's solution consists in finding the coefficient of $1/x$ in a certain development, it is clear that the solution can be expressed in terms of residues. It seems, however, that residues were first mentioned in England in 1837, when Gregory published his exposition [3]. See also [4].

REFERENCES

1. D. Morduhaj-Boltovskoj: 'On the Inversion of Definite Integrals with the Help of the Theory of Residues', (Russian), *Učen. Zap. Universiteta, Rostov n/D.* 1 (1934), 111–118.
2. R. Murphy: 'On the Inverse Method of Definite Integrals, with Physical Applications', *Trans. Camb. Phil. Soc.* 4 (1833), 353–408.
3. D. F. Gregory: 'On the Residual Calculus', *The Cambridge Mathematical Journal* 1 (1837), 145–155. Second edition 1846, 158–169.
4. E. Rouché: 'Mémoire sur le calcul inverse des intégrales définies', *Compt. Rend. Acad. Sci. Paris* 51 (1860), 126–128.

7.4. APPLICATIONS TO PARTIAL DIFFERENTIAL EQUATIONS

7.4.1. *An Auxilliary Result*

In 4.6.2 we proved the following general theorem, due to Cauchy:

THEOREM 1. *Suppose that the function f satisfies Dirichlet's conditions on the segment* $[t_0, t_1]$, *and let p, q, and r be entire functions of the complex variable z such that:*

$1°$ *r has a countable number of zeros which do not coincide with the zeros of p and q;*

$2°$ $p(z) + q(z) = r(z);$ $3°$ $\lim\limits_{|z| \to +\infty} \dfrac{p(z)}{r(z)}\,e^{z(t-t_0)} = 0;$

$4°$ $\lim\limits_{|z| \to +\infty} \dfrac{p(-z)}{r(-z)} = 1;$ $5°$ $\lim\limits_{|z| \to +\infty} \dfrac{q(z)}{r(z)} = 1;$

$6°$ $\lim\limits_{|z| \to +\infty} \dfrac{q(-z)}{r(-z)}\,e^{z(t_1-t)} = 0.$

Then, for $t \in (t_0 . t_1)$ we have

$$f(t) = \sum_z \mathrm{Res}\, \frac{q(z)}{r(z)} \int_{t_0}^{t_1} e^{z\,(t-\xi)} f(\xi)\, \mathrm{d}\xi,$$

where the sum is taken over all the zeros of r.

Consider now a special case of this theorem. Let F be a polynomial, b an arbitrary constant and $a > \frac{1}{4}\, (t_1 - t_0)$, and put

$$p(z) = F(b - z)\, e^{a\,(b-z)}, \quad q(z) = F(z)\, e^{az}, \quad r(z) = F(z)\, e^{az} + F(b - z)\, e^{a\,(b-z)}.$$

It is easily verified that all the conditions of the previous theorem are satisfied, provided that $t_1 - 2a < t < t_0 + 2a$. Therefore, if $t \in (t_1 - 2a,\ t_0 + 2a)$, the following expansion is valid

$$f(t) = \sum \mathrm{Res}\, \frac{F(z)\, e^{az}}{F(z)\, e^{az} + F(b-z)\, e^{a\,(b-z)}} \int_{t_0}^{t_1} e^{z\,(t-\xi)} f(\xi)\, \mathrm{d}\xi,$$

where the sum is taken over all the roots of the equation

$$F(z)\, e^{az} + F(b - z)\, e^{a\,(b-z)} = 0.$$

7.4.2. Solution of a Linear Equation with Initial and Boundary Conditions

Consider the second order linear partial differential equation

$$(1) \qquad A u_{tt} + B u_t = a u_{xx} + b u_x + c u,$$

where A, B, a, b, c are given constants, together with

$$(2) \qquad u_x(0, t) + h u(0, t) = 0; \quad u_x(l, t) + k u(l, t) = 0;$$
$$(3) \qquad u(x, 0) = p_0(x); \quad u_t(x, 0) = p_1(x),$$

where p_0 and p_1 are given functions and k, h, l are given constants.

We look for a solution of (1) in the form

$$(4) \qquad u(x, t) = X(x)\, T(t),$$

where X and T are unknown twice differentiable functions.

Substituting (4) into (1), we find

$$(AT'' + BT') X = (aX'' + bX' + cX) T,$$

i.e.

(5) $$\frac{AT'' + BT'}{T} = \frac{aX'' + bX' + cX}{X} = a,$$

where α is independent of x and t.

The Equations (5) can be represented in the form

(6) $$AT'' + BT' - aT = 0,$$

(7) $$aX'' + bX' + (c - a) X = 0.$$

The above equations are ordinary differential equations with constant coefficients. The corresponding characteristic equations are

(8) $$A\mu^2 + B\mu = a,$$

(9) $$a\lambda^2 + b\lambda + c = a.$$

The general solution of (7) is

$$X(x) = C_1 e^{\lambda_1 x} + C_2 e^{\lambda_2 x},$$

where C_1 and C_2 are arbitrary constants, and λ_1, λ_2 are solutions (by hypothesis distinct) of the Equation (9).

Since the Equation (9) contains an arbitrary constant α, the solutions λ_1 and λ_2 will be functions of α, i.e.

$$\lambda_1 = \varphi_1 (\alpha), \qquad \lambda_2 = \varphi_2 (\alpha),$$

which implies that one solution can be expressed as a function of the other, for instance $\lambda_2 = \omega(\lambda_1)$.

Put $\lambda_1 = \rho$. Then

(10) $$a\varrho^2 + b\varrho + c = \alpha,$$

and the Equation (9) becomes $a\lambda^2 + b\lambda + c = a\rho^2 + b\rho + c$, i.e.

(11) $$(\lambda - \varrho) (a (\lambda + \varrho) + b) = 0.$$

The solutions of (11) are $\lambda_1 = \rho$ and $\lambda_2 = -(b/a) - \rho = r - \rho$, where $r = -b/a$.

Hence, the general solution of (7) reads

$$X(x) = C_1 e^{\varrho x} + C_2 e^{(r-\varrho)x},$$

where C_1 and C_2 are arbitrary constants, and r and ρ have the same meaning as above.

We now determine the constants C_1 and C_2 so that the conditions (2) are fulfilled. We find

(12) $C_1(\varrho + h) + C_2(r - \varrho + h) = 0, \; C_1(\varrho + k) e^{\varrho l} + C_2(r - \varrho + k) e^{(r-\varrho)l} = 0.$

The system (12) will have a nontrivial solution in C_1 and C_2 if its determinant is equal to zero, i.e. if

(13) $\Delta(\varrho) \equiv (\varrho + h)(r - \varrho + k) e^{(r-\varrho)l} - (\varrho + k)(r - \varrho + h) e^{\varrho l} = 0.$

If the condition (13) is fulfilled, we can take, for instance

$$C_1 = -(r - \varrho + h), \qquad C_2 = \varrho + h,$$

so that the solution of (7) which satisfies the boundary value conditions becomes

(14) $X(x) = (\varrho + h) e^{(r-\varrho)x} - (r - \varrho + h) e^{\varrho x}.$

In view of (10), the Equation (8) becomes

(15) $A\mu^2 + B\mu = a\varrho^2 + b\varrho + c.$

Denote by $\mu_1 = \sigma$ and $\mu_2 = \tau$ the solutions of (15). They are in fact, functions of ρ. The solution of (6) is thus given by

$$T(t) = D e^{\sigma t} + E e^{\tau t}$$

(D and E arbitrary constants.)

Therefore, the solution of (1) which satisfies (2) can be represented in the form

(16) $u(x, t) = \sum_\varrho (D_\varrho e^{\sigma t} + E_\varrho e^{\tau t}) X(x),$

where $X(x)$ is given by (14), and the sum is taken over all the values of ρ which satisfy (13).

Cauchy writes the solution (16) in the form

$$u(x, t) = \sum_{\varrho} \text{Res} \left(\sum_{\mu} \text{Res} \frac{f(\mu, \varrho) e^{\mu t}}{F(\mu, \varrho)} \frac{X(x)}{\varDelta(\varrho)} \right),$$

where

$$F(\mu, \varrho) = A \mu^2 + B \mu - a \varrho^2 - b \varrho - c,$$

and f is an arbitrary function.

The function f is chosen so that the conditions (3) are also satisfied. Put

$$f(\mu, \varrho) = (\varrho + k) \int_{c}^{l} \frac{F(\varrho, \mu) - F(\varrho, p)}{\mu - p} e^{\varrho(l - x)} dx,$$

where after the indicated division, p^0 and p^1 should be replaced by $p_0(x)$ and $p_1(x)$, respectively.

Indeed, using a similar reasoning as in Section 7.2.3, we find that in this case

$$(17) \qquad u(x, 0) = \sum_{\varrho} \text{Res} \left(\frac{\varrho + k}{\varDelta(\varrho)} X(x) \int_{0}^{l} e^{\varrho(l - x)} p_0(x) dx \right),$$

$$(18) \qquad u_t(x, 0) = \sum_{\varrho} \text{Res} \left(\frac{\varrho + k}{\varDelta(\varrho)} X(x) \int_{0}^{l} e^{\varrho(l - x)} p_1(x) dx \right).$$

It remains to prove that the expressions on the right-hand sides of (17) and (18) are equal to $p_0(x)$ and $p_1(x)$, respectively.

We have

$$\sum_{\varrho} \text{Res} \frac{\varrho + k}{\varDelta(\varrho)} X(x) \int_{0}^{l} e^{\varrho(l - x)} p_0(x) dx$$

$$= \sum_{\varrho} \text{Res} \frac{(\varrho + k)((\varrho + h) e^{(r - \varrho)x} - (r - \varrho + h) e \varrho x)}{(\varrho + h)(r - \varrho + k) e^{(r - \varrho)l} - (\varrho + k)(r - \varrho + k) e \varrho l} \int_{0}^{l} e^{\varrho(l - \xi)} p_0(\xi) d\xi$$

$$= \sum_{\varrho} \text{Res} \frac{(\varrho + k)(r - \varrho + h) e \varrho l}{(\varrho + k)(r - \varrho + k) e \varrho l - (\varrho + h)(r - \varrho + k) e^{(r - \varrho)l}} \int_{0}^{l} e^{\varrho(x - \xi)} p_0(\xi) d\xi$$

$$- \sum_{\varrho} \text{Res} \frac{(\varrho + k)(\varrho + h) e \varrho l e^{rx}}{(\varrho + k)(r - \varrho + k) e \varrho l - (\varrho + h)(r - \varrho + h) e^{(r - \varrho)l}} \int_{0}^{l} e^{-\varrho(x + \xi)} p_0(\xi) d\xi.$$

If we put $F(z) = (z+k)(r-z+h)$, $a=l$, $b=r$, $t_0 = 0$, $t_1 = l$ into the formula (1) of Section 7.5.1, we immediately conclude that

$$\sum_{\varrho} \text{Res} \; \frac{(\varrho+k)(r-\varrho+h)e\varrho^l}{(\varrho+k)(r-\varrho+k)e\varrho^l-(\varrho+h)(r-\varrho+k)e^{(r-\varrho)l}} \int_0^l e^{\varrho(x-\xi)} p_0(\xi)\,d\xi = p_0(x).$$

On the other hand, in proving the result of Section 4.6.2 we saw that, for every $x > 0$,

$$\sum_{\varrho} \text{Res} \; \frac{(\varrho+k)(\varrho+h)e\varrho^l e^{rx}}{(\varrho+k)(r-\varrho+k)e\varrho^l-(\varrho+h)(r-\varrho+k)e^{(r-\varrho)l}} \int_0^l e^{-\varrho(x+\xi)} p_0(\xi)\,d\xi = 0.$$

Therefore, we proved that the expression on the right-hand side of (17) is really equal to $p_0(x)$. In a similar manner, it can be proved that expression on the right-hand side of (18) is equal to $p_1(x)$.

Thus, we arrived at the following result:

THEOREM 1. *The solution of the problem defined by* (1), (2), (3) *is*

$$u(x, t) = \sum_{\varrho} \text{Res} \left(\sum_{\mu} \text{Res} \; \frac{f(\mu, \varrho)e^{\mu t}}{F(\mu, \varrho)} \frac{X(x)}{\Delta(\varrho)} \right),$$

where $X(x)$ *is given by* (14), $\Delta(\rho)$ *by* (13), *and where*

$$F(\mu, \varrho) = A\mu^2 + B\mu - a\varrho^2 - b\varrho - c,$$

$$f(\mu, \varrho) = (\varrho+k) \int_0^l \frac{F(\varrho, \mu) - F(\varrho, p)}{\mu - p} e^{\varrho(l-\xi)}\,d\xi,$$

where after the division under the integral sign, p^0 *should be replaced by* $p_0(\xi)$ *and* p^1 *by* p_1 (ξ).

EXAMPLE 1. Consider the wave equation $u_{tt} - u_{xx} = 0$, with the conditions

$$u_x(0, t) = u_x(1, t) = 0, \qquad u(x, 0) = u_t(x, 0) = e^x.$$

In this particular case the solutions of (11) are $\lambda_1 = \rho$ and $\lambda_2 = -\rho$, and so

$$X(x) = \varrho(e^{\varrho x} + e^{-\varrho x}) = 2\varrho \, \text{ch} \, \varrho x.$$

We also get

$$\Delta(\varrho) = \varrho^2(e^\varrho - e^{-\varrho}) = 2\varrho^2 \, \text{sh} \, \varrho.$$

Furthermore, we have $F(\mu, \rho) = \mu^2 - \rho^2$, and

$$f(\mu, \varrho) = \varrho \int_0^1 \frac{\mu^2 - p^2}{\mu - p} \, e^{\varrho(1-\xi)} \, d\xi$$

$$= \varrho \, (1 + \mu) \int_0^1 e^{\varrho(1-\xi)+\xi} \, d\xi = \frac{\varrho \, (1 + \mu)}{1 - \varrho} \, (e - e^{\varrho}).$$

In virtue of Theorem 1, we conclude that the solution of this problem is given by

$$u\,(x,\,t) = \sum_{\varrho} \mathrm{Res} \left(\sum_{\mu} \mathrm{Res} \, \frac{(1+\mu)\,(e-e^{\varrho})\,e^{\mu t}}{(1-\varrho)\,(\mu^2-\varrho^2)} \, \frac{\mathrm{ch}\,\varrho x}{\mathrm{sh}\,\varrho} \right)$$

$$= \sum_{\varrho} \mathrm{Res} \, \frac{(e-e^{\varrho})}{\varrho\,(1-\varrho)} \, \frac{\mathrm{ch}\,\varrho x}{\mathrm{sh}\,\varrho} \, (\mathrm{sh}\,\varrho t + \varrho\,\mathrm{ch}\,\varrho t).$$

The function

$$\varrho \mapsto A\,(\varrho, x, t) \equiv \frac{(e-e^{\varrho})}{\varrho\,(1-\varrho)} \, \frac{\mathrm{ch}\,\varrho x}{\mathrm{sh}\,\varrho} \, (\mathrm{sh}\,\varrho t + \varrho\,\mathrm{ch}\,\varrho t)$$

has only simple poles at $\rho = k\pi i$ $(k \in \mathbb{Z})$. We have

$$\mathrm{Res}_{\varrho=0} A\,(\varrho, x, t) = (e-1)\,(t+1);$$

$$\mathrm{Res}_{\varrho=k\pi i} A\,(\varrho, x, t) = \frac{(-1)^k e-1}{(1-k\pi i)\,k\pi} \, (\cos k\pi x)\,(\sin k\pi t + k\pi \cos k\pi t) \quad (k = \pm 1, \pm 2, \ldots),$$

and hence

$$u\,(x,\,t) = (e-1)\,(t+1) + \sum_{\substack{k=-\infty \\ k \neq 0}}^{+\infty} \frac{(-1)^k e-1}{(1-k\pi i)\,k\pi} \, (\cos k\pi x)\,(\sin k\pi t + k\pi \cos k\pi t)$$

$$= (e-1)\,(t+1) + 2 \sum_{k=1}^{+\infty} \frac{(-1)^k e-1}{k\pi\,(1+k^2\pi^2)} \, (\sin k\pi t + k\pi \cos k\pi t) \, \cos k\pi x.$$

The exposed method can be extended to more general problems for partial differential equations. Namely, the following result is valid.

Let F_1 and F_2 be polynomials of degree n and m, respectively, and let f_{k0}, f_{k1} $(k = 1, \ldots, n)$ be polynomials of degree $\leq n - 1$. Consider the equations

$$(19) \qquad F_1\left(\frac{\partial}{\partial x}\right) u + F_2\left(\frac{\partial}{\partial t}\right) u = 0,$$

together with the boundary value conditions

(20) $f_{k0}\left(\dfrac{\partial}{\partial x}\right) u\bigg|_{x=x_0} + f_{k1}\left(\dfrac{\partial}{\partial t}\right) u\bigg|_{x=x_1} = 0$ $(k = 1, \ldots, n)$

and the initial conditions

(21) $\dfrac{\partial^k u}{\partial t^k}\bigg|_{t=0} = p_k(x)$ $(k = 0, 1, \ldots, m)$.

Suppose that $X(x, \lambda)$ is the solution of the equation

$$F_1\left(\frac{\mathrm{d}}{\mathrm{d}x}\right) X = F_1(\lambda) X,$$

which also satisfies

$$f_{k0}\left(\frac{\mathrm{d}}{\mathrm{d}x}\right) X\bigg|_{x=x_0} + f_{k1}\left(\frac{\mathrm{d}}{\mathrm{d}x}\right) X\bigg|_{x=x_1} = -f_{k0}(\lambda) \qquad (k = 1, \ldots, n).$$

The solution of the problem (19)–(20)–(21) is

$$u(x, t) = \sum_\lambda \operatorname{Res}\left(\sum_\mu \operatorname{Res} \frac{e^{\mu t} X(x, \lambda)}{F_1(\lambda) + F_2(\mu)} \int_{x_0}^{x_1} e^{-\lambda(\xi - x_0)} \frac{F_1(\lambda) - F_1(p)}{\lambda - p} \, \mathrm{d}\xi \right),$$

where after the division under the integral, p^k should be replaced by $p_k(\xi)$.

Cauchy applied Calculus of residues to partial differential equations for the first time in [1]. Somewhat later Tortolini in three papers ([2], [3], [4]) gave a detailed survey of Cauchy's methods and also considered some systems of partial differential equations. Cauchy's methods are also given in book [5], and they are compared with classical methods, e.g. with Poisson's method.

Cauchy's residue method has a number of disadvantages. Tamarkin [6] mentions the following:

1° the method cannot be extended to equations with functional coefficients;

2° the convergence of series which figure in the proof of Cauchy's formula is not strictly proved.

Rasulov [7] adds that:

3° the method cannot be applied to the case when (1) contains the mixed derivative u_{xt}, even when the coefficients are constant.

Further developments of Cauchy's method were mainly attempts to remove those disadvantages. The first results in connection with that were obtained by Birkhoff ([8], [9]) who leaned on Poincaré's representation of functions by residues [10]. Papers by Tamarkin [11], Geppert [12], and Keldyš [13] are also interesting, and special attention to this problem was paid by Rasulov, who published over 15 papers dealing with the applications of residues to partial differential equations. His monograph [14] (also translated into English; see [15]) contains all the results in this domain published up to 1964.

REFERENCES

1. A.-L. Cauchy: 'Mémoire sur l'application du calcul des résidus à la solution des problèmes de physique mathématique', Paris 1827, 56 pp. $\equiv$ *Oeuvres* (2) 15, 90–137.
2. B. Tortolini: 'Memoria sull'applicazione del calcolo dei residui all'integrazione dell'equazioni lineari a derivate parziali', *Giornale Arcad.* 93 (1842), 3–41.
3. B. Tortolini: 'Seconda memoria sull'applicazione del calcolo dei residui all'integrazione dell'equazioni lineari a derivate parziali', *Giornale Arcad.* 94 (1842), 58–121.
4. B. Tortolini: 'Continuazione e fine della seconda memoria sull'applicazione de calcolo ec., inserita nell'antededente', *Giornale Arcad.* 95 (1842), 3–66.
5. A. N. Krilov: *On Some Differential Equations of Technical Physics*, (Serbian), Beograd 1952, pp. 227–234.
6. J. D. Tamarkin: *On Some General Problems in the Theory of Ordinary Linear Differential Equations and Expansions of Arbitrary Functions in Series*, (Russian), Petrograd 1917.[1]
7. M. L. Rasulov: 'The Residue Method for Solution of Mixed Problems for Differential Equations and a Formula for Expansion of an Arbitrary Vector Function in Fundamental Functions of a Boundary Problem with a Parameter', (Russian), *Mat. Sb.* 48 (90) (1959), 277–310.
8. G. D. Birkhoff: 'On the Asymptotic Character of the Solutions of Certain Linear Differential Equations, Containing a Parameter', *Trans. Am. Math. Soc.* 9 (1908), 219–231.
9. G. D. Birkhoff: 'Boundary Value and Expansion Problems of Ordinary Linear Differential Equations', *Trans. Am. Math. Soc.* 9 (1908), 373–395.
10. H. Poincaré: 'Sur les équations de la physique mathématique', *Rend. Circ. Mat. Palermo* 8 (1894), 57–156.
11. J. D. Tamarkin: 'Some General Problems of the Theory of Ordinary Linear Differential Equations and Expansion of an Arbitrary Function in Series of Fundamental Functions', *Math. Z.* 27 (1928), 1–54.
12. H. Geppert: 'Entwicklungen willkürlicher Funktionen nach funktionentheoretischen Methoden', *Math. Z.* 20 (1924), 29–94.
13. M. V. Keldyš: 'On the Characteristic Values and Characteristic Functions of Certain Classes of Non-Self-Adjoint Equations', (Russian), *Dokl. Akad. Nauk SSSR* 77 (1951), 11–14.
14. M. L. Rasulov: *The Contour Integral Method, and its Application to the Investigation of Problems for Differential Equations*, (Russian), Moscow 1964.
15. M. L. Rasulov: *Methods of Contour Integration*, Amsterdam 1967.

[1] References [6] and [8]–[13] are taken from the paper [7].

Chapter 8

Applications of Calculus of Residues to Special Functions

8.1. GAMMA AND BETA FUNCTIONS

8.1.1. *Definition and Some Properties of the Gamma Function*

The gamma function is usually defined by one of the following formulas:

$$\Gamma(z) = \lim_{n \to +\infty} \frac{n! \, n^z}{z(z+1)\cdots(z+n)},$$

(1)
$$\frac{1}{\Gamma(z)} = z e^{Cz} \prod_{k=1}^{+\infty} \left(1 + \frac{z}{k}\right) e^{-z/k} \qquad (C \text{ is Euler's constant})$$

$$\Gamma(z) = \int_0^{+\infty} e^{-t} t^{z-1} \, dt \qquad (\mathrm{Re}\, z > 0).$$

It can be proved that for $\mathrm{Re}\, z > 0$, the above definitions are equivalent to each other.

The gamma function satisfies the following functional equations

(2) $\Gamma(z+1) = z \, \Gamma(z),$

(3) $\Gamma(z) \, \Gamma(1-z) = \dfrac{\pi}{\sin \pi z}.$

Using the above definitions we can prove that the gamma function possesses derivatives of all orders. In particular, starting with (1) we obtain

$$\frac{\Gamma'(z)}{\Gamma(z)} = -C - \frac{1}{z} + \sum_{k=1}^{+\infty} \left(\frac{1}{k} - \frac{1}{z+k}\right),$$

which imples

(4) $\dfrac{d^n}{dz^n} \log \Gamma(z) = \displaystyle\sum_{k=0}^{+\infty} \frac{(-1)^n (n-1)!}{(z+k)^n} \qquad (n \geq 2).$

286

8.1.2. *Asymptotic Expansion of the Logarithm of the Gamma Function*

A particular case of the formula (4) from 8.1.1, i.e.

$$(1) \qquad \frac{d^2}{dz^2} \log \Gamma(z) = \sum_{k=0}^{+\infty} \frac{1}{(z+k)^2}$$

will be the starting point.

We now apply Theorem 2 from 6.4.2, with $m = 0$, to the function $z \mapsto 1/((t+z)^2)$, and we get

$$\sum_{n=0}^{+\infty} \frac{1}{(t+n)^2} = \frac{1}{2t^2} + \int_0^{+\infty} \frac{1}{(t+x)^2} \, dx + 4 \int_0^{+\infty} \frac{ty}{(t^2+y^2)^2 (e^{2\pi y}-1)} \, dy.$$

Hence, in virtue of (1),

$$\frac{d^2}{dz^2} \log \Gamma(z) = \frac{1}{z} + \frac{1}{2z^2} + \int_0^{+\infty} \frac{4zy}{(z^2+y^2)^2 (e^{2\pi y}-1)} \, dy,$$

and after integration we find

$$(2) \qquad \frac{d}{dz} \log \Gamma(z) = K + \log z - \frac{1}{2z} - 2 \int_0^{+\infty} \frac{y}{(z^2+y^2)^2 (e^{2\pi y}-1)} \, dy,$$

where K is an arbitrary real constant, since all other terms are real when z is real.

Integrating again, we obtain

$$\log \Gamma(z) = K_1 + Kz + \left(z - \frac{1}{2}\right) \log z - z + J(z),$$

where K_1 is an arbitrary real constant, and

$$J(z) = 2 \int_0^{+\infty} \frac{1}{e^{2\pi y}-1} \arctg \frac{y}{z} \, dy = \frac{1}{\pi} \int_0^{+\infty} \frac{z}{z^2+y^2} \log \frac{1}{1-e^{-2\pi y}} \, dy.$$

Applying the functional equation (2) from 8.1.1, we get

$$\log \Gamma (x+1) = K_1 + K(x+1) + \left(x+\frac{1}{2}\right)\log (x+1) - (x+1) + J(x+1)$$

$$= K_1 + Kx + \left(x+\frac{1}{2}\right)\log x - x + J(x),$$

which implies

(3) $\qquad K = -\left(x+\frac{1}{2}\right)\log\left(1+\frac{1}{x}\right) + 1 + J(x) - J(x+1).$

Since $\lim_{x \to +\infty} J(x) = 0$, if $x \to +\infty$ in (3), we obtain $K = 0$, and so

(4) $\qquad \log \Gamma (z) = K_1 + \left(z-\frac{1}{2}\right)\log z - z + J(z).$

In order to evaluate K_1, apply the functional equation (3) from 8.1.1.
Putting $z = \frac{1}{2} + iu$, we find

$$\Gamma\left(\frac{1}{2}+iu\right)\Gamma\left(\frac{1}{2}-iu\right) = \frac{2\pi e^{-\pi u}}{1+e^{-2\pi u}},$$

and further

(5) $\qquad \mathrm{Re} \log \Gamma\left(\frac{1}{2}+iu\right) = \log\sqrt{2\pi} - \frac{\pi u}{2} - \frac{1}{2}\log (1+e^{-2\pi u}).$

On the other hand, if we put $z = \frac{1}{2} + iu$ into (4), we get

$$\mathrm{Re} \log \Gamma\left(\frac{1}{2}+iu\right) = K_1 - u \,\mathrm{arctg}\, 2u - \frac{1}{2} + \mathrm{Re} J\left(\frac{1}{2}+iu\right),$$

which together with (5) implies

(6) $\qquad K_1 = \log\sqrt{2\pi} + u\left(\mathrm{arctg}\, 2u - \frac{\pi}{2}\right) + \frac{1}{2}$

$$-\frac{1}{2}\log (1+e^{-2\pi u}) - \mathrm{Re} J\left(\frac{1}{2}+iu\right).$$

Now, if p and q are positive, we have

$$\operatorname{Re} J(p+iq) = \frac{1}{\pi} \int_0^{+\infty} \frac{p(p^2+q^2+y^2)}{(p^2+q^2+y^2+2qy)(p^2+(q-y)^2)} \log \frac{1}{1-e^{-2\pi y}} \, dy$$

$$< \frac{p}{(p^2+q^2/4)} \int_0^{q/2} \log \frac{1}{1-e^{2\pi y}} \, dy + \log \frac{1}{1-e^{-\pi q}} \int_{q/2}^{+\infty} \frac{p}{p^2+(q-y)^2} \, dy$$

$$< \frac{4p}{\pi q^2} \int_0^{+\infty} \log \frac{1}{1-e^{2\pi y}} \, dy + \log \frac{1}{1-e^{-\pi q}} \int_{-\infty}^{+\infty} \frac{p}{p^2+y^2} \, dy,$$

and so

$$\lim_{u \to +\infty} \operatorname{Re} J\left(\frac{1}{2}+iu\right) = 0.$$

We also have

$$\lim_{u \to +\infty} u\left(\operatorname{arctg} 2u - \frac{\pi}{2}\right) = -\frac{1}{2}.$$

Therefore, if $u \to +\infty$ in (6), we obtain $K_1 = \log \sqrt{2\pi}$.
Hence, the asymptotic expansion of the function $z \mapsto \log \Gamma(z)$ reads

$$\log \Gamma(z) = \log \sqrt{2\pi} + \left(z-\frac{1}{2}\right)\log z - z + J(z),$$

where

$$J(z) = \frac{1}{\pi} \int_0^{+\infty} \frac{z}{z^2+y^2} \log \frac{1}{1-e^{-2\pi y}} \, dy.$$

REMARK 1. The following asymptotic formula can also be proved

$$(7) \qquad \log \Gamma(z) \sim \log \sqrt{2\pi} + \left(z-\frac{1}{2}\right)\log z - z + \sum_{k=1}^{+\infty} \frac{(-1)^{k-1} B_k}{2k(2k-1)z^{2k-1}},$$

where B_k are Bernoulli's numbers, which holds for $|z|$ sufficiently large and $|\arg z| < \pi$. Using this formula, we obtain the asymptotic formula for the gamma function

$$\Gamma(z) \sim \sqrt{2\pi}\, e^{-z} z^{z-\frac{1}{2}} \left(1 + \frac{1}{12z} + \frac{1}{288 z^2} + \cdots\right),$$

which holds under the same conditions as (7).

8.1.3. $\Gamma(z)$ and $1/\Gamma(z)$ as Contour Integrals

Consider the contour integral $\int_{C'} e^t t^{-z}\, dt$.

The function $t \mapsto e^t t^{-z}$ has a branch-point at the origin, but each branch is a regular one-valued function in any region of the complex plane which is cut from $-\infty$ to 0. We choose the branch for which

$$e^t t^{-z} = e^{t - z \log t},$$

where we take the principal value of the logarithm.

Let C' be a contour in the cut plane which consists of the line segment from $-R$ to $-r$ on the lower edge of the cut, positively oriented circle $\{t \mid |t| = r\}$ and the line segment from $-r$ to $-R$ on the upper edge of the cut (see Figure 8.1.3).

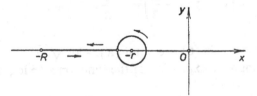

Fig. 8.1.3.

On the upper edge of the cut we have $t = u e^{\pi i}$, where $u > 0$, and so the function $t \mapsto e^t t^{-z}$ takes the value

$$e^{-u - z \log u - z \pi i} = e^{-u} u^{-z} e^{-z \pi i}.$$

Similarly, on the lower edge of the cut we have $t = u e^{-\pi i}$ and the integrand is $e^{-u} u^{-z} e^{z\pi i}$.

Hence

$$(1) \qquad \int_{C'} e^t t^{-z}\, dt = (e^{z\pi i} - e^{-z\pi i}) \int_r^R e^{-u} u^{-z}\, du + \int_{-\pi}^{\pi} e^{r e^{\theta i}} i r^{1-z} e^{(1-z)\theta i}\, d\theta.$$

But if $z = x + iy$, then

$$\left| \int_{-\pi}^{\pi} e^{re^{\theta i}} i r^{1-z} e^{(1-z)\theta i} d\theta \right| \leq \int_{-\pi}^{\pi} r^{1-x} e^{r\cos\theta + y\theta} d\theta \leq 2\pi r^{1-x} e^{r+\pi|y|},$$

and so, for $x < 1$, we conclude that

$$\lim_{r \to 0} \int_{-\pi}^{\pi} e^{re^{\theta i}} i r^{1-z} e^{(1-z)\theta i} d\theta = 0.$$

In other words, if $x < 1$, i.e. $\operatorname{Re}(1 - z) > 0$, and if $r \to 0$ in (1), we obtain

$$\int_{C'} e^t t^{-z} dt = 2i \sin \pi z \int_0^R e^{-u} u^{-z} du.$$

Let C denote the contour obtained from C' by letting $R \to + \infty$. We then have

$$\int_C e^t t^{-z} dt = 2i \sin \pi z \int_0^{+\infty} e^{-u} u^{-z} du = 2i \sin \pi z \, \Gamma(1 - z),$$

which implies

$$\Gamma(1-z) = \frac{1}{2i \sin \pi z} \int_C e^t t^{-z} dt, \qquad (\operatorname{Re}(1 - z) > 0)$$

i.e.

(2) $$\Gamma(z) = \frac{1}{2i \sin \pi z} \int_C e^t t^{z-1} dt. \qquad (\operatorname{Re} z > 0)$$

Formula (2) is proved under the supposition $\operatorname{Re} z > 0$. However, by the theory of analytic continuation, it holds for all $z \notin \mathbf{Z}$.

From (2) and the functional equation (3) from 8.1.1, we find

$$(3) \qquad \frac{1}{\Gamma(z)} = \frac{i}{2\pi i} \int_C e^t t^{-z} dt,$$

for all z.

REMARK 1. It is stated in [1] that formula (3) was proved by Hankel in 1864. It was rediscovered by Heine in 1880 (see [2]) who used it for obtaining the potential series expansion of $1/\Gamma(z)$, which converges for all z.

REFERENCES
1. Copson, pp. 225–227.
2. E. Heine: 'Einige Anwendungen der Residuenrechnung von Cauchy', *J. Reine Angew. Math.* **89** (1880), 19–39.

8.1.4. *Legendre's Duplication Formula*

Using Liouville's theorem, we shall now prove Legendre's duplication formula for the gamma function, i.e. we shall prove that

$$(1) \qquad \frac{2^{z-1}}{\sqrt{\pi}} \Gamma\left(\frac{z}{2}\right) \Gamma\left(\frac{z+1}{2}\right) = \Gamma(z).$$

Consider the function f defined by

$$f(z) = \frac{2^{2z} \Gamma(z) \Gamma\left(z + \frac{1}{2}\right)}{\Gamma(2z)}.$$

The only possible singularities of f are the poles of the numerator and the zeros of the denominator. The numerator has simple poles at $z = 0, -\frac{1}{2}, -1, \ldots$, but those points are also simple poles of the denominator and therefore they are not singularities of f. Moreover, since the denominator never vanishes, we conclude that f is a regular function in the finite plane.

Furthermore, f is a periodic function. Indeed, we have

$$\frac{f(z+1)}{f(z)} = \frac{2^{2z+2} \Gamma(z+1) \Gamma\left(z+\frac{3}{2}\right) \Gamma(2z)}{\Gamma(2z+2) 2^{2z} \Gamma(z) \Gamma\left(z+\frac{1}{2}\right)} = \frac{4z\left(z+\frac{1}{2}\right)}{(2z+1)2z} = 1$$

because

$$\frac{\Gamma(z+1)}{\Gamma(z)}=z, \quad \frac{\Gamma\left(z+\dfrac{3}{2}\right)}{\Gamma\left(z+\dfrac{1}{2}\right)}=z+\frac{1}{2},$$

$$\frac{\Gamma(2z)}{\Gamma(2z+2)}=\frac{\Gamma(2z)}{\Gamma(2z+1)}\cdot\frac{\Gamma(2z+1)}{\Gamma(2z+2)}=\frac{1}{2z}\cdot\frac{1}{2z+1}.$$

Hence, if we prove that f is bounded in the region $\{z \mid \mathrm{Re}\, z \geq 1\}$, it will follow, by periodicity, that f is bounded in the whole z-plane. But, using the asymptotic formula for the gamma function (see 8.1.2) we have

$$(2) \qquad f(z)=2\sqrt{\frac{\pi}{e}}\left(1+\frac{1}{2z}\right)^z\left(1+O\left(\frac{1}{|z|}\right)\right)=2\sqrt{\pi}\left(1+O\left(\frac{1}{z}\right)\right)$$

for sufficiently large $|z|$ and $|\arg z| < \pi$, and therefore f is bounded for $\mathrm{Re}\, z \geq 1$.

We have thus proved that f is regular and bounded as $|z| \to +\infty$, and by Liouville's theorem (see 4.1) we conclude that $f(z)$ is constant. However, from (2) follows

$$\lim_{|z|\to+\infty} f(z)=2\sqrt{\pi}$$

and so $f(z) = 2\sqrt{\pi}$, i.e.

$$(3) \qquad \Gamma(2z)=\frac{2^{2z-1}}{\sqrt{\pi}}\,\Gamma(z)\,\Gamma\left(z+\frac{1}{2}\right).$$

Replacing z by $z/2$ from (3) we obtain Legendre's duplication formula (1).

REFERENCE
Copson, pp. 224–225.

8.1.5. *Definition of the Beta Function*

The beta function $(p, q) \mapsto B(p, q)$ is a function in two complex variables p and q, defined by

$$(1) \qquad B(p, q)=\int_0^1 t^{p-1}(1-t)^{q-1}\,dt,$$

where

$$t^{p-1} = e^{(p-1)\log t}, \qquad (1-t)^{q-1} = e^{(q-1)\log(1-t)},$$

and in both cases the principal value of the logarithm is taken.

The integral on the right-hand side of (1) is convergent for Re $p > 0$ and Re $q > 0$. For other values of p and q, the beta function can be defined by means of analytic continuation.

It can be proved that for Re $p > 0$, Re $q > 0$, we have

$$(2) \qquad B(p, q) = \frac{\Gamma(p)\,\Gamma(q)}{\Gamma(p+q)}.$$

For other values of p and q, (2) can be taken as the definition of the beta function.

8.1.6. *Beta Function as a Contour Integral*

The beta function can be written in the form of a contour integral along an unusual contour. This representation was obtained by Pochhammer [1].

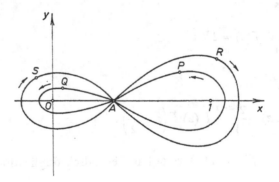

Fig. 8.1.6.1.

Consider the integral of

$$z \mapsto f(z) = z^{p-1}(1-z)^{q-1}$$

along the closed contour C shown on Figure 8.1.6.1. It is composed of:
1° a loop APA round $z = 1$ in the positive direction;
2° a loop AQA round $z = 0$ in the positive direction;

3° a loop ARA round $z = 1$ in the negative direction;

4° a loop ASA round $z = 0$ in the negative direction.

The point A is the starting and the end point of the contour C.

The function f is multiform, but at every point of C the branch integrated is uniform and continuous.

The value of the integral $\oint_C f(z)\,dz$ will be unaltered if the contour C is deformed in the following way. Instead of the loop APA we take the contour shown on Figure 8.1.6.2, consisting of: the segment joining A to $1 - r$ on the x-axis; positively oriented circle $\{z \,|\, |z - 1| = r\}$, and the segment joining $1 - r$ to A.

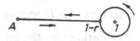

Fig. 8.1.6.2.

Similarly, loops AQA, ARA, ASA can be replaced by the contours shown on Figure 8.1.6.3.

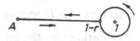

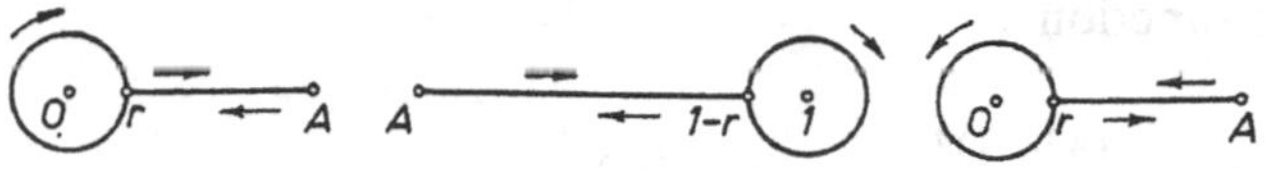

Fig. 8.1.6.3.

As z describes the circular path in positive direction round $z = 1$, the value of $f(z)$ changes from $f(x)$ to $f(x)e^{2q\pi i}$. Similarly the descriptions of the circular paths of the other three circles give $f(z)$ the values $f(x)e^{2(p+q)\pi i}$, $f(x)e^{2p\pi i}$ and $f(x)$, respectively.

We now let $r \to 0$, and for $p > 0$ and $q > 0$ we get

$$\oint_C z^{p-1}(1-z)^{q-1}\,dz = (1 - e^{2q\pi i} + e^{2(p+q)\pi i} - e^{2p\pi i})\int_0^1 x^{p-1}(1-x)^{q-1}\,dx$$

$$= (1 - e^{2p\pi i})(1 - e^{2q\pi i})\,B(p, q),$$

which implies

(1) $$B(p, q) = \frac{1}{(1 - e^{2p\pi i})(1 - e^{2q\pi i})}\oint_C z^{p-1}(1-z)^{q-1}\,dz.$$

Since the functions on both sides of (1) are regular for all p and q, the equality does not hold only for $p > 0$, $q > 0$, but for all values of those variables.

REFERENCES
1. L. Pochhammer: 'Zur Theorie der Euler'schen Integrale', *Math. Annalen* **35** (1890), 495–526.
2. T. M. Macrobert: *Functions of a Complex Variable*, Fifth edition, London 1966, pp. 144–145.

8.2. SOME SPECIAL FUNCTIONS AS CONTOUR INTEGRALS

8.2.1. *Legendre's Polynomials as Contour Integrals*

Legendre's polynomial P_n can be defined by the equality

$$P_n(z) = \frac{1}{2^n n!} \frac{d^n}{dz^n} (z^2 - 1)^n \qquad (n \in \mathbf{N}_0).$$

We shall express the polynomial P_n in the form of a contour integral. Let $z \neq \pm 1$. The function

$$w \mapsto \frac{(w^2-1)^n}{(w-z)^{n+1}} \qquad (n \in \mathbf{N}_0)$$

has only one singularity, a pole of order $n + 1$ at $w = z$. Hence, if $C = \{w \mid |w - z| = r\}$, we have

$$\oint_C \frac{(w^2-1)^n}{(w-z)^{n+1}}\, dw = 2\pi i \operatorname*{Res}_{w=z} \frac{(w^2-1)^n}{(w-z)^{n+1}} = \frac{2\pi i}{n!} \lim_{w \to z} \frac{d^n}{dw^n} (w^2-1)^n$$

$$= \frac{2\pi i}{n!} \frac{d^n}{dz^n} (z^2-1)^n$$

and therefore

(1) $$P_n(z) = \frac{1}{2\pi i} \oint_C \frac{(w^2-1)^n}{2^n (w-z)^{n+1}}\, dw.$$

Formula (1) is called Schläfli's formula.

REMARK 1. For associated Legendre's functions, or Ferrers' functions P_n^m, defined by

$$P_n^m(x) - (1-x^2)^{m/2} \frac{d^m}{dx^m} P_n(x),$$

where $m, n \in \mathbb{N}$, $m \leq n$, $|x| < 1$, the following formula is valid

$$P_n^m(x) = \frac{(n+m)!}{2^n n!} (1-x^2)^{m/2} \frac{1}{2\pi i} \oint_C \frac{(w^2-1)^n}{(w-x)^{n+m+1}} \, dw,$$

which is analogous to Schläfli's formula for Legendre's polynomials.

8.2.2. Laguerre's Polynomials as Contour Integrals

Laguerre's polynomials L_n, defined by

$$L_n(x) = e^x \frac{d^n}{dx^n} (x^n e^{-x}) \qquad (n = 0, 1, \ldots),$$

can be expressed as contour integrals.

Indeed, if C is a closed contour which encircles the point x, then

$$\oint_C \frac{e^{-w} w^n}{(w-x)^{n+1}} \, dw = 2\pi i \operatorname*{Res}_{w=x} \frac{e^{-w} w^n}{(w-x)^{n+1}}.$$

The point x is a pole of order $n+1$ of the function

$$w \longmapsto \frac{e^{-w} w^n}{(w-x)^{n+1}},$$

and so

$$\operatorname*{Res}_{w=x} \frac{e^{-w} w^n}{(w-x)^{n+1}} = \frac{1}{n!} \lim_{w \to x} \frac{d^n}{dw^n} (e^{-w} w^n) = \frac{1}{n!} \frac{d^n}{dx^n} (e^{-x} x^n).$$

Hence, we obtain

$$L_n(x) = \frac{n! e^x}{2\pi i} \oint_C \frac{e^{-w} w^n}{(w-x)^{n+1}} \, dw.$$

REMARK 1. For generalized Laguerre's polynomials L_n^s, defined by

$$L_n^s(x) = x^{-s} e^x \frac{d^n}{dx^n} (e^{-x} x^{n+s}),$$

we have the following analogous formula

$$L_n^s(x) = \frac{n! \, x^{-s} e^x}{2\pi i} \oint_C \frac{e^{-w} w^{s+n}}{(w-x)^{n+1}} \, dw.$$

8.2.3. *Hermite's Polynomials as Contour Integrals*

Hermite's polynomials H_n, defined by

$$H_n(x) = (-1)^n e^{x^2} \frac{d^n}{dx^n} (e^{-x^2}) \qquad (n = 0, 1, \ldots),$$

can be expressed as contour integrals.

Suppose that C' is a closed contour which encircles the point $z = x$. Then the function

$$z \mapsto \frac{e^{-z^2}}{(z-x)^{n+1}}$$

has only one singularity in the region int C', namely a pole of order $n + 1$ at $z = x$. Hence

$$\operatorname*{Res}_{z=x} \frac{e^{-z^2}}{(z-x)^{n+1}} = \frac{1}{n!} \lim_{z \to x} \frac{d^n}{dz^n} (e^{-z^2}) = \frac{1}{n!} \frac{d^n}{dx^n} (e^{-x^2}).$$

However, from Cauchy's residue theorem we have

$$\oint_{C'} \frac{e^{-z^2}}{(z-x)^{n+1}} \, dz = 2\pi i \operatorname*{Res}_{z=x} \frac{e^{-z^2}}{(z-x)^{n+1}},$$

and hence

$$\oint_{C'} \frac{e^{-z^2}}{(z-x)^{n+1}} \, dz = \frac{2\pi i}{n!} \frac{d^n}{dx^n} (e^{-x^2}) = \frac{2\pi i}{n!} (-1)^n e^{-x^2} H_n(x).$$

Thus, we proved the formula

(1) $$H_n(x) = \frac{(-1)^n n!}{2\pi i} e^{x^2} \oint_{C'} \frac{e^{-z^2}}{(z-x)^{n+1}} \, dz.$$

Finally, if make the substitution $z = x - t$ in (1), we get

$$H_n(x) = \frac{n!}{2\pi i} \oint_C \frac{e^{2xt - t^2}}{t^{n+1}} \, dt,$$

where C is a closed contour which encircles the point $t = 0$.

8.2.4. Definition of Bessel's Functions

Bessel's functions J_n $(n \in Z)$ are defined by

$$e^{\frac{z}{2}\left(t - \frac{1}{t}\right)} = \sum_{n=-\infty}^{+\infty} J_n(z) t^n.$$

From Laurent's theorem follows

(1) $$J_n(z) = \frac{1}{2\pi i} \oint_C \frac{1}{t^{n+1}} \exp\left(\frac{z}{2}\left(t - \frac{1}{t}\right)\right) dt,$$

where $C = \{t \,|\, |t| = 1\}$.

Starting with (1) and applying Cauchy's residue theorem, we obtain a formula which expresses the Bessel function J_n as an infinite potential series. Indeed, if we put $t = 2u/z$ in (1), we get

(2) $$J_n(z) = \frac{1}{2\pi i} \left(\frac{z}{2}\right)^n \oint_{C_1} \frac{1}{u^{n+1}} \exp\left(u - \frac{z^2}{4u}\right) du,$$

where $C_1 = \{u \,|\, |u| = \frac{1}{2} |z|\}$.

The function

$$u \mapsto f(u, z) = \frac{1}{u^{n+1}} \exp\left(u - \frac{z^2}{4u}\right)$$

has only one singularity in int C_1, an essential singularity at $u = 0$, and so

$$(3) \qquad \oint_{C_1} f(u, z)\, du = 2\pi i \operatorname*{Res}_{u=0} f(u, z).$$

The residue of f at $u = 0$ is equal to the coefficient of $1/u$ in the expansion of that function in a Laurent's series at $u = 0$, or equivalently to the coefficient of u^n in the expansion of the function

$$u \mapsto \exp\left(u - \frac{z^2}{4u}\right)$$

in a Laurent's series at $u = 0$.
  Since

$$e^u = \sum_{k=0}^{+\infty} \frac{1}{k!} u^k, \qquad \exp\left(-\frac{z^2}{4u}\right) = \sum_{k=0}^{+\infty} (-1)^k \frac{1}{k!} \frac{1}{u^k} \left(\frac{z}{2}\right)^{2k},$$

we have

$$\exp\left(u - \frac{z^2}{4u}\right) = \left(\sum_{k=0}^{+\infty} \frac{1}{k!} u^k\right)\left(\sum_{k=0}^{+\infty} (-1)^k \frac{1}{k!\, u^k} \left(\frac{z}{2}\right)^{2k}\right),$$

and the coefficient of u^n is

$$\sum_{k=0}^{+\infty} (-1)^k \frac{1}{(k+n)!} \frac{1}{k!} \left(\frac{z}{2}\right)^{2k}.$$

  Hence

$$(4) \qquad \operatorname*{Res}_{u=0} \frac{1}{u^{n+1}} \exp\left(u - \frac{z^2}{4u}\right) = \sum_{k=0}^{+\infty} (-1)^k \frac{1}{(k+n)!} \frac{1}{k!} \left(\frac{z}{2}\right)^{2k}.$$

From (2), (3), and (4) follows

$$J_n(z) = \left(\frac{z}{2}\right)^n \sum_{k=0}^{+\infty} (-1)^k \frac{1}{k!\,(n+k)!} \left(\frac{z}{2}\right)^{2k}.$$

8.2.5. *Polylogarithms*

For $n \in N_0$ define the function L^n by

$$L^n(z) = \frac{1}{2\pi i} \int\limits_{C-i\infty}^{C+i\infty} \frac{\Gamma(s)\,\Gamma(-s)}{(-s)^{n-1}} (-z)^s \, ds.$$

It can be shown, by contour integration, that

$$L^n(z) = \sum_{k=1}^{+\infty} \frac{z^k}{k^n} \quad (|z|<1); \qquad L^n(z) = -\sum_{k=1}^{+\infty} \frac{z^{-k}}{(-k)^n} - r_n(z) \qquad (|z|>1),$$

where

$$r_n(z) = \operatorname*{Res}_{s=0} \frac{\Gamma(s)\,\Gamma(-s)}{(-s)^{n-1}} (-z)^s.$$

The function L^n is called polylogarithm of order n. In particular, for $n = 1$, we have

$$L^1(z) = -\log(1-z) \quad \text{and} \quad r_1(z) = \log(-z).$$

A number of definite integrals can be evaluated by means of polylogarithms. We give a few examples.

$$1° \qquad \int\limits_0^1 \frac{1}{u} \log^{n-2} u \, \log(1-uz) \, du = (-1)^{n-1}(n-2)! \, L^n(z),$$

where $n = 2, 3, \ldots$ and $|\arg(1-z)| < \pi$;

$$2° \qquad \int\limits_1^{+\infty} \frac{x \log u}{(u-x)u} \, du = -L^2\!\left(\frac{1}{x}\right) - \frac{1}{2}\log^2 x + \frac{1}{3}\pi^2 \qquad (x>1);$$

$$3° \qquad \int\limits_1^{+\infty} \frac{x \log^2 u}{(u-x)u} \, du = L^3\!\left(\frac{1}{x}\right) - \frac{1}{6}\log^3 x + \frac{1}{3}\pi^2 \log x \qquad (x>1).$$

REFERENCE
O. I. Maričev: *A Method for Evaluating Integrals of Special Functions*, (Russian), Minsk 1978, pp. 90–93.

8.3. PERIODIC AND ELLIPTIC FUNCTIONS

8.3.1. *Periods and Residues*

As known, a function f is said to be periodic if there exists a nonzero constant Ω, called a period of f, such that the equation

$$f(z + \Omega) = f(z)$$

holds for all values of z. Clearly, if $n \in \mathbf{Z}$, then $n\Omega$ is also a period of f. The number Ω will be called the fundamental period if no submultiple of it is a period, i.e. if it is the period nearest to the origin.

A periodic function which has only one fundamental period is said to be simply-periodic, and a function which possesses more than one fundamental period is said to be multiply-periodic. For example, a function f which has two fundamental periods Ω_1 and Ω_2 is said to be doubly-periodic. For such a function we have

$$f(z + \Omega_1) = f(z), \qquad f(z + \Omega_2) = f(z), \qquad f(z + m\Omega_1 + n\Omega_2) = f(z)$$

$(m, n \in \mathbf{Z})$ for all values of z.

Suppose that f is a uniform analytic function in a simply-connected region D, and define the function F by

$$F(z) = \int_{z_0}^{z} f(z) \, dz.$$

If f is regular in D, then the function F will also be uniform and regular in D, provided that the path of integration lies entirely within D. If, however, the region D contains one or more singularities of f, F may be a multiform function, each branch being a regular function in a simply-connected region containing no singularities of f. Of course, the path of integration must not pass through a singularity of f.

Suppose now that the equation $w = F(z)$ defines the inverse function $w \mapsto z = G(w)$, which is called the inversion of the integral (1). We will show that the existence of a singularity of f implies in general that G is a periodic function, and conversely, that if f is regular in D, than G is not a periodic function.

Suppose that a is a pole of f and consider the integral

$$w = F(z) = \int_{z_0}^{z} f(z)\, dz$$

along the segment L joining z_0 to z, and along the path C shown on Figure 8.3.1. Since the integrands along l cancel, we conclude that

$$\int_C f(z)\, dz = \int_L f(z)\, dz + 2\pi i \operatorname*{Res}_{z=a} f(z).$$

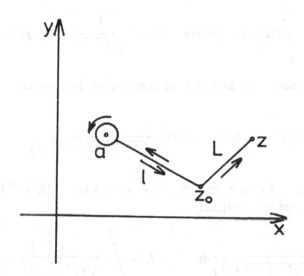

Fig. 8.3.1.

In other words, for a given value of z, we obtain two different values: w and $w + 2\pi i \operatorname{Res}_{z=a} f(z)$. Hence, for the inverse function G we have

$$G(w + 2\pi i \operatorname*{Res}_{z=a} f(z)) = G(z),$$

i.e. it is periodic with period $2\pi i \operatorname{Res}_{z=a} f(z)$.

EXAMPLE 1. The function $z \mapsto 1/z$ has a simple pole at $z = 0$ with residue 1. Therefore, the inverse function of the function F, defined by

$$F(z) = \int_1^z \frac{1}{z}\,dz$$

is periodic and has period $2\pi i$.

EXAMPLE 2. The function $z \mapsto z^{-2}$ has a pole at $z = 0$ with residue 0. Hence, the integral $\int_1^z z^{-2}\,dz$ defines a uniform function, and its inverse function is not periodic.

In the case when f is a multiform function, we do not have such a simple connection between the residues and the periods of the inversion of the integral. Nevertheless, it can be shown, by contour integration, that the inversion of the integral

$$w = \int_0^z f(z)\,dz, \quad \text{where} \quad f(z) = \frac{1}{\sqrt{1-z^2}}, \quad f(0) = 1$$

is periodic with period 2π, and that the inversion of the integral

$$w = \int_0^z f(z)\,dz, \quad \text{where} \quad f(z) = \frac{1}{\sqrt{(1-z^2)(1-k^2 z^2)}}, \quad f(0) = 1$$

where $0 < k < 1$, is doubly-periodic with periods $\Omega_1 = 4K$ and $\Omega_2 = 2iK'$, where K and K' are real integrals

$$K = \int_0^1 \frac{1}{\sqrt{(1-u^2)(1-k^2 u^2)}}\,du, \quad K' = \int_1^{1/k} \frac{1}{\sqrt{(u^2-1)(1-k^2 u^2)}}\,du.$$

8.3.2. Poles and Zeros of Elliptic Functions

A doubly-periodic analytic function whose only possible singularities in the finite plane are poles is called an elliptic function. If Ω_1 and Ω_2 are two periods of a doubly-periodic function such that every other period is a sum of multiples of Ω_1 and Ω_2, we say that Ω_1 and Ω_2 form a pair of primitive periods.

Let f be an elliptic function with a pair of primitive periods Ω_1 and Ω_2 and let a be an arbitrary fixed complex number. Denote by A_{mn} the point in the complex plane which corresponds to the number $a + m\Omega_1 + n\Omega_2$ (m, $n \in Z$), and let P_{mn} be the open parallelogram with vertices A_{mn}, $A_{m+1,n}$, $A_{m+1,n+1}$, $A_{m,n+1}$ such that the sides $A_{mn}A_{m+1,n}$ and $A_{mn}A_{m,n+1}$ and the vertex A_{mn} also belong to P_{mn}. The complex plane is completely covered by this system of nonoverlapping parallelograms, i.e. each point belongs to one and only one parallelogram. Such a parallelogram is called a period-parallelogram.

We shall now be concerned with the number of poles and zeros of an elliptic function in a given period-parallelogram P, and we suppose that the poles (or zeros) do not lie on its boundary. There is no loss in generality in this assumption, since we can always translate the parallelograms (i.e. change the point a) until no pole or zero lies on their boundaries.

THEOREM 1. *For an arbitrary elliptic function f we have*

$$\oint_{\partial P} f(z)\,dz = 0.$$

Proof. This follows on account of periodicity of f. Indeed,

$$\oint_{\partial P} f(z)\,dz = \int_0^{\Omega_1} f(z)\,dz + \int_{\Omega_1}^{\Omega_1+\Omega_2} f(z)\,dz + \int_{\Omega_1+\Omega_2}^{\Omega_2} f(z)\,dz + \int_{\Omega_2}^{0} f(z)\,dz.$$

However, putting $z = u - \Omega_2$ we find

$$\int_0^{\Omega_1} f(z)\,dz = \int_{\Omega_2}^{\Omega_1+\Omega_2} f(u - \Omega_2)\,du = \int_{\Omega_2}^{\Omega_1+\Omega_2} f(u)\,du,$$

i.e.

$$\int_0^{\Omega_1} f(z)\,dz + \int_{\Omega_1+\Omega_2}^{\Omega_2} f(z)\,dz = 0.$$

Similarly,

$$\int\limits_{\Omega_1}^{\Omega_1+\Omega_2} f(z)\,dz + \int\limits_{\Omega_2}^{0} f(z)\,dz = 0.$$

THEOREM 2. *The sum of residues of an elliptic function at its poles in any period-parallelogram is zero.*

This is a direct consequence of Theorem 1.

The number of poles of an elliptic function in any period-parallelogram, each counted according to its multiplicity, is called the order of the function.

THEOREM 3. *The order of an elliptic function is at least 2.*

Proof. If f is an elliptic function of order 1, than it has only one simple pole in a period-parallelogram of residue zero, which is impossible.

THEOREM 4. *An elliptic function of order n has n zeros in each period-parallelogram.*

Proof. If f is an elliptic function of order n, then f' is also an elliptic function with the same periods as f, and therefore so is also $z \mapsto f'(z)/f(z)$. Let m be the number of zeros of f in a period-parallelogram P. Then

$$\frac{1}{2\pi i}\oint\limits_{\partial P} \frac{f'(z)}{f(z)}\,dz = m - n = \sum \mathrm{Res}\frac{f'(z)}{f(z)}.$$

Since $z \mapsto f'(z)/f(z)$ is an elliptic function, by Theorem 2, the last expression is zero, and hence $m = n$.

THEOREM 5 (Liouville). *An integral nonconstant function cannot be an elliptic function.*

Proof. Suppose that f is an integral nonconstant elliptic function. Then the function $z \mapsto f(z) - f(z_0)$ has zeros but has no poles, which is in contradiction with Theorem 4.

REMARK. Notice an important difference between simply and doubly-periodic functions. We have just proved that there are no integral nonconstant doubly-periodic functions. On the other hand, functions such as $z \mapsto e^z$, $z \mapsto \sin z$, etc. are integral nonconstant simply-periodic functions.

REFERENCES
1. Copson, pp. 345–420.
2. É. Picard: *Traité d'analyse*, tome II, 2ème édition, Paris 1905.
3. É. Goursat: *Cours d'analyse mathématique*, tome II, 5ème édition, Paris 1929, pp. 168–200.
4. L. Čakalov: *An Introduction to the Theory of Analytic Functions*, (Bulgarian), Sofia 1972, pp. 360–417.

Chapter 9

Calculus of Finite Differences

9.1. CALCULUS WITH THE OPERATOR Δ

9.1.1. *Introduction*

DEFINITION 1. The finite difference operator Δ is defined by

$$\Delta f(x) = f(x+h) - f(x),$$

where h is fixed.

The operator Δ is linear, i.e. we have

$$\Delta(C_1 f_1(x) + C_2 f_2(x)) = C_1 \Delta f_1(x) + C_2 \Delta f_2(x),$$

where C_1, C_2 do not depend on x.

DEFINITION 2. The operator Δ^n is defined recursively by

$$\Delta^1 = \Delta, \quad \Delta^{n+1} = \Delta(\Delta^n) \quad (n \in \mathbf{N}),$$

i.e.

$$\Delta^{n+1} f(x) = \Delta^n f(x+h) - \Delta^n f(x).$$

DEFINITION 3. The operator Δ^{-1}, inverse to Δ, is defined by

$$\Delta^{-1}(\Delta f(x)) = \Delta(\Delta^{-1} f(x)) = f(x).$$

In other words, the equalities

$$\Delta f(x) = g(x) \quad \text{and} \quad \Delta^{-1} g(x) = f(x)$$

are equivalent.

DEFINITION 4. The operator Δ^{-n} $(n \in \mathbb{N})$ is defined as $(\Delta^{-1})^n$ $(n \in \mathbb{N})$.

It is easily shown that Δ^n and Δ^{-n} $(n \in \mathbb{N})$ are inverse to each other, i.e. that the equalities $\Delta^n f(x) = g(x)$ and $\Delta^{-n} g(x) = f(x)$ are equivalent.

9.1.2. Evaluation of the Expression $\Delta^m x^n$ $(m, n \in \mathbb{N}; m \leq n)$

THEOREM 1. *Let m and n be positive integers such that $m \leq n$. Then*

$$(1) \qquad \Delta^m x^n = \frac{n!}{(n-m)!} M_0 h^m x^{n-m} + \frac{n!}{(n-m-1)!} M_1 h^{m+1} x^{n-m-1} + \cdots$$

$$+ \frac{n!}{0!} M_{n-m} h^n$$

$$= \sum_{k=0}^{n-m} \frac{n!}{(n-m-k)!} M_k h^{m+k} x^{n-m-k},$$

where the coefficients $M_0, M_1, \ldots$ are defined by the expansion

$$(2) \qquad \left(1 + \frac{t}{2!} + \frac{t^2}{3!} + \cdots \right)^m = M_0 + M_1 t + M_2 t^2 + \cdots$$

Proof. We have

$$x^n = n! \operatorname*{Res}_{z=0} \frac{e^{xz}}{z^{n+1}} \quad (n \in \mathbb{N}),$$

and so

$$(3) \qquad \Delta^m x^n = n! \operatorname*{Res}_{z=0} \frac{\Delta^m e^{xz}}{z^{n+1}} \quad (m, n \in \mathbb{N}).$$

We now prove the formula

$$(4) \qquad \Delta^m e^{xz} = (e^{hz} - 1)^m e^{xz} \quad (m \in \mathbb{N}).$$

Since

$$\Delta e^{xz} = e^{(x+h)z} - e^{xz} = (e^{hz} - 1) e^{xz},$$

formula (4) is valid for $m = 1$. Suppose that it is valid for some $m \in \mathbb{N}$, and apply the operator Δ. We find

$$\Delta^{m+1} e^{xz} = \Delta (e^{hz} - 1)^m e^{xz} = (e^{hz} - 1)^m \Delta e^{xz} = (e^{hz} - 1)^{m+1} e^{xz},$$

which means that (4) is true for all $m \in \mathbf{N}$. Hence, (3) can be written in the form

$$(5) \qquad \Delta^m x^n = n! \operatorname*{Res}_{z=0} \frac{(e^{hz}-1)^m e^{xz}}{z^{n+1}} \qquad (m, n \in \mathbf{N}).$$

The residue which appears on the right-hand side of (5) is equal to the coefficient of z^{-1} in the expansion of the function

$$z \mapsto \frac{(e^{hz}-1)^m e^{xz}}{z^{n+1}}$$

into Laurent's series, or equivalently, to the coefficient of z^{n-m} in the expansion of the function

$$z \mapsto h^m e^{xz} \left(1 + \frac{hz}{2!} + \frac{h^2 z^2}{3!} + \cdots\right)^m.$$

However, from (2) follows

$$h^m e^{xz} \left(1 + \frac{hz}{2!} + \frac{h^2 z^2}{3!} + \cdots\right)^m$$

$$= h^m \left(1 + \frac{xz}{1!} + \frac{x^2 z^2}{2!} + \cdots\right)(M_0 + M_1 hz + M_2 h^2 z^2 + \cdots),$$

and hence

$$(6) \qquad \operatorname*{Res}_{z=0} \frac{(e^{hz}-1)^m e^{xz}}{z^{n+1}} = \sum_{k=0}^{n-m} \frac{1}{(n-m-k)!} M_k h^{m+k} x^{n-m-k}.$$

From (5) and (6) we obtain (1).
The first few coefficients in the expansion (2) are:

$$M_0 = 1, \quad M_1 = \frac{m}{2}, \quad M_2 = \frac{m(3m+1)}{24}, \quad M_3 = \frac{m^2(m+1)}{48}, \ldots$$

From formula (1), as special cases, we have

$$\Delta^m x^m = m! \, h^m,$$

$$\Delta^m x^{m+1} = (m+1)! \, h^{m+1} \left(\frac{x}{h} + \frac{m}{2} \right),$$

$$\Delta^m x^{m+2} = (m+2)! \, h^{m+2} \left(\frac{x^2}{2h^2} + \frac{mx}{2h} + \frac{m(3m+1)}{24} \right),$$

$$\Delta^m x^{m+3} = (m+3)! \, h^{m+3} \left(\frac{x^3}{6h^3} + \frac{mx^2}{4h^2} + \frac{m(3m+1)x}{24h} + \frac{m^2(m+1)}{48} \right),$$
$$\vdots$$

REFERENCE

A.-L. Cauchy: 'Sur les différences finies des puissances entières d'une seule variable', *Exercices de mathématique*, troisième année. Paris 1828. ≡ *Oeuvres* (2) 8, 180–182.

9.1.3. Solution of the Equation $\Delta f(x) = x^n$ $(n \in N)$

We shall determine a particular solution of the difference equation

(1) $\Delta f(x) = x^n$ $(n \in N)$

where f is the unknown function.
  Since

(2) $x^n = n! \, \operatorname*{Res}_{z=0} \dfrac{e^{xz}}{z^{n+1}}$ $(n \in N)$,

from (1) and (2) follows

(3) $f(x) = n! \, \operatorname*{Res}_{z=0} \dfrac{\Delta^{-1} e^{xz}}{z^{n+1}}$.

  However, since

$$\Delta e^{xz} = e^{(x+h)z} - e^{xz} = (e^{hz} - 1) \, e^{xz},$$

we have

$$\Delta^{-1} (\Delta e^{xz}) = (e^{hz} - 1) \, \Delta^{-1} e^{xz},$$

which implies

$$\Delta^{-1} e^{xz} = \frac{e^{xz}}{e^{hz} - 1}.$$

Therefore, from (3) we get

$$(4) \qquad f(x) = n! \; \operatorname*{Res}_{z=0} \frac{e^{xz}}{(e^{hz}-1) \, z^{n+1}}.$$

Let

$$(5) \qquad \left(1 + \frac{t}{2!} + \frac{t^2}{3!} + \cdots \right)^{-1} = A_0 + A_1 t + A_2 t^2 + \cdots$$

In order the evaluate the residue which appears in (4), it is enough to determine the coefficient of z^{n+1} in Laurent's expansion of the function

$$z \mapsto \frac{1}{h} \, e^{xz} \left(1 + \frac{hz}{2!} + \frac{h^2 z^2}{3!} + \cdots \right)^{-1}.$$

However, in view of (5) we have

$$\frac{1}{h} \, e^{xz} \left(1 + \frac{hz}{2!} + \frac{h^2 z^2}{3!} + \cdots \right)^{-1}$$

$$= \frac{1}{h} \left(1 + \frac{xz}{1!} + \frac{x^2 z^2}{2!} + \cdots \right) (A_0 + A_1 hz + A_2 h^2 z^2 + \cdots),$$

and so

$$(6) \qquad \operatorname*{Res}_{z=0} \frac{e^{xz}}{(e^{hz}-1) \, z^{n+1}} = \sum_{k=0}^{n+1} A_k \frac{h^{k-1} \, x^{n-k+1}}{(n-k+1)!}.$$

From (4) and (6) we obtain the solution of (1) in the form

$$f(x) = \sum_{k=0}^{n+1} \frac{n!}{(n-k+1)!} \, A_k \, h^{k-1} \, x^{n-k+1},$$

where the coefficients $A_0, \ldots, A_{n+1}$ are defined by (5).

REMARK 1. The general solution of (1) is

$$f(x) = H(x) + \sum_{k=0}^{n+1} \frac{n!}{(n-k+1)!} \, A_k \, h^{k-1} \, x^{n-k+1},$$

where H is an arbitrary h-periodic function, i.e. $H(x+h) = H(x)$.

REMARK 2. Since Bernoulli's numbers B_n ($n = 0, 1, \ldots$) are defined by the expansion

$$\frac{z}{e^z-1} = 1 - \frac{1}{2}z + \sum_{k=1}^{+\infty} \frac{B_{2k}}{(2k)!} z^{2k} \qquad (|z|<2\pi),$$

and since

$$\frac{z}{e^z-1} = \left(1 + \frac{z}{2!} + \frac{z^2}{3!} + \cdots\right)^{-1},$$

we conclude that $k!A_k = B_k$. In particular, $A_{2n+1} = 0$ ($n = 1, 2, \ldots$)

REFERENCE
A.-L. Cauchy: 'Sur les intégrales aux différences finies des puissances entières d'une seule variable', *Exercices de mathématiques*, troisième année. Paris 1828. $\equiv$ *Oeuvres* (2) 8, 183–188.

9.1.4. *Solution of the Equation* $\Delta^m f(x) = x^n$ ($m, n \in \mathbf{N}$)

Consider the difference equation

(1) $\Delta^m f(x) = x^n$ ($m, n \in \mathbf{N}$)

where f is the unknown function. Using a procedure similar to that of 9.1.3, we find a solution (1) in the form

$$f(x) = n! \operatorname*{Res}_{z=0} \frac{1}{(e^{hz}-1)^m} \frac{e^{xz}}{z^{n+1}},$$

i.e.

(2) $$f(x) = \sum_{k=0}^{m+n} \frac{n!}{(m+n-k)!} M_k h^{k-m} x^{m+n-k},$$

where the coefficients $M_0, M_1, \ldots$ are defined by

$$\left(1 + \frac{t}{2!} + \frac{t^2}{3!} + \cdots\right)^{-m} = M_0 + M_1 t + M_2 t^2 + \cdots$$

In particular we have

$$M_0 = 1, \quad M_1 = -\frac{m}{2}, \quad M_2 = \frac{m(3m-1)}{24}, \quad M_3 = -\frac{m^2(m-1)}{48}, \quad \ldots$$

REMARK 1. The general solution of (1) is

$$f(x) = \sum_{k=1}^{m} x^{m-k} H_k(x) + \sum_{k=0}^{m+n} \frac{n!}{(m+n-k)!} M_k h^{k-m} x^{m+n-k},$$

where $H_1, \ldots, H_m$ are arbitrary h-periodic functions.

REMARK 2. For $m = 1$ we obtain the result of 9.1.3.

REFERENCE
A.-L. Cauchy: 'Sur les intégrales aux différences finies des puissances entières d'une seule variable', *Exercices de mathématiques*, troisième année. Paris 1828. ≡ *Oeuvres* (2) 8, 183–188.

9.1.5. *Solution of the Equation* $P(\Delta)f(x) = F(x)$

Consider the difference equation

$$(1) \qquad a_n \Delta^n f(x) + a_{n-1} \Delta^{n-1} f(x) + \cdots + a_1 \Delta f(x) + a_0 f(x) = F(x)$$

where f is the unknown function, which also satisfies the initial conditions

$$(2) \qquad f(x_0) = \Delta f(x_0) = \cdots = \Delta^{n-1} f(x_0) = 0.$$

We shall describe here a formal method for determination of f which satisfies (1) and (2).

If we put $P(z) = a_n z^n + a_{n-1} z^{n-1} + \cdots + a_1 z + a_0$, then the Equation (1) can be formally written as

$$P(\Delta)f(x) = F(x).$$

Putting $R(z) = 1/P(z)$ into Cauchy's formula (see 4.7.1)

$$R(z) = \sum_{k=1}^{n} \operatorname*{Res}_{t=t_k} \frac{R(t)}{z-t},$$

where $t_1, \ldots, t_n$ are poles of the rational function R, we get

$$(3) \qquad \frac{1}{P(z)} = \sum_{k=1}^{n} \operatorname*{Res}_{t=t_k} \frac{1}{z-t} \frac{1}{P(t)},$$

where $t_1, \ldots, t_n$ are the zeros of P.

By a formal replacement of z by the operator Δ, from (3) follows

$$(4) \qquad \frac{1}{P(\Delta)} = \sum_{k=1}^{n} \operatorname*{Res}_{t=t_k} \frac{1}{\Delta-t} \frac{1}{P(t)}.$$

and finally, from (4) we obtain

$$(5) \qquad \frac{1}{P(\Delta)} F(x) = \sum_{k=1}^{n} \operatorname*{Res}_{t=t_k} \frac{F(x)}{\Delta-t} \frac{1}{P(t)}.$$

In the formula (5), $F(x)/(\Delta - t)$ denotes the solution of the equation $(\Delta - t)f(x) = F(x)$ with respect to the unknown function f. Therefore

$$\frac{F(x)}{\Delta-t} = (1+t)^{\frac{x}{h}-1} \sum_{s=x_0}^{x} (1+t)^{-\frac{s}{h}} F(s),$$

and (5) becomes

$$\frac{1}{P(\Delta)} F(x) = \sum_{k=1}^{n} \operatorname*{Res}_{t=t_k} \frac{1}{P(t)} (1+t)^{\frac{x}{h}-1} \sum_{s=x_0}^{x} (1+t)^{-\frac{s}{h}} F(s)$$

$$= \sum_{k=1}^{n} \operatorname*{Res}_{t=t_k} \frac{1}{P(t)} \sum_{s=x_0}^{x} (1+t)^{\frac{x-s}{h}-1} F(s).$$

This solution was found by Tortolini [1], and was quoted by Cauchy in [2].
Tortolini [3] extended this method to partial difference equations. Namely if P is a polynomial, Tortolini considered the equation

$$P(\Delta_x, \Delta_y) u = 0,$$

where

$$\Delta_x u = u(x+h, y) - u(x, y), \quad \Delta_y u = u(x, y+k) - u(x, y),$$

and obtained the solution

$$(6) \qquad u(x, y) = \sum \operatorname{Res} \frac{f(t) e^{\theta y} (1+t)^{x/h}}{P(e^{\theta k}-1, t)},$$

where θ is arbitrary, f is an arbitrary regular function and the sum is taken over all the solutions of the equation in t: $F(e^{\theta k} - 1, t) = 0$.

EXAMPLE 1. Let $h = k = 1$. Using Tortolini's formula (6) we obtain the following solution

$$u(x, y) - 3^x f(x+y) + 2^x g(x+y),$$

for the equation

$$(\Delta_x^2 - 3\Delta_x - 5\Delta_x\Delta_y + 7\Delta_y + 6\Delta_y^2 + 2)u - 0$$

where f and g are arbitrary functions.

REFERENCES
1. B. Tortolini: 'Trattato del calcolo dei residui', *Giornale Arcad.* 63 (1834–35), 86–138.
   – See, in particular, p. 113.
2. A.-L. Cauchy: 'Mémoire sur l'emploi des équations symboliques dans le Calcul infinitésimal et dans le calcul aux différences finies', *Compt. Rend. Acad. Sci. Paris* 17 (1843), 449–458. $\equiv$ *Oeuvres* (1) 8, 28–38.
3. B. Tortolini: 'Seconda memoria sull'applicazione del calcolo dei residui all'integrazione delle equazioni lineari a differenze finite', *Giornale Arcad.* 91 (1842), 3–67.

9.1.6. *Systems of Difference Equations*

Tortolini [1] applied calculus of residues to systems of difference equations. Since the method is more or less the same as the method given in 9.1.5, we shall only give an example, which is also taken from [1].

Let f, g, and h be unknown functions which satisfy the system of difference equations

$$\Delta f(x) + f(x) + 15 h(x) = 0, \qquad f(x) + 5 g(x) - \Delta g(x) + 7 h(x) = 0,$$

$$\Delta h(x) - 2g(x) - 11 h(x) = 0,$$

and the initial conditions $f(1) = 0, g(1) = 1, h(1) = 0$, where Δ is defined by $\Delta F(x) = F(x + 1) - F(x)$.

The solution of this problem is given by

$$f(x) = \sum \text{Res} \frac{(1+t)^{x-1} A(t)}{P(t)}, \quad g(x) = \sum \text{Res} \frac{(1+t)^{x-1} B(t)}{P(t)},$$

$$h(x) = \sum \text{Res} \frac{(1+t)^{x-1} C(t)}{P(t)},$$

where $A(t), B(t), C(t)$ are nontrivial solutions of the system

$$(1 + t) A - 14 B = 0, \quad A - (t - 5) B + 7 C = 0, \quad -2B + (t - 11) C = 0,$$

and $P(t)$ is the determinant of this system. The sums are taken over all the zeros of P.

It is easily shown that

$$A(t) = 14t - 154, \quad B(t) = (t+1)(t-11),$$

$$C(t) = 2(t+1), \quad P(t) = (t+3)(t-5)(t-13),$$

and hence the solution is given by

$$f(x) = \frac{1}{64}(14 \cdot 6^x + 14^x + 49(-2)^x), \quad g(x) = \frac{1}{64}(6 \cdot 6^x + 14^x - 7(-2)^x),$$

$$h(x) = \frac{1}{64}(14^x - 2 \cdot 6^x + (-2)^x).$$

REFERENCE

1. B. Tortolini: 'Memoria sull'applicazioni del calcolo dei residui all'integrazione delle equazioni lineari a differenze finite', *Giornale Arcad.* 9 (1842), 84–113.

9.2. CALCULUS OF DIVIDED DIFFERENCES

9.2.1. *Divided Differences*

DEFINITION 1. Suppose that the function f is defined at the points x_0, $x_1, \ldots, x_n$ which are distinct. The divided difference of order n of the function f with respect to $n+1$ points $x_0, x_1, \ldots, x_n$, denoted by

$$[x_0, x_1, \ldots, x_n; f]$$

is defined recursively

$$[x_0; f] = f(x_0),$$

$$[x_0, x_1, \ldots, x_n; f] = \frac{[x_0, x_1, \ldots, x_{n-1}; f] - [x_1, x_2, \ldots, x_n; f]}{x_0 - x_n} \quad (n \in \mathbf{N}).$$

From Definition 1 follows

$$[x_0, x_1; f] = \frac{f(x_0) - f(x_1)}{x_0 - x_1},$$

$$[x_0, x_1, x_2; f] = \frac{f(x_0)}{(x_0 - x_1)(x_0 - x_2)} + \frac{f(x_1)}{(x_1 - x_0)(x_1 - x_2)} + \frac{f(x_2)}{(x_2 - x_0)(x_1 - x_0)},$$

$$\vdots$$

(1) $$[x_0, x_1, \ldots, x_n; f] = \sum_{k=0}^{n} \frac{f(x_k)}{\prod\limits_{\substack{s=0 \\ s \neq k}}^{n} (x_k - x_s)}.$$

Notice the equality $[x, x + h; f] h = \Delta f(x)$.

9.2.2. *Divided Difference as a Contour Integral*

The nth order divided difference can be expressed as a contour integral. This representation is often more useful than the representation (1) from 9.2.1.

Suppose that f is a regular function in the region G and on its boundary, which is a closed contour $\Gamma = \partial G$, and suppose that distinct numbers x_0, $x_1, \ldots, x_n$ belong to G. The function F defined by

$$F(z) = \frac{f(z)}{(z - x_0)(z - x_1) \cdots (z - x_n)},$$

has in G only simple poles at $x_0, x_1, \ldots, x_n$. From Cauchy's residue theorem we get

$$\oint_\Gamma F(z)\, dz = \sum_{k=0}^{n} \operatorname*{Res}_{z = x_k} F(z).$$

However,

$$\operatorname*{Res}_{z = x_k} F(z) = \lim_{z \to x_k} (z - x_k) F(z) = \frac{f(x_k)}{\prod\limits_{\substack{s=0 \\ s \neq k}}^{n} (x_k - x_s)}$$

and so

(1) $$\oint_\Gamma \frac{f(z)}{(z - x_0)(z - x_1) \cdots (z - x_n)}\, dz = \sum_{k=0}^{n} \frac{f(x_k)}{\prod\limits_{\substack{s=0 \\ s \neq k}}^{n} (x_k - x_s)}.$$

From (1), and formula (1) from 9.2.1, we immediately obtain

(2) $$[x_0, x_1, \ldots, x_n; f] = \oint_\Gamma \frac{f(z)}{(z - x_0)(z - x_1) \cdots (z - x_n)}\, dz.$$

Formula (2) is often taken as a definition of nth order divided difference, even in the case when $x_0, x_1, \ldots, x_n$ are not distinct.

We give a simple application of (2).

Let f be a regular function in the region G and on the contour $\Gamma = \partial G$ of length L. If $x_0, x_1, \ldots, x_n \in G$, then

$$|[x_0, x_1, \ldots, x_n; f]| \leq \frac{L}{2\pi} \frac{\max\limits_{z \in \Gamma} |f(z)|}{\min\limits_{z \in \Gamma} \left| \prod\limits_{k=0}^{n} (z - x_k) \right|}.$$

9.2.3. Hermite's Formula

THEOREM 1. *Let*

$$(1) \qquad x_0, x_1, \ldots, x_n$$

be a finite sequence of complex numbers, where some of them may be equal. Let $z_0, z_1, \ldots, z_m$ be the subsequence of (1) consisting of distinct numbers from (1), i.e. $z_i \neq z_j$ for $i \neq j$. Denote by p_v the number of appearances of z_v in the sequence (1).

Let f be a regular function in the region G and on its boundary, the closed contour $\Gamma = \partial G$. If $z_v \in G$ ($v = 0, 1, \ldots, m$), then

$$(2) \quad [x_0, x_1, \ldots, x_n; f] = \sum_{v=0}^{m} \sum_{r=0}^{p_v - 1} \frac{f^{(p_v - r - 1)}(z_v)}{r!\,(p_v - r - 1)!} \lim_{z \to z_v} \frac{d^r}{dz^r} \left(\frac{(z - z_v)^{p_v}}{Q(z)} \right),$$

where

$$Q(z) = \prod_{v=0}^{m} (z - z_v)^{p_v}.$$

Proof. In virtue of the equality (2) from 9.2.2 which, by definition, holds also in the case when $x_0, x_1, \ldots, x_n$ are not distinct, we have

$$[x_0, x_1, \ldots, x_n; f] = \frac{1}{2\pi i} \sum_{v=0}^{m} \oint_{\gamma_v} \frac{f(z)}{(z - z_0)^{p_0} \cdots (z - z_n)^{p_n}}\, dz,$$

where γ_ν is the circle such that int γ_ν contains the point z_ν and none of the points z_k $(k \neq \nu)$. Introduce the notation

$$I_\nu = \frac{1}{2\pi i} \oint_{\gamma_\nu} \frac{f(z)}{Q(z)} \, dz = \frac{1}{2\pi i} \oint_{\gamma_\nu} (z - z_\nu)^{p_\nu} \frac{f(z)}{Q(z)} \frac{1}{(z - z_\nu)^{p_\nu}} \, dz,$$

so that

$$(3) \qquad [x_0, x_1, \ldots, x_n; f] = \sum_{\nu=0}^{m} I_\nu.$$

The function

$$z \mapsto H(z) = (z - z_\nu)^{p_\nu} \frac{f(z)}{Q(z)}$$

is regular in int γ_ν, and by Cauchy's integral theorem we have

$$I_\nu = \frac{1}{(p_\nu - 1)!} \lim_{z \to z_\nu} \frac{d^{p_\nu - 1}}{dz^{p_\nu - 1}} H(z).$$

After an easy calculation we find

$$(4) \qquad I_\nu = \sum_{r=0}^{p_\nu - 1} \frac{f^{(p_\nu - r - 1)}(z_\nu)}{r! \, (p_\nu - r - 1)!} \lim_{z \to z_\nu} \frac{d^r}{dz^r} \left(\frac{(z - z_\nu)^{p_\nu}}{Q(z)} \right).$$

From (3) and (4) follows (2).

Hermite's interpolation formula can be obtained by the use of Theorem 1. We give it in the following form.

THEOREM 2. *Under the conditions of Theorem 1 we have*

$$(5) \qquad f(x) = \sum_{\nu=1}^{m} \sum_{r=0}^{p_\nu - 1} \sum_{s=0}^{r} \frac{f^{(p_\nu - r - 1)}(z_\nu)}{(r-s)! \, (p_\nu - r - 1)!} \lim_{z \to z_\nu} \frac{d^{r-s}}{dz^{r-s}} \left(\frac{(z - z_\nu)^{p_\nu}}{R(z)} \right) \frac{R(x)}{(x - z_\nu)^{s+1}}$$

$$+ R(x)[x, x_1, \ldots, x_n; f],$$

where $x \in G$ and

$$R(z) = \prod_{\nu=1}^{n} (z - z_\nu)^{p_\nu}.$$

Proof. Put $p_0 = 1$, $z_0 = x_0 = x$, $Q(z) = (z - x)R(z)$ in (2). By an application of Leibniz' rule for differentiation of a product, we obtain

$$[x, x_1, \ldots, x_n; f]$$

$$= \frac{f(x)}{R(x)} - \sum_{\nu=1}^{m} \sum_{r=0}^{p_\nu-1} \frac{f^{(p_\nu - r - 1)}(z_\nu)}{(p_\nu - r - 1)!} \sum_{s=0}^{r} \frac{1}{(r-s)!} \lim_{z \to z_\nu} \frac{d^{r-s}}{dz^{r-s}} \left(\frac{(z-z_\nu)^{p_\nu}}{R(z)} \right) \frac{1}{(x-z_\nu)^{s+1}},$$

and this implies (5).

REFERENCE

A. O. Gel'fond: *Calculus of Finite Differences*, (Russian), Moscow 1967, pp. 43–46.

REMARK. Residues were widely applied to finite and divided differences in the book:
   N. E. Nörlund: *Vorlesungen über Differenzenrechnung*, Berlin 1924; see, in particular, pp. 81–94.

Augustin-Louis Cauchy, at three different periods of his life.

An excerpt from paper [7] (see references on p. 337) where it can be seen
how Cauchy wrote the sign **E** which denotes the sum of all the residues.

Chapter 10

Augustin-Louis Cauchy

10.1. BIOGRAPHY

French mathematician Augustin-Louis Cauchy was born in Paris on 21 August 1789, less than six weeks after the fall of the Bastille. In order to avoid risks which seem to be inevitable in a post-revolutionary period, the family escaped to the village of Arcueil, where they were neighbours of Laplace and Berthollet, so that Cauchy, as a young boy, became acquainted with famous scientists. In 1800 Cauchy's father returned to Paris where he was appointed secretary to the Senate. Cauchy often sat in his father's office and there he met Lagrange who used to visit the Senate on business. Lagrange soon became interested in the boy (it is said that he forcasted Cauchy's scientific genius) and gave his father useful advices regarding his education.

After having completed his elementary education, Cauchy attended the *École Centrale du Panthéon*, and in 1805 was admitted to *École Polytechnique*. In 1807 he entered the *École des Ponts et Chaussées*.

In March 1810 Cauchy was appointed military engineer at the harbour of Cherbourg which held a central position in Napoleon's plans for the invasion of England. At Cherbourg, where he remained until 1813, Cauchy wrote his first mathematical papers, and immediately drew attention of the leading French mathematicians.

In 1815 Cauchy began to lecture in Mathematical Analysis at the *École Polytechnique*. Next year, when the republican and Bonapartist Gaspard Monge was expelled from the *Académie des Sciences*, Cauchy was appointed (not elected) a member. Soon after he became Full Professor and lectured in Algebra at the *Sorbonne*, Mathematical Physics at the *Collège de France* and Mechanics at the *École Polytechnique*.

All his life Cauchy was a devoted, almost fanatic Catholic and a sincere supporter of the Bourbon dynasty. After the July Revolution of 1830, he refused to take the oath of allegiance to Louis-Philippe, the Orleans "Citizen-King", lost his chairs and went into exile.

In the period up to 1830 Cauchy published a large number of papers dealing with various branches of Mathematics. Among other things, he laid

323

down the foundations of Combinatorics, which led him to the Theory of
Finite Groups, he made fundamental contributions to Analysis (both real
and complex) and to Mathematical Physics (notably Hydrodynamics and
Elasticity). From 1826 to 1830 he published a kind of a private journal
entitled *Exercices de mathématiques*, and its twelve issues a year were filled
by Cauchy himself. The journal was selling well and had great influence on
contemporary mathematicians.

From 1830 till 1832 Cauchy lived in Torino where he held the chair of
Mathematical Physics, created especially for him by Charles Albert, king of
Sardinia. However, the ex-king Charles X, who settled in Prague, called him
to assist in the education of the crown prince, and for six years Cauchy spent
a lot of valuable time in teaching the young Duc de Bordeau all the prepar-
atory school subjects and in looking after his behaviour. As a reward, he was
created baron. Cauchy was very proud of the title received from the lawful
king. Although the life at the exiled court took much of his time, Cauchy
managed to devote some of his energies to research. From that period dates
his famous treatise on the propagation of light.

In 1838 Cauchy returned to Paris and at first lectured in Higher Mathematics at various institutions under the control of the Church. He also took
his seat at the Academy, but when he was unanimously elected Professor at
the *Collège de France*, he again refused to take the oath of allegiance, and so
could not take that chair. He was then coopted by the *Bureau des Longitudes*
and he worked there 'illegally' until the fall of Louis-Philippe in 1848, since
the government refused to confirm this cooption.

The Second Republic of 1848 repealed the act requiring the oath of allegiance, and Cauchy resumed the chair of Mathematical Astronomy at the
Sorbonne. The oath was reestablished by Napoleon III in 1852, but Cauchy
was exempted from taking it and so retained this chair until his death.

Cauchy was an extremely fertile scientist. He published 789 papers (some
of them are extensive studies of several hundred pages) from all branches of
Mathematics, and also Mechanics, Optics and Astronomy. Only two mathematicians (Euler and Cayley) published more papers than Cauchy, though
when total volumes are compared Cauchy comes before Cayley. Besides,
Cauchy published eight articles on general topics in which he defends his
religious convictions, discusses the problems of education, criminal, etc. His
unbelievable productivity led to the restriction of the Academy (which still
holds) that in the weekly *Comptes rendus hebdomadaires de l'Académie des
Sciences* papers longer than four pages cannot be printed. The collected
papers of Cauchy are published in 27 volumes of over 13000 pages in $4°$.

It is interesting to note that Cauchy rarely entered into polemics. His views on the subject are best illustrated by his reply [4] to Poinsot: "Je regarde comme perdu pour la science tout le temps que les savants emploient à guerroyer les uns contre les autres, et je crois qu'il vaut beaucoup mieux résoudre une question de plus que de se livrer à cette polémique".

Cauchy lived in an extremely turbulent and unstable period of the French history. Born during the reign of Louis XVI, his childhood and youth fall in the period of Robespierre's Terror and Napoleon's conquests. As a young man of 27 he rejoiced at the return of the Bourbons (1815). Then came their fall, the bourgeois monarchy of Louis-Philippe (1830–1848), The Second Republic (1848–1852) and the Second Empire of Napoleon III. Nevertheless, Cauchy lived a peaceful and a quiet life. He was happily married from 1818, and had two daughters. Cauchy's stubborn devotion to the unpopular Bourbons and his fanatical Catholicism seem to be a sort of paravane which enabled him to stand apart from the social changes he was witnessing and to devote all his time to Mathematics.

Cauchy died on 23 May 1857 in a small village Sceaux, near Paris, where he retired to recover from pneumonia.

REFERENCES

1. C.-A. Valson: *La vie et les travaux du Baron Cauchy*, Paris 1868: t. I: XXIV + 290 pp., t. II: XXIII + 178 pp. Second edition: Paris 1970.
2. E. T. Bell: *Men of Mathematics*, New York 1937, 593 pp.
3. *Cauchy, Baron Augustin-Louis*, Article in Nouveau Larousse illustré, pp. 575–576.
4. A.-L. Cauchy: 'Note', *Bulletin des sciences mathématiques* (*Bulletin de Férussac*) 7 (1827), 333–337. ≡ *Oeuvres* (2) **15**, 138–140.

10.2. SCIENTIFIC CONTRIBUTIONS

To describe Cauchy's contributions to Mathematics and Mathematical Physics would require a considerable amount of space. In this short note we shall only roughly mention his main results and shall give some statistical data.

Cauchy's first papers were devoted to Geometry and Algebra. He started his mathematical career in 1811 by solving Poinsot's problem in connection with the existence of regular polyhedrons. Next year he published his first paper on symmetric functions, and he often returned to this topic. Cauchy developed the theory of substitutions and founded the theory of finite groups which Galois later successfully applied to general algebraic equations.

Since 1814 Cauchy paid special attention to Mathematical Analysis and Mathematical Physics. In that year he wrote his treatise of definite integrals that was later to become the basis of the theory of complex functions. Next year he solved an open problem of Fermat from the number theory, and immediately afterwards, in 1816, we won the Grand Prix of the French Academy for his theory of the propagation of waves at the surface of a liquid. This paper, of over 300 pages in print, became a classic in Hydrodynamics.

In 1821 Cauchy published his lectures held at the *École Polytechnique*. He gave strict definitions of limits and continuity and developed the "$\epsilon - \delta$" technique. His method of exposition is even to-day present in many text-books on analysis. Cauchy was particularly concerned with infinite series and their convergence, which were up to then used formally. His rigorous approach frightened Laplace who used series, without investigating their convergence, in his Celestial Mechanics. Aware of the fact that Cauchy's approach was correct, Laplace had made a thorough revision of his life work, and fortunately found that his conclusions remain valid.

Later Cauchy had great success in applying the series to differential equations, and using series he proved the fundamental theorems on the existence of solutions of those equations.

Calculus of Residues, a method developed within the framework of Complex Analysis, was a discovery extremely valued by Cauchy. The notion of the residue is not so interesting in itself, but the applications found by Cauchy are amazing. He applied Calculus of Residues to the evaluation of definite integrals, the summation of series, to ordinary and partial differential equations, difference and algebraic equations, to the theory of symmetric functions and finally to mathematical physics. Cauchy's work on the Calculus of Residues will be considered in more detail in 10.3.

As we mentioned before, Cauchy published 789 papers, 501 on Pure Mathematics and 288 on Mathematical Physics. Mathematical papers can be classified as follows:

    (i) Arithmetic, Number theory — 69 papers,
   (ii) Geometry — 39 papers,
  (iii) Foundations of Analysis — 72 papers,
  (iv) Residues and Definite Integrals — 81 papers,
   (v) Symmetric Functions and the Theory of Substitutions — 40 papers,
  (vi) Series — 73 papers,
 (vii) Theory of Equations — 48 papers,
(viii) Inverse Periodic Functions — 39 papers,
  (ix) Differential Equations — 84 papers.

Papers from Mathematical Physics can be classified to papers from Mechanics (113), Optics (102), and Astronomy (73).

The greatest part of his writings Cauchy published in the journal *Comptes rendus hebdomadaires des séances de l'Académie des sciences de Paris* — altogether 549 papers. We also mention that his pupil Abbé François Moigno edited three books ([1], [2], [3]) totaling 1770 pages which are based on Cauchy's work.

Besides that, Cauchy was also referee for a large number of manuscripts of other authors, and he often presented detailed reports in writing. In Vol. 15 of the 2nd series of Cauchy's collected works, pp. 515–580, a list of 128 manuscripts which Cauchy was to referee in the period 1816–1830 is given. For those manuscripts Cauchy gave 51 written reports and 18 oral reports. Other papers were not refereed by Cauchy.

However, Cauchy was not always prompt in refereeing, as shown by the case of Abel's important work, which was the foundation of the theory of elliptic functions.[1]

During his lifetime, as well as later on, Cauchy was criticised for his enormous production. True, he sometimes published works of little or no merit, but it seems that his papers which gave the greatest contributions to Mathematics of that time were purposely overlooked. Nevertheless, Cauchy was always acknowledged as a great mathematician, as can be seen from his biography. Only eleven years after his death, Professor C.-A. Valson published a book, in two volumes, on Cauchy's life and work ([4], [5]). This book was reprinted in 1970.

A lasting monument to Cauchy's work are his Collected works which were, under the title *Oeuvres complètes d'Augustin Cauchy*, published in Paris from 1882 to 1974. They are divided into two series.

The first series which consists of 12 volumes in 4° was published in the period 1882–1911, and it contains Cauchy's writings published in the editions

---

[1] The history of that manuscript is one more tragic episode in Abel's tragic life. Namely, on 30th October 1826 the Academy sent the manuscript to Cauchy and Legendre for examination. Cauchy took it first but he never looked at it. Abel, however, knew the value of his work, and was curious of its reception. Only when Jacobi enquired about the paper (he found a reference to it in a published paper by Abel), Cauchy recommended it for publication on 29th June 1829. Unfortunately, Abel was already dead (he died on 6th April 1829). However, this is not the end of the story. The 1830 revolution and the general inertia in the Academy led to it that the paper was published 12 years after Abel's death in the jorunal *Mémoires présentés par divers savants* 7 (1841), 176–264. Hence, it is not present in the 1839 edition of Abel's works. Libri, a member of the Academy with a reputation that he was apt to steel books and manuscripts, was instructed to see the paper through the press, and he 'lost' the manuscript just after the printing. It was rediscovered in the 1950's in a Florence library.

of the Academy. Vol. 1 contains the papers written by Cauchy before he became a member of the Academy; they are published in the journal *Mémoires présentés par divers savants*. The papers published in the journal *Mémoires de l'Académie des Sciences* are reprinted in Vols. 2 and 3. Finally, Vol. 4 – Vol. 12 contain Cauchy's papers published, since 1836, in the *Comptes rendus*. At the end of Vol. 12 there is a table of contents for the first series.

The second series consists of 15 volumes in 4°. The first two volumes contain Cauchy's writings not published in the editions of the Academy, but in *Journal de l'École Polytechnique* (Vol. 1) and *Journal de Mathématiques pures et appliquées, Bulletin des sciences mathématiques* (*Bulletin de Férussac*), *Bulletin de la Société philomatique, Annales de Mathématiques* and *Correspondance sur l'École Polytechnique* (Vol. 2). Text-books written by Cauchy are reprinted in Vols. 3–5. They are: *Cours d'analyse de l'École Polytechnique, 1ère partie: Analyse algébrique*, Paris 1821 (Vol. 3); *Resumé des leçons données à l'École Polytechnique sur le calcul infinitésimal*, Paris 1823, and *Leçons sur le calcul différentiel*, Paris 1829 (Vol. 4); *Leçons sur les applications du calcul infinitésimal à la géometrié*, Paris 1826 and 1828 (Vol. 5). Volumes 6–9 contain all 51 issues of Cauchy's privite journal *Exercices des mathématiques*, Paris 1826–1830; Vol. 10 contains *Resumés analytiques*, Torino 1833, and *Nouveaux exercices de mathématiques*, Prague 1835–1836; and Vols. 11–14 the 48 issues of Cauchy's journal *Exercices d'analyse et de physique mathématique*, Paris 1839–1853.

The last volume, published as late as 1974, largely consists of works which Cauchy published as pamphlets or litographs, though there are also a few papers which were missed out of Vol. 2. It includes two handwritten litographs (submitted to the Academy in Turin) which are reprinted photographically. This volume also contains various bibliographies for Cauchy, a bibliography of secondary literature on Cauchy and a table of contents for the second series.

A number of articles were written on the publication of this last volume of Cauchy's works. A detailed survey is given in [6]. In reviews [7] and [8] Grattan-Guinness made some critical remarks, the most important of which seems to be that translations of Cauchy's books were not mentioned. Grattan-Guinness gives in [7] a short list of translations in German and Russian.[2]

---

[2] It is interesting to note the enormous difference in the reviews by Dugac [6] and by Grattan-Guinness [8]. We give a characteristic example.

Dugac: La très riche "Bibliographie des travaux concernant la vie et l'oeuvre de Cauchy" (p. 607–609) . . . .

Grattan-Guinness: . . . pp. 607–611 contain an extremely patchy bibliography of secondary literature on Cauchy, and a rather incomplete list of manuscripts.

REFERENCES

1. *Leçons de calcul différentiel et de calcul intégral redigées d'après les méthodes et les ouvrages publiés ou inédits de M. A.-L. Cauchy,* par M. l'abbé Moigno, tome I, Calcul différentiel. Paris 1840, XXXVI + 531 pp.
2. *Leçons de calcul différentiel et de calcul intégral rédigées principalement d'après les méthodes de M. A.-L. Cauchy et étendues aux travaux les plus récents des géomètres,* par l'abbé Moigno, tome II, Calcul intégral. Première partie. Paris 1844, XLVII + 783 pp.
3. *Leçons de calcul différentiel et de calcul intégral d'après les méthodes et les ouvrages publiés ou inédits de A.-L. Cauchy,* par M. l'abbé Moigno, tome IV, 1re fascicule, Calcul des variations. Rédigé en collaboration avec M. Lindelöf. Paris 1861, XX + 352 pp.
4. C.-A. Valson: *La vie et les travaux du baron Cauchy,* tome I, partie historique. Paris 1868, XXIV + 290 pp.
5. C.-A. Valson: *La vie et les travaux du baron Cauchy,* tome II, partie scientifique. Paris 1868, XXIV + 178 pp.
6. P. Dugac: 'Sur la publication du dernier volume des oeuvres d'Augustin Cauchy', *Rev. Hist. Sci.* 27/1 (1975), 75–83.
7. I. Grattan-Guinness: 'On the Publication of the Last Volume of the Works of Augustin Cauchy', *Janus* 62 (1975), 179–191.
8. I. Grattan-Guinness: 'Oeuvres complètes d'Augustin Cauchy', IIe série, tome XV. (Review). *Historia Mathematica* 3 (1976), 481–485.

10.3. HISTORICAL DEVELOPMENT OF THE CALCULUS OF RESIDUES

10.3.1. *Cauchy's Work on the Calculus of Residues*

> C'est à Cauchy que revient la gloire d'avoir fondé la théorie des intégrales prises entre des limites imaginaires.
>
> H. Poincaré, [11] from 10.3.2.

Although complex functions were in use during the 18th century, the derivative and the integral of a complex function were not strictly defined. The first written document in which the question of complex integration is treated was a letter written in 1811 by Gauss to his friend Bessel. In that letter Gauss states that the integral $\int_0^c f(z)\,dz$ has the same value independently of the path of integration which joins the points $z = 0$ and $z = c$. However, Gauss published his investigations much later, in 1832.

In the meantime Cauchy published a series of papers devoted to complex integrals. His first paper from this field [1] dates from 1814. In that paper Cauchy also arrived at the notion of the residue. Investigating the interchange

of the order of integration for a function in two real variables (at that time
the notation $z = x + iy$ was not in general use), Cauchy defined the so-called
singular integral. It is, in fact, the limit

$$A = \lim_{\varepsilon \to 0} \int_{-\varepsilon}^{\varepsilon} f(a + \xi)\,d\xi,$$

where a is a pole of the function f. For some simple functions Cauchy gave
the formula

$$A = 2\pi i \lim_{\varepsilon \to 0} \varepsilon f(a + \varepsilon),$$

where a has the same meaning as before.

It is easily verified that

$$\text{(1)} \qquad \lim_{\varepsilon \to 0} \varepsilon f(a + \varepsilon) = \operatorname*{Res}_{z=a} f(z),$$

under the condition that a is a simple pole of f. However, if a is a pole of
order higher than 1, formula (1) ceases to hold, which is easily seen by considering an example, such as: $f(z) = (z + 1)z^{-2}$, $a = 0$.

Later on, in 1826, Cauchy [2] gave a formula for the evaluation of residues in the case of nth order poles. He wrote it in the form:

*The residue of the function $f(z)$, such that $(z - a)^n f(z) = g(z)$ is finite and
different from zero, at $z = a$, is*

$$\frac{g^{(n-1)}(a)}{1 \cdot 2 \cdot 3 \cdots (n-1)}.$$

In fact, this formula coincides with the formula (1) of 2.1.2.

Cauchy's first text-book [3] from 1821 contains an extensive and a rather
complete theory of explicit algebraic and elementary transcendent functions
of a complex variable, but not the general definition of a complex function,
nor the integral of a complex function.

A decisive step forward Cauchy made in his pamphlet [4] from 1825,
where he defined the integral between complex limits as the limit of the
corresponding sum. Integrals over infinite regions are also considered, but the
conditions for their existence are not given. This paper was, in a way, a reply
to Poisson who several times sharply criticised complex integration. A well-
written and a thorough survey of the evolution of complex integration, and
the polemics between Cauchy and Poisson, is given in paper [5] (Poisson's
remarks are given on pp. 117–118).

Next year, in 1826, Cauchy began to publish his *Exercices de mathématiques* [2]. He used there, for the first time, the term *résidu* for the limit (1), and the term *résidu intégral* for the sum of all residues in a certain region. Special notations are also introduced. So, for example,

$$\mathbf{E}\left((f(z))\right)$$

denotes[1] the *résidu intégral* of the function f, i.e. the sum of all residues of f. The notation

$$\mathbf{E}\,\frac{g(z)}{((h(z)))}$$

means that the residues are taken with respect to the zeros of h, and not the poles of g. Finally,

$${}_a^c\mathbf{E}_b^d\left((f(z))\right)$$

denotes the sum of all residues "taken with respect to the zeros of the equation $1/f(z) = 0$, but so that the real parts lie between a and c, and the imaginary parts between b and d, where c and d could be $+\infty$, and a and b could be $-\infty$".

In the same paper Cauchy considered in detail the case when a pole lies on the boundary of the considered rectangular region, and came to the conclusion that if a pole is on the side of the rectangle, then the residue must be multiplied by $\frac{1}{2}$, and if it is at a vertex, then it must be mutiplied by $\frac{1}{4}$. Besides, Cauchy states that if the pole is at the vertex, then the integration should be carried along both sides which meet in that vertex, but up to a distance ϵ from it, and then ϵ should be made to approach zero. It is interesting to note that many modern text-books do not treat the case when a singularity lies on the contour of integration (this fact was particularly stressed by Freudenthal [6]).

Cauchy [4] proceeded to investigate infinite regions, and proved a number of theorems on the evaluation of real integrals. For instance, Cauchy proved the theorem:

---

[1] Cauchy and his contemporaries (as well as Lindelöf [27] in 1905) did not use the letter E, but the sign which can be seen, in Cauchy's handwriting, on page 322, formula (28).

If $f(x + iy) = 0$ for every positive y and for $x = \pm\infty$, and if for every x we have $\lim_{y\to+\infty} (x + iy)f(x + iy) = F$, *then*

$$\int\limits_{-\infty}^{+\infty} f(x)\,\mathrm{d}x = 2\pi i \left({}_{-\infty}^{+\infty}\mathbf{E}_0^{+\infty}\left((f(z))\right) - \frac{1}{2}F\right).$$

In the same pamphlet Cauchy proved the following important theorem:
The sum of residues of a uniform function in the extended plane is zero.

He finally considered regions which are not rectangular, but especially annular sectors, or regions bounded by two curves and two line segments.

Using residues Cauchy deduced the formula which is now called Cauchy's integral formula. In the first version, from 1826, the formula reads

$$f(z) = \frac{1}{2\pi}\int\limits_{-\pi}^{\pi} \frac{f(e^{ix})}{1 - ze^{-ix}}\,\mathrm{d}x,$$

while later, e.g. in 1831, Cauchy writes it in the form

$$(2)\qquad f(z) = \frac{1}{2\pi}\int\limits_{-\pi}^{\pi} \frac{\zeta f(\zeta)}{\zeta - z}\,\mathrm{d}x \qquad (\zeta = re^{ix}),$$

which closely resembles the modern form of that formula.

It should be noted that Cauchy, at that time, was not quite sure of the conditions which ensure the validity of his results. So, for example, he states that (2) will hold provided that the function f is 'finite and continuous' for $|z| < r$. Later (in 1840) Cauchy states that the same conditions must be imposed also on the derivative f'.

A detailed and a complete survey of complex integration and residues Cauchy gave in paper [7] which he read to the Academy in Turin on 27th November 1831. The complete paper was published for the first time in 1974, in Vol. 15 of the second series of Cauchy's works, where the original manuscript was reproduced photographically. A summary of results from this paper is published in 1831 in *Bulletin des sciences mathématiques* (*Bulletin de Férussac*).

Cauchy's paper [8] is also interesting. He proved the following important theorems:

(1) *If*

$$\lim_{z\to\infty} zf(z) = A, \qquad then \qquad \mathbf{E}\left((f(z))\right) = A.$$

(2) *If the function f is finite for finite and infinite values of z, then f is constant.*

(3) *If the function $f(z)$ is always finite and continuous for finite values of z and if* lim $z^{-m} f(z) = 0$ *(m positive integer), as $z \to \infty$ along any direction, then $f(z)$ is a polynomial of degree $m-1$.*

Clearly, from (3) for $m = 1$ follows (2), and this theorem is usually called Liouville's theorem. In fact, Liouville proved the theorem for doubly periodic functions, and Cauchy generalized it to analytic functions. In modern form, this theorem was first stated and proved by Jordan in his textbook [9].

In the next period Cauchy continued to develop the theory of complex functions. He occupied himself by the expansions of those functions in series, he proved Laurent's theorem by means of residues, and he also arrived at some important properties which are fundamental for the theory of analytic continuation, as for example (see [10]): A functional equation holds in the annulus of convergence if it holds on a circle which belongs to that annulus.

During the last years of his life, almost 40 years after the publication of his first paper on complex integration, Cauchy returned to his theory and tried to make it more rigorous. On that occasion he introduced certain definitions which were in use for a long time. A function which has a derivative independent of the direction of the tangent is called monogenic, uniform functions are called monodrome, and a function which is monodrome, monogenic and finite is called a sinectic function.

Despite his rigorous approach to the theory of functions, Cauchy never mentioned essential singularities, nor did he consider the behaviour of a function in the neighbourhood of an essential singularity. For him a singularity is always a pole. Only once (see [11]) Cauchy implicitly mentioned an essential singularity. Namely, he stated that tg t takes the values $+i$ or $-i$ for $t = \infty$, depending on the sign of Im t.

In paper [12] Cauchy gave a new definition of the residue. He now defined it not as a limit, but as the 'mean-value':

$$\int \zeta f(z + \zeta) \, d\zeta.$$

Cauchy's last paper on residues [13] was published on 2 March 1857, less than three months before his death. In that paper Cauchy describes his first steps in the field, made 30 years ago: *C'est dans le premier volume des Exercices de Mathématiques, publié en 1826, que j'ai, pour la première fois, exposé les principes du Calcul des résidus* (Cauchy is referring to paper [2]). In [13] he also pointed out his papers [4] and [7] as important contributions to residues.

In this short survey of Cauchy's work on residues we mentioned only his principal contributions. Naturally, Cauchy published a large number of less important papers in which he mainly applied residues to various branches of Mathematics and Mathematical Physics. Some of those applications are given in preceeding chapters.

Cauchy's calculus of residues attracted attention of his contemporaries. The theory of residues soon crossed the boarders of France, and already in 1834 (see [14] − [15]) in Italy appeared two surveys of the calculus of residues written by Piola and Tortolini (Tortolini's review is particularly complete and well-written). Soon after, the English mathematicians became acquainted with residues owing to a note by Gregory [16]. Much later, in 1879, an account of the calculus of residues [17] appeared in America. Judging by the style, it seems that [17] is the first American text on the subject. Some original contributions also appeared. So, for example, in [18] and [19] the authors used different methods to arrive at the same results which Cauchy obtained by residues ([18] is concerned with differential equations, and [19] with definite integrals). Conversely, in [20] some definite integrals are written as sums of residues. Moreover, in [21] some general formulas involving residues are derived. Functions in several variables are also considered, but the residue is not defined for higher dimensions. Residues were particularly well applied by Tortolini, who in the period 1834–1850, besides the mentioned survey [15], published several articles containing original contributions, notably to difference equations. His papers are mentioned earlier in this book.

On the other hand, Cauchy's results were not immediately accepted as correct. For instance, G. Radike [22] criticized Cauchy's results, but his criticism (see Burkhardt [23]) was founded on misunderstanding, mainly in connection with uniform and multiform functions. Namely, Radike pointed out that the definition of a residue for an nth order pole at the point a as the coefficient of $1/\epsilon$ in the expansion of $f(a + \epsilon)$ is not always equivalent to the formula

$$\frac{d^{n-1}}{dz^{n-1}} (z-a)^n f(z)\bigg|_{z=a} .$$

To illustrate this statement Radike takes the function f defined by

$$f(z) = 7z + \frac{5}{z-2} + \frac{1+z}{(z-2)^{3/2}} + \frac{z}{(z-2)^3} ,$$

where, by the first definition, the residue at $z = 2$ is 5, and by the second is infinite. Of course, this 'contradiction' takes place because $z = 2$ is not a pole of the function f.

Falk's paper [24] from 1877 is worth mentioning. The author believes that Cauchy's definition of the residue is not natural, and he tries to introduce the residue without the use of integrals or Laurent's series. The problem he sets himself is as follows:

Let F be a regular function in a region D, except that $a \in D$ is an mth order pole of f. Determine the function G, regular in D, so that the function $z \mapsto F(z) - G(z)$ is finite at $z = a$.

Falk, quite correctly, believes that this procedure will lead him to the residue of F at $z = a$. Indeed, if F is written in the form $F(z) = (z - a)^{-m} f(z)$, where f is regular in D, then the solution of this problem is given by

$$G(z) = \sum_{k=0}^{m-1} \frac{1}{k!} \frac{f^{(k)}(a)}{(z-a)^{m-k}} = \frac{1}{(m-1)!} \lim_{u \to a} \frac{\partial^{m-1}}{\partial u^{m-1}} \left(\frac{(u-a)^m F(u)}{z-u} \right).$$

In fact, the function G is the principal part of Laurent's expansion of F, and Falk thus arrived at the same conclusion as Cauchy, though he used a different method.

In the mentioned paper [23] Burkhardt gave a critical historical survey of the development of the calculus of residues. His paper contains an extensive bibliography, although he did not mention all the papers published up to 1904; for instance the papers [14], [15], [17], [21].

Cauchy was interested in the work of other mathematicians on the calculus of residues, and he often began his papers by quoting their names (though without bibliographical date, as was the custom of that time). So, for example, in [25] he says: *Parmi les travaux relatifs à cet object, on peut citer ceux de MM. Ostrogradski et Bouniakowski, de l'Académie de Saint-Pétersbourg, et ceux de M. Tortolini, professeur au Collège Romain ... On peut citer encore un Mémoire où M. Richelot a démontré ...* In paper [11] Cauchy mentions Blanchet and again Tortolini, etc. The present authors identified all the papers of other mathematicians quoted by Cauchy, except those by Ostrogradski and Bouniakowski. See Note A9, p. 351.

In connection with this we mention that according to the rather accurate book *The History of National Mathematics* (Russian), Vol. 2, Kiev 1967, p. 256, the first Russian paper on complex functions was written in 1868 by Yu. V. Sohockiĭ. It was his Master's thesis entitled: *Theory of Integral Residues with Some Applications* (Russian). In the preface Sohockiĭ writes: "In further text I give the general priciples of evaluation of integral residues and I demonstrate some applications, namely those which I either did not find in Cauchy's work, or which were presented by him in a form which is not as simple and obvious as the form which I give them." In fact, Sohockiĭ showed that the theory of residues can be used as a method in solving various problems from the theory of special functions.

Still, Cauchy did not mention all the papers on the calculus of residues which appeared during his life-time. For example, papers [14], [16], [21], or [22] are not quoted by Cauchy.

It is interesting to note that Academy journal *Comptes rendus* ofter placed calculus of residues (*calcul des résidus*) as an independent mathematical discipline.

Ever since the middle of the last century, every textbook on Complex Analysis contains a chapter (or at least a section) on the Calculus of Residues.[1] However, there exist books devoted entirely to residues. Soon after Cauchy's death, H. Laurent [27] published the first monograph on the Calculus of Residues. After him, monographs were written by Lindelöf [28] and Watson [29]. Gel'fond [30] published a short text book on residues, and Mitrinović [31] a short collection of problems and exercices.

Each one of those books treats one particular aspect of the theory of residues. Laurent is mainly concerned with definite integrals and infinite series; the chief preoccupations of Lindelöf are various summation formulas; Watson treats only definite integrals, Gel'fond summation formulas, and Mitrinović definite integrals and infinite series.

This monograph covers all known applications of the Calculus of Residues.

REMARK. Residues can also be defined on Riemann surfaces. In that case we do not define the residue of an analytic function, but the residue of an analytic differential. For details, consult [32].

REFERENCES

1. A.-L. Cauchy: 'Mémoire sur la théorie des intégrales définies', *Recueil des mémoires des savants étrangers présentés à l'Académie des sciences* 1 (1827), 599–799 ≡ *Oeuvres* (1) 1, 319–506. This paper was presented to the Academy on 22 August 1814.
2. A.-L. Cauchy: 'Sur une nouveau genre de calcul analogue au calcul infinitésimal',[2] *Exercices de mathématiques*, Paris 1826. ≡ *Oeuvres* (2) 6, 23–37.
3. A.-L. Cauchy: 'Cours d'analyse de l'École Polytechnique, 1ère partie: Analyse algébrique', Paris 1821. ≡ *Oeuvres* (2) 3, 153–301.

[1] In connection with this, notice that Bertrand in his textbook [26] rarely mentions residues, though he gives the usual applications of Cauchy's theory.

[2] The beginning of this paper is reproduced, in English, in the book: *A Source Book in Classical Analysis*, edited by G. Birkhoff (with the assistance of Uta Merzbach), Cambridge (Mass.) 1973, pp. 40–44.

4. A.-L. Cauchy: *Mémoire sur les intégrales définies prises entre des limites imaginaires*, Paris 1825, 68 pp. ≡ *Oeuvres* (2) **15**, 41–89. A short version of this work, under the title 'Sur les intégrales définies prises entre des limites imaginaires', was published in *Bulletin des sciences mathématiques (Bulletin de Férussac)* **3** (1825), 214–221. It is completely reproduced in *Bull. Sci. Math.* **7** (1874), 265–304 and **8** (1875), 43–55 and 148–159.

5. P. Stäckel: 'Integration durch imaginäres Gebiet', *Bibliotheka Mathematica* (3) **1** (1900), 109–128.

6. H. Freudenthal: 'Cauchy, Augustin-Loius', *Dictionary of scientific biography* **3**, New York 1971, pp. 131–148.

7. A.-L. Cauchy: 'Mémoire sur les rapports qui existent entre le calcul des résidus et le calcul ᴅ s limites, et sur les avantages qu'offrent ces deux nouveaux calculs dans la résolution des équations algébriques ou transcendantes', ≡ *Oeuvres* (2) **15**, 182–261.

8. A.-L. Cauchy: 'Mémoire sur quelques propositions fondamentales du calcul des résidus et sur la théorie des intégrales singulières', *Compt. Rend. Acad. Sci. Paris* **19** (1844), 1337–1346. = *Oeuvres* (1) **8**, 366–375.

9. C. Jordan: *Cours d'analyse de l'École Polytechnique*, t. **2**, Paris 1913.

10. A.-L. Cauchy: 'Note sur divers conséquences du théorème relatif aux valeurs moyennes des fonctions', *Compt. Rend. Acad, Sci. Paris* **20** (1845), 119–127. ≡ *Oeuvres* (1) **8**, 414–422.

11. A.-L. Cauchy: 'Sur les rapports differentiels des quantités géométriques, et sur les intégrales synectiques des équations différentielles', *Compt. Rend. Acad, Sci. Paris* **40** (1855), 445–455. ≡ *Oeuvres* (1) **12**, 225–235.

12. A.-L. Cauchy: 'Considérations nouvelles sur les résidus', *Compt. Rend. Acad. Sci. Paris* **41** (1855), 41–42. ≡ *Oeuvres* (1) **12**, 300–301.

13. A.-L. Cauchy: 'Théorie nouvelle des résidus', *Compt. Rend. Acad. Sci. Paris* **44** (1857), 406–417. ≡ *Oeuvres* (1) **12**, 433–444.

14. G. Piola: 'Sui principi e sugli usi del calcolo dei residui (da varie memorie del Sig. A. L. Cauchy)', *Opuscoli Mat. Fiz. Milano* **2** (1834), 237–260.

15. B. Tortolini: 'Trattato del calcolo dei residui', *Giornale arcadico di science, lettere ed arti. Roma* **63** (1834–35), 86–138.

16. D. F. Gregory: 'On the Residual Calculus', *The Cambridge Math. Journal* **1** (1837), 145–155.

17. C. H. Kummell: 'An Account of Cauchy's Calcul des résidus', *The Analyst (J. Pure Appl. Math.), De Moines, Iowa* **6** (1879), 1–9, 41–46, 173–176.

18. P.-H. Blanchet: 'Mémoire sur la propagation et la polarisation du mouvement dans un milieu élastique indéfini, cristallisé d'une manière quelconque', *J. Math. Pures Appl.* **5** (1840), 1–30.

19. B. Boncompagni: 'Recherches sur les intégrales définies', *J. Reine Angew. Math.* **25** (1843), 74–96.

20. F. Richelot: 'De integralibus quibusdam definitis, quorum summa ad quadraturam divisionemque circuli revocatur', *J. Reine Angew. Math.* **21** (1840), 293–327.

21. E. Roselli: 'Alcune formole sul calcolo dei residui e loro applicazione', *Giornale arcadico di science, lettere ed arti. Roma* **114** (1848), 1–32, 121–168 i **115** (1848), 3–24.

22. G. Radike: 'Einige Bemerkungen über die Principien der Cauchyschen Residuenrechnung', *J. Reine Angew. Math.* **25** (1843), 216–239.

23. H. Burkhardt: *Exkurs betr. die Entwicklungsgeschichte von Cauchys Residuentheorie.* Encyklopädie der Mathematischen Wissenschaften, zweiter Band, erster Teil, zweite Hälfte, Leipzig 1904–1916, pp. 1001–1032.

24. M. Falk: *Sur les fonctions imaginaires, a l'égard spécial du calcul des résidus.* Nova Acta Regiae Societatis Scientiarium Upsaliensis. 1887, No. 7, 32 pp.

25. A.-L. Cauchy: 'Mémoire sur des formules générales qui se déduisent du calcul des résidus et qui paraissent devoir concourir notablement aux progrès de l'Analyse infinitésimale', *Compt. Rend. Acad. Sci. Paris* 12 (1841), 871–878. ≡ *Oeuvres* (1) 6, 149–158.

26. J. Bertrand: *Traité de calcul différentiel et de calcul intégral. Calcul intégral. Intégrales définies et indéfinies,* Paris 1870, pp. 291–330.

27. H. Laurent: *Théorie des résidus,* Paris 1865, VIII + 176 pp.

28. E. Lindelöf: *Calcul des résidus et ses applications à la théorie des fonctions,* Paris 1905, VIII + 144 pp. Reprinted 1947.

29. G. N. Watson: *Complex Integration and Cauchy's Theorem.* Cambridge 1914, VIII + 78 pp. Reprinted New York (without the year of publication).

30. A. O. Gel'fond: *Residues and their Applications,* (Russian), Moscow 1966, 112 pp. There is a Spanish edition of this book, also published in Moscow under the title A. Guelfond: *Los residuos y sus aplicaciones,* Moscu 1968, 112 pp.

31. D. S. Mitrinović: *Calculus of Residues,* Groningen 1966, 88 pp.

32. G. Springer: *Introduction to Riemann Surfaces,* Reading, Mass., 1957, VIII + 307 pp., Russian translation: Moscow 1960, 343 pp.

10.3.2. *Residues of Functions in Several Variables*

M. Marie first tried to extend the definitions of the integral and the residue to functions in several variables. In 1853 he submitted his paper *Sur les périodes des intégrales simples et doubles* to the Academy in Paris, and next year Cauchy and Sturm reported favourably on it. Almost twenty years later, Marie published a series of papers ([1]–[10]) where he again expressed his ideas, thinking that his earlier results were, in a way, overlooked.

However, as shown by Poincaré, Marie's results are not correct. In his long paper [11] Poincaré analysed in detail the reasons which led Marie to fallacious results. In the same paper Poincaré stated that Jacobi obtained important results on that topic, but that he did not publish them. On the other hand, Južakov [12] quotes Jacobi's paper [13] in which he used multidimensional residues as coefficients of powers of degree -1 in multiple Laurent expansions. Južakov also mentions Didon's paper [14] which is devoted to residues of functions in two variables. Poincaré stated in [11] that Stieltjes sent an interesting paper to Hermite in which he generalized various Cauchy's and Lagrange's formulas. Unfortunately, certain questions in that paper were not clear enough, and Stieltjes could not clear them up satisfactorily, and so he did not publish his work. Poincaré [11] analysed Stieltjes's results and showed how they can be proved rigorously.

We also mention Appell [15] who derived some formulas for the residue of a function in two variables, and Picard [16], [17], who studied some questions which are indirectly relevant to this topic.

In paper [18] Poincaré defined the integral of a complex function in two complex variables, reducing it to integrals of real functions. His definition reads:

Let $z_1 = x_1 + iy_1$ and $z_2 = x_2 + iy_2$ ($x_1, x_2, y_1, y_2 \in \mathbf{R}$) be two complex variables, and let F be a function of those variables, whose real and imaginary parts are P and Q, namely: $F(z_1, z_2) = P + iQ$. Suppose that the region of integration is defined by the equations

$$x_1 = f_1(u, v), \quad y_1 = g_1(u, v), \quad x_2 = f_2(u, v), \quad y_2 = g_2(u, v)$$

where u and v are real parameters. Then

$$\iint F(z_1, z_2)\, dz_1\, dz_2 = \iint \Bigg((P + iQ) \frac{\partial(x_1, x_2)}{\partial(u, v)} + (iP - Q) \frac{\partial(x_1, y_2)}{\partial(u, v)}$$

$$+ (iP - Q) \frac{\partial(y_1, x_2)}{\partial(u, v)} - (P + iQ) \frac{\partial(y_1, y_2)}{\partial(u, v)} \Bigg) du\, dv,$$

where $\partial(x, y)/\partial(u, v)$ is the Jacobian, i.e.

$$\frac{\partial(x, y)}{\partial(u, v)} = \frac{\partial x}{\partial u} \frac{\partial y}{\partial v} - \frac{\partial x}{\partial v} \frac{\partial y}{\partial u}.$$

Poincaré's paper [11] is, in fact, an extension and a development of the paper [18]. At the same time Picard [19] arrived at some important results for the integrals of complex functions in two variables.

Južakov [12] mentions that Lefschetz [20], de Rham [21], Segre [22], Severi [23] also obtained some results for the residues of functions in several variables, while Wirtinger [24], Caccioppoli [25], Martinelli [26], Bergman [27], and Sorani [28] considered generalizations of logarithmic residues to functions in two or more variables.

In connection with the generalizations of Cauchy's results to higher dimensions, we also mention Carvallo [29] who by geometric reasoning (by the use of vectors) deduced a formula which is the space analog of Cauchy's integral formula.

In the case of functions in n variables, the role of an isolated singularity is played by analytic sets (geometrically, $(2n-2)$-dimensional subspaces of $2n$-dimensional real space) which may have very complicated structures. Hence, in the study of the integrals of analytic functions in several variables over closed surfaces, we are faced with serious difficulties of topological nature. However, the development of algebraic topology led in the last twenty years to an essential progress in the theory of multidimensional residues. Leray (see [30], [31], and his extensive paper [32]) developed the theory of residues on complex analytic varieties. He defined the form-residue, the class-residue, generalized the concept of the residue to closed differential forms with singularities on analytic subvarieties and proved a residue formula which reduces the order of the integral. Further generalizations were given by Norguet [33]. We also notice that Šabat [34] states that some important integrals of theoretical physics were evaluated in [35] by Leray's method.

An exposition of the Leray-Norguet theory of meromorphic differential forms and its possible homological generalisations is given by Dolbeault [36] and [37].

Finally, we mention Aronszajn's paper [38] which treats the question of the so-called Cauchy kernels. Necessary conditions for a function K, holomorphic in $\mathbf{C}^n \times \mathbf{C}^n$, to be a Cauchy kernel is provided, and it is shown that, in certain cases, this condition can also be sufficient.

In the U.S.S.R., a group of mathematicians (L. A. Aizenberg, G. P. Egoryčev, A. P. Južakov, etc.) at the University of Krasnoyarsk made valuable contributions to multidimensional residues and especially to their applications. A few of those were mentioned earlier in this book. Their work up to now is summarized in two well written monographs [39] and [40]. Monograph [39] is entirely devoted to the evaluation of combinatorial sums by means of residues, while in the monograph [40] residues are applied to: (i) expansions of implicit functions in potential and functional series; (ii) systems of nonlinear equations; (iii) combinatorial sums; (iv) the theory of functions in several complex variables. Both monographs contain very detailed bibliographies, and [40] some very useful historical comments.

REFERENCES

1. M. Marie: 'Théorie élémentaire des intégrales simples et de leurs périodes', *Compt. Rend. Acad. Sci. Paris* 75 (1872), 524–527.
2. M. Marie: 'Théorie élémentaire des intégrales doubles et de leurs périodes', *Compt. Rend. Acad. Sci. Paris* 75 (1872), 576–579.

3. M. Marie: 'Théorie élémentaire des intégrales doubles et de leurs périodes (suite)', *Compt. Rend. Acad. Sci. Paris* 75 (1872), 614–616.

4. M. Marie: 'Théorie élémentaire des intégrales doubles et de leurs périodes (fin)', *Compt. Rend. Acad. Sci. Paris* 75 (1872), 660–663.

5. M. Marie: 'Théorie des résidus des intégrales doubles', *Compt. Rend. Acad. Sci. Paris* 75 (1872), 695–698.

6. M. Marie: 'Théorie des résidus des intégrales doubles (suite et fin)', *Compt. Rend. Acad. Sci. Paris* 75 (1872), 751–755.

7. M. Marie: 'Extension de la méthode de Cauchy à l'étude des intégrales doubles, ou théorie des contours élémentaires dans l'espace', *Compt. Rend. Acad. Sci. Paris* 75 (1872), 865–868.

8. M. Marie: 'Théorie élémentaire des intégrales d'ordre quelconque et de leurs périodes', *Compt. Rend. Acad. Sci. Paris* 75 (1872), 1078–1081.

9. M. Marie: 'Théorie élémentaire des intégrales d'ordre quelconque et de leurs périodes (suite et fin)', *Compt. Rend. Acad. Sci. Paris* 75 (1872), 1247–1250.

10. M. Marie: 'Théorie des résidus des intégrales d'ordre quelconque', *Compt. Rend. Acad. Sci. Paris* 75 (1872), 1475–1479.

11. H. Poincaré: 'Sur les résidus des intégrales doubles', *Acta Mathematice* 9 (1887), 321–380.

12. A. P. Južakov: *Elements of the Theory of Multidimensional Residues*, (Russian), Krasnoyarsk 1975.[1]

13. C. G. Jacobi: 'De resolutions aequationum per series infinitas', *J. Reine Angew. Math.* 6 (1830), 257–286.

14. M. F. Didon: 'Note sur une formule de calcul intégral', *Ann. École Norm. Sup.* (2) 2 (1873), 31–48.

15. P. Appell: 'Relations entre les résidus d'une fonction d'un point analytique (x, y) qui se reproduit, multipliée par une constante, quand le point (x, y) décrit un cycle', *Compt. Rend. Acad. Sci. Paris* 95 (1882), 714–717.

16. É. Picard: 'Sur une classe de fonctions de deux variables indépendantes', *Compt. Rend. Acad. Sci. Paris* 96 (1883), 320–324.

17. É. Picard: 'Sur les intégrales de différentielles totales de second espèce', *Compt. Rend. Acad. Sci. Paris* 102 (1886), 250–253.

18. H. Poincaré: 'Sur les résidus des intégrales doubles', *Compt. Rend. Acad. Sci. Paris* 102 (1886), 202–204.

19. É. Picard: 'Sur les périodes des intégrales doubles', *Compt. Rend. Acad. Sci. Paris* 102 (1886), 349–350, 410–412.

20. S. Lefschetz: 'On the Residues of Double Integrals to an Algebraic Surfaces', *Quart. J. Math. Oxford* 47 (1916), 333–343.

---

[1] This book is not easy to find (only 300 copies were published) and so we give a brief description of its contents. The book has 182 pages and contains, besides a very well written historical introduction, four chapters: Integration of differential forms (pp. 12–59), Elements of the general theory of multidimensional residues (pp. 60–97), Residues of some rational and meromorphic functions (pp. 98–127) and Multidimensional analogs of the logarithmic residue with some applications (pp. 128–174). At the end of each chapter there are a few exercices, and 112 references are quoted.

21. G. de Rham: 'Relations entre la topologie et la théorie des intégrales multiples', *Enseignement Math.* **35** (1936), 213–228.

22. B. Segre: *Sull'estensioni della formula di Cauchy e sui residue degli integrali n-pli nella teoria della funzioni di n variabili complesse.* Atti del primo Congresso dell'Unione matematica Italiana, Bologna 1938, pp. 174–180.

23. F. Severi: *Funzioni analitiche e forme differenziali.* Atti del quarto Congresso dell'Unione matematica Italiana, t. 1. Roma 1953, pp. 125–140.

24. W. Wirtinger: 'Ein Integralsatz über analytische Gebilde im Gebeite von mehreren komplexen Veränderlichen', *Monat. Math. Phys.* **45** (1937), 418–431.

25. R. Caccioppoli: 'Residui di integrali doppi e intersezioni di curve analitiche', *Ann. Mat. Pura Appl.* (4) **29** (1949), 1–14.

26. E. Martinelli: 'Sulle estenzioni della formula integrale di Cauchy alle funzioni analitiche di piu variabili complesse', *Ann. Mat. Pura Appl.* (4) **34** (1953), 277–347.

27. S. Bergmann: 'The Number of Intersection Points of Two Analytic Surfaces in the Space of Two Complex Variables', *Math. Z.* **72** (1960), 294–306.

28. G. Sorani: 'Sull'indicatore logaritmico per le funzioni di piu variabili complesse', *Rend. Mat. Appl.* (5) **19** (1960), 130–142.

29. E. Carvallo: 'Généralisation et extension à l'espace du théorème des résidus de Cauchy', *Bull. Soc. Math. France* **24** (1896), 180–184.

30. J. Leray: 'La théorie des résidus sur une variété analytique complexe', *Compt. Rend. Acad. Sci. Paris* **247** (1958), 2253–2257.

31. J. Leray: 'Le calcul différentiel et intégral sur une variété analytique complexe', *Comp. Rend. Acad. Sci. Paris* **248** (1959), 22–28.

32. J. Leray: 'Le calcul différentiel et intégral sur une variété analytique complexe (Problème de Cauchy, III)', *Bull. Soc. Math. France* **87** (1959), 81–180. – This paper is translated into Russian, and published as a book.

33. F. Norguet: 'Sur la théorie des résidus', *Compt. Rend. Acad. Sci. Paris* **248** (1959), 2057–2059.

34. B. V. Šabat: *Introduction to Complex Analysis*, (Russian), Moscow 1969, p. 501.

35. R. Hua and V. Toeplitz: *Homology and Feinman Integrals*, (Russian), Moscow 1969.

36. P. Dolbeault: *Theory of Residues and Homology*, Séminaire Pierre Lelang (Analyse), Année 1969, pp. 152–163. Lecture Notes in Math., Vol. 116, Berlin 1970.

37. P. Dolbeault: *Valeurs principales et résidus sur les espaces analytiques complexes d'apres M. Herrera et D. Liebermann*, Séminaire Pierre Lelong (Analyse), Année 1970–1971, pp. 14–26, Lecture Notes in Math., Vol. 275, Berlin 1972.

38. N. Aronszajn: 'Calculus of Residues and General Cauchy Formulas in C^n, *Bull. Sci. Math.* (2) **101** (1977), 319–352.

39. G. P. Egoryčev: *Integral Representation and Evaluation of Combinatorial Sums*, (Russian), Novosibirsk 1977.

40. L. A. Aizenberg and A. P. Južakov: *Integral Representations and Residues in Multidimensional Complex Analysis*, (Russian), Novosibirsk 1979.

NOTES ADDED IN PROOF

A1. The calculus of residues is usually applied to the exact evaluation of integrals. However, if the singularities cannot be determined exactly, then calculus of residues can be used to obtain an approximate value of the integral. We illustrate such a possibility here, and in order to simplify the exposition, we suppose that poles are the only singularities.

Let $z = a\ (\neq \infty)$ be a pole of order k of an analytic function f. Then f can be written in the form

$$f(z) = \frac{g(z)}{(z-a)^k} ,$$

where g is a regular function at $z = a$ and $g(a) \neq 0$, provided that $z \in \{z \mid 0 < < |z - a| < R\}$. Suppose that the pole $z = a$ cannot be determined exactly, but with an error Δa, so that the approximate value is $\alpha = a + \Delta a$. Let

$$f^*(z) = \frac{g(z)}{(z - a - \Delta a)^k} ,$$

and let $\Gamma = \{z \mid |z - a| = r < R\}$, so that $\alpha \in \text{int } \Gamma$, i.e. $|\Delta a| < r$. Then in virtue of the Cauchy residue theorem, we have

$$I = \oint_\Gamma f(z)\, dz = 2\pi i \operatorname*{Res}_{z=a} f(z) = \frac{2\pi i}{(k-1)!}\, g^{(k-1)}(a)$$

and

$$I^* = \oint_\Gamma f^*(z)\, dz = 2\pi i \operatorname*{Res}_{z=\alpha} f^*(z) = \frac{2\pi i}{(k-1)!}\, g^{(k-1)}(a + \Delta a).$$

Let M be a constant such that

$$|g^{(k-1)}(a + \Delta a) - g^{(k-1)}(a)| \leq M\, |\Delta a|,$$

343

i.e.

$$\left| g^{(k)}(a) + \frac{1}{2!} g^{(k+1)}(a) \Delta a + \cdots \right| \leq M.$$

Then we have

(1) $$|I - I^*| \leq \frac{2\pi M}{(k-1)!} |\Delta a|,$$

which means that if we want to evaluate the integral I with an error not greater than ϵ, it is sufficient to determine the pole a, so that

$$|\Delta a| \leq \frac{(k-1)!}{2\pi M} \epsilon.$$

Suppose now that the function f is analytic in the domain $G = \operatorname{int} \Gamma$ and continuous on Γ. Let $a_1, \ldots, a_n$ be the poles of f of orders $k_1, \ldots, k_n$ respectively belonging to G, and suppose that they are determined with the errors $\Delta a_1, \ldots, \Delta a_n$, so that $\alpha_s = a_s + \Delta a_s \ (\in G)$.

In order to determine $I = \oint_\Gamma f(z) \, dz$ approximately, define the circles $\Gamma_s = \{z \,|\, |z - a_s| = r_s \}$, so that $\alpha_s \in \operatorname{int} \Gamma_s$, and $a_1, \ldots, a_{s-1}, a_{s+1}, \ldots, a_n$, $\alpha_1, \ldots, \alpha_{s-1}, \alpha_{s+1}, \ldots, \alpha_n \notin \operatorname{int} \Gamma_s$. According to Cauchy's theorem we have

$$I = \oint_\Gamma f(z) \, dz = \sum_{s=1}^{n} \oint_{\Gamma_s} f(z) \, dz = \sum_{s=1}^{n} I_s.$$

If I_s is approximated by

$$I_s^* = \oint_{\Gamma_s} f^*(z) \, dz, \quad \text{where} \quad f^*(z) = \frac{g_s(z)}{(z - \alpha_s)^{k_s}},$$

and if

$$I^* = \sum_{s=1}^{n} I_s^*,$$

we have

$$I - I^* = \sum_{s=1}^{n} (I_s - I_s^*).$$

Let now

$$|g_s^{(k_s)}(a_s) + \frac{1}{2!} g_s^{(k_s+1)}(a_s)\,\Delta a_s + \cdots| \le M_s \qquad (s = 1, \ldots, n)$$

and

$$\max(|\Delta a_1|, \ldots, |\Delta a_n|) = |\Delta a|.$$

Then, in virtue of (1) we have

$$|I - I^*| \le \sum_{s=1}^{n} |I_s - I_s^*| \le 2\pi \sum_{s=1}^{n} \frac{M_s}{(k_s - 1)!}\,|\Delta a_s|,$$

i.e.

$$|I - I^*| \le 2\pi M\,|\Delta a| \qquad \left(M = \sum_{s=1}^{n} \frac{M_s}{(k_s - 1)!}\right).$$

EXAMPLE. Consider the process of evaluating the integral

$$(2) \qquad \text{v.p.} \int_{-\infty}^{+\infty} \frac{1}{x^3 + x - 5}\,dx.$$

The equation $z^3 + z - 5 = 0$ has only one real root $a \in (1, 2)$. Take an approximate value $\alpha_1 = 1.515$ of a (all digits are exact). Then an approximate factorization gives

$$z^3 + z - 5 \cong (z - 1.515)(z^2 + 1.515z + 3.300).$$

In domain $\operatorname{Im} z > 0$ exists the pole

$$\alpha_2 = \frac{-1.515 + \sqrt{1.515^2 - 4 \cdot 3.300}}{2} \cong \frac{1}{2}(-1.515 + 3.302i).$$

By calculus of residues we find

$$I^* = \frac{2\pi i}{3\alpha_2^2 + 1} + \frac{\pi i}{3\alpha_1^2 + 1} \cong \pi i\,\{(-0.127 + 0.174i) + 0.127\},$$

i.e. $I^* \cong -0.546$. All these calculations are done by rounding to three decimal places.

Notice that the value of (2) can also be determined exactly. Namely, since $a^3 + a - 5 = 0$ we have $(z^3 + z - 5) : (z - a) = z^2 + az + 1 + a^2$, so the poles of the function $f(z) = 1/(z^3 + z - 5)$ are·

$$a_1 = a, \qquad a_2 = \frac{1}{2}(-a + i\sqrt{3a^2 + 4}), \qquad a_3 = \bar{a}_2.$$

Since

$$\operatorname*{Res}_{z=a_2} f(z) = \frac{3ai\sqrt{3a^2 + 4} - (3a^2 + 4)}{2(6a^2 + 45a + 4)},$$

by residual calculus we find

$$I = \text{v.p.} \int_{-\infty}^{+\infty} \frac{1}{x^3 + x - 5}\, dx = \pi i\left(2 \operatorname*{Res}_{z=a_2} f(z) + \operatorname*{Res}_{z=a} f(z)\right)$$

$$= \pi i\, \frac{3ai\sqrt{3a^2 + 4}}{6a^2 + 45a + 4},$$

i.e.

$$I = -\frac{3\pi a}{\sqrt{679}},$$

with use of the equality $a^3 = 5 - a$. By Cardano's formulas we easily find that

$$a = \sqrt[3]{(\sqrt{2037} + 45)/18} - \sqrt[3]{(\sqrt{2037} - 45)/18} \cong 1.515\,980\,228.$$

Hence, $I = -0.548\,314\,558 \ldots$. Note that $|I - I^*| < 3 \times 10^{-3}$.

REMARK. This text was written by G. Milovanović, who also gave the above example. It is, in fact, an improved version of a result from [1].

REFERENCE
1. K. Adžievski: *Approximative Integration with Applications to Approximative Calculus of Residues* (Macedonian), M.A. thesis. Mathematical Faculty, Skopje 1981.

A2.   Many series involving binomial coefficients can be summed by using the following integral representations:

$$(1) \quad \binom{x}{n} = \frac{1}{2\pi i} \int_{C_1} z^{-n-1}(1 + z)^x\, dz = \frac{1}{2\pi i} \int_{C_2} z^{-n-1}(1 - z)^{-x+n-1}\, dz,$$

where x is any complex number, n any positive integer, C_1 and C_2 being the positive paths along any simple closed Jordan curves enclosing $z = 0$ (the plane is cut along real z from -1 to $-\infty$ for C_1 and from 1 to $+\infty$ for C_2 to exclude branch points, the powers taking their principal values). Formulas are immediate consequences of the Cauchy residue theorem. We quote two particular cases from the general result which is obtained by an application of (1):

$$\sum_{k=1}^{+\infty} \binom{1+3k}{k} \left(\frac{2}{27}\right)^k = \frac{1}{2}, \qquad \sum_{k=1}^{+\infty} \binom{4k}{2k} \left(\frac{6}{25}\right)^{2k} = \frac{13}{7}.$$

REFERENCE

G. P. M. Heselden: 'The Sum of a Certain Series Involving Binomial Coefficients', *Math. Gaz.* 41 (1957), 280–282.

A3. Let $P(z) = z^n + a_{n-1}z^{n-1} + \cdots + a_1 z + a_0$, and suppose that $z_1, \ldots, z_n$ are distinct complex numbers. The following iteration scheme for factorizing P has been suggested by Kerner [1]:

$$\hat{z}_i = z_i - \frac{P(z_i)}{\displaystyle\prod_{\substack{j=1 \\ j \neq i}}^{n} (z_i - z_j)} \qquad (i = 1, \ldots, n).$$

Laurie [2] proposed the following problem: If $\sum_{i=1}^{n} z_i = -a_{n-1}$, prove that $\sum_{i=1}^{n} \hat{z}_i = -a_{n-1}$.

The following solution, communicated to us by G. V. Milovanović, shows that $\sum_{i=1}^{n} \hat{z}_i = -a_{n-1}$, regardless of the value of $\sum_{i=1}^{n} z_i$. Indeed, let $Q(z) = \Pi_{j=1}^{n} (z - z_j)$. Then

$$\hat{z}_i = z_i - \frac{P(z_i)}{Q'(z_i)} \qquad (i = 1, \ldots, n)$$

and so:

$$(1) \qquad \sum_{i=1}^{n} \hat{z}_i = \sum_{i=1}^{n} z_i - \sum_{i=1}^{n} \frac{P(z_i)}{Q'(z_i)}.$$

The function P/Q has only simple poles. Hence, if $r > \max_i |z_i|$, $C = \{z \,|\, |z| = r\}$, then

$$I = \frac{1}{2\pi i} \oint_C \frac{P(z)}{Q(z)} \, dz = \sum_{i=1}^{n} \operatorname*{Res}_{z=z_i} \frac{P(z)}{Q(z)} = \sum_{i=1}^{n} \frac{P(z_i)}{Q'(z_i)}.$$

On the other hand, if we introduce the substitution $w = 1/z$, and let $C' = \{w \mid |w| = 1/r\}$, we see that

$$I = \frac{1}{2\pi i} \oint_{C'} \frac{1}{w^2} \frac{P(1/w)}{Q(1/w)} \, dw = \lim_{w \to 0} \frac{d}{dw} \left(\frac{w^n P(1/w)}{w^n Q(1/w)} \right) = a_{n-1} + \sum_{i=1}^{n} z_i.$$

Therefore,

$$\sum_{i=1}^{n} \frac{P(z_i)}{Q'(z_i)} = a_{n-1} + \sum_{i=1}^{n} z_i,$$

and (1) becomes

$$\sum_{i=1}^{n} \hat{z}_i = \sum_{i=1}^{n} z_i - \left(a_{n-1} + \sum_{i=1}^{n} z_i \right) = -a_{n-1}.$$

REFERENCES

1. I. Kerner: 'Ein Gesamtschrittverfahren zur Berechung der Nullstellen von Polynomen', *Numer. Math.* 8 (1966), 290–294.
2. D. P. Laurie: 'Problem H-315', *The Fibonacci Quarterly* 18 (1980), 190 and 19 (1981), 476.

A4. In paper [1] Čebyšev stated the following theorem:

Suppose that $\varphi(x)/\psi(x)$ is one of the 'fractions convergentes (réduites)' of the integral

$$\int_a^b \frac{f(z)}{x - z} \, dz,$$

where f is positive for $z \in [a, b]$. If

$$\varphi(x) = \prod_{\nu=1}^{n} (x - x_\nu),$$

where $a < x_1 < \cdots < x_n < b$, then

$$(1) \qquad \int_{x_k}^{x_m} f(x) \, dx < \sum_{\nu=k}^{m} \frac{\varphi(x_\nu)}{\psi'(x_\nu)} < \int_{x_{k-1}}^{x_{m+1}} f(x) \, dx, \quad (1 < k < m < n).$$

This theorem was proved by Markov [2].

However, Čebyšev returned to this problem in paper [3] and gave a proof of (1), and of some other related inequalities, such as

$$(2) \qquad \sum_{\nu=k+1}^{m-1} \frac{\varphi(x_\nu)}{\psi'(x_\nu)} < \int_{x_k}^{x_m} f(x)\, dx < \sum_{\nu=k}^{m} \frac{\varphi(x_\nu)}{\psi'(x_\nu)} \,.$$

In his proof Čebyšev uses the notion of residues. Indeed, if certain conditions are fulfilled the middle term of (1) is equal to

$$\sum_{\nu=k}^{m} \operatorname*{Res}_{z=x_\nu} \frac{\varphi(z)}{\psi(z)} \,,$$

and similar inequalities hold for left-hand and right-hand terms of (2).

REMARK. Formulas (1) and (2) are important in the theory of numerical integration.

REFERENCES

1. P. Tchébycheff: 'Sur les valeurs limites des intégrales', *J. Math. Pures Appl.* (2) 19 (1874), 151–160.
2. A. Markov: 'Démonstration de certaines inégalités de M. Tchébychef', *Math. Annalen* 24 (1884), 172–178.
3. P. Tchébycheff: 'Sur la représentation des valeurs limites des intégrales par des résidus intégraux', *Acta Mathematica* 9 (1887), 35–56.

A5. If n, k, l are positive integers, define $I_n(k, l)$ by

$$I_n(k, l) = \frac{1}{\pi i} \oint \left(\frac{w - w^k}{1 - w^{k+1}} \right)^l \frac{1}{z^{n+1}}\, dz,$$

where the path of integration is a small circle with radius $< \frac{1}{4}$ around the origin. The letter w stands for

$$w = \frac{1 - \sqrt{1 - 4z}}{1 + \sqrt{1 - 4z}}$$

and for $|z| < \frac{1}{4}$ we take the principal value. Using contour integration de Bruijn [1] proved the following result:

If $x, y \in \mathbf{R}$ are fixed, and if $k \sim y\sqrt{2n}$, $l \sim x\sqrt{2n}$ $(n \to +\infty)$, then

$$\lim_{n \to +\infty} \frac{\pi i n}{4n} I_n(k, l) = \int_{1-i\infty}^{1+i\infty} u \exp\left(-2xu \operatorname{cth} uy + \frac{1}{2} u^2 \right) du.$$

This is an answer to a question proposed by E. Csáki and I. Vincze in probability theory.

REFERENCE
N. G. de Bruijn: 'An Asymptotic Problem', *Proceedings of the Prague Symposium on Asymptotic Statistics*, 3–6 September 1973, pp. 31–35.

A6. It is known that every meromorphic and doubly periodic function f is a quotient of two entire functions G_1 and G_2. Petrovitch [1] proved that each of the functions G_1 and G_2 can be expressed in the form

$$\frac{1}{2\pi i} \int_C \Phi(t, z)P(t - z)\, dt,$$

where Φ is a meromorphic function in t and z, fixed for all functions f which have the same elementary periods; P is a polynomial depending on f; C is a contour which encircles the parallelogram of periods which contains the point z.

This result is proved by means of residues.

REFERENCE
1. M. Petrovitch: 'Un mode général de représentation des fonctions elliptiques', *Compt. Rend. Acad. Sci. Paris* 198 (1934), 698–700.

A7. The integrals

$$I(z) = \int_{-\infty}^{+\infty} \frac{\operatorname{ch} zu - 1}{\operatorname{sh} \pi u \, \operatorname{sh} \pi t u}\, du; \qquad J(z) = \int_{-\infty}^{+\infty} \frac{\operatorname{ch} zu - 1}{\operatorname{sh} \pi u \, \operatorname{sh} \pi t u} \frac{1}{1 + u^2}\, du$$

both absolutely convergent when $|\operatorname{Re} z| < \pi(1 + |\operatorname{Re} t|)$, were considered by Brosa [1]. His paper has arisen from a study of nuclear collective motion in connection with Mehler–Fock transform. The integrals $I(z)$ and $J(z)$ appear when Parseval's relation for this transform of order 0 or -1 is used.

Using calculus of residues Brosa proved that if t is a rational number then $I(z)$ can be expressed in terms of elementary functions, and $J(z)$ by elementary functions and the Clausen integral $\operatorname{Cl}(z)$, defined by

$$\operatorname{Cl}(z) = -\int_0^z \log\left(2 \sin\frac{t}{2}\right) dt = \sum_{k=1}^{+\infty} \frac{\sin kz}{k^2}, \qquad (0 \leqslant z \leqslant \pi).$$

So, for example, for $t = 1$, Brosa established the equalities:

$$I(z) = \int_{-\infty}^{+\infty} \frac{\operatorname{ch} zu - 1}{\operatorname{sh}^2 \pi u} \, du = \frac{2}{\pi} - \frac{z}{\pi} \operatorname{ctg} \frac{z}{2};$$

$$J(z) = \int_{-\infty}^{+\infty} \frac{\operatorname{ch} zu - 1}{\operatorname{sh}^2 \pi u} \frac{1}{1 + u^2} \, du = \frac{\pi}{3}(1 - \cos z) + \frac{z^2}{2\pi} \cos z -$$

$$- \frac{\sin z}{\pi} \Big(z \log(2(1 - \cos z)) + 2\operatorname{Cl}(z) \Big), \qquad (|\operatorname{Re} z| \leq 2).$$

REFERENCE

1. U. Brosa: 'Two Types of Integrals Containing Two Hyperbolic Sine Functions in the Denominator of the Integrand', *Simon Stevin* **54** (1980), 213–221.

A8. 'A general solution' of an integro-differential equation with constant coefficients by the Cauchy method of residues is obtained in [1], but we have not been able to find that paper.

REFERENCE

1. R. A. Kušpeleva: 'An Application of the Theory of Residues to the Solution of Integrodifferential Equations with Constant Coefficients (Russian)', *Studies in Integro-differential Equations in Kirgizia*, No. 8, Frunze 1971, pp. 256–267.

A9. We were informed in August 1983 by Professor A. P. Juškevič (Moscow) about the following facts.

Cauchy knew about Ostrogradski's work on integrals and residues because they worked together in Paris; however, Ostrogradski's papers remained unpublished. A. P. Juškevič found them in 1963 in the archives of the French Academy and he published a short summary of those papers in his work: 'On Early Unpublished Works of M. V. Ostrogradski', (Russian), *Istor.-Mat. Issled.* **16** (1965), 11–48.

Among all the unpublished papers of Ostrogradski two of them deal with integrals and residues. They are:

(1) Mémoire sur la difficulté qui se rencontre dans le calcul des intégrales définies, lorsque la fonction à intégrer est discontinue entre les limites de l'intégration — dated 24 July, 1824;

(2) Remarques sur les intégrales définies — dated 7 August, 1824.

In both papers Ostrogradski examines improper integrals of the form

$\int_a^b f(x)\,\mathrm{d}x$, where f is not continuous on $[a,\,b]$. The first paper suggests a treatment similar to Cauchy's principal value, while in the second paper residues are explicitely used, without mentioning the word 'residue'. Namely, in certain cases Ostrogradski used the expression

$$\frac{1}{(n-1)!}\,\frac{\mathrm{d}^{n-1}}{\mathrm{d}z^{n-1}}\,(z-a)^n f(z),$$

which is the residue of f with respect to an nth order pole at $z=a$. He applied his results to the integral

$$\int_0^\pi \frac{x\sin x}{\cos t - \cos x}\,\mathrm{d}x$$

and he used the value of this integral to evaluate the integrals

$$\int_0^{\pi/2} (x/\mathrm{tg}\,x)\,\mathrm{d}x \quad \text{and} \quad \int_0^{\pi/2} \log\sin x\,\mathrm{d}x.$$

It should be mentioned that Ostrogradski was not quite satisfied with his results and that is why they remained in the archives instead of going to the printers.

Regarding Bouniakovski, according to the same communication by Prof. A. P. Juškevič, he became acquainted with residues by listening to Cauchy's lectures at the Collège de France, and he applied them in the second part of his Ph.D. thesis entitled *Détermination du rayon vecteur dans le mouvement elliptique des planètes*, on pp. 15–17, which he defended in May 1825 in Paris.

A10. Let $|a_{rs}|$ denote the determinant of order n, the element in the rth row and sth column being equal to a_{rs}. If for all $s=1,\ldots,n$ we have $y_s \neq z_s$, then

$$\left| \frac{x_r + z_s}{x_r + y_s} \right| = \frac{XY \displaystyle\prod_{s=1}^{n} (z_s - y_s)}{\displaystyle\prod_{r,\,s=1}^{n} (x_r + y_s)} \left(1 + \sum \operatorname{Res} \frac{(y+x_1)\cdots(y+x_n)}{q(y)\,(y-\varphi(y))} \right),$$

where

$$X = \prod_{1 \leqslant s < r \leqslant n} (x_r - x_s), \quad Y = \prod_{1 \leqslant s < r \leqslant n} (y_r - y_s), \quad q(y) = \prod_{r=1}^{n} (y - y_r),$$

$\varphi(y)$ is a polynomial such that $z_s = \varphi(y_s)$, $s = 1, \ldots, n$, and the sum is taken over all the roots of $y = \varphi(y)$ and the point $y = \infty$.

This result was proved in [1], and a number of applications were also given, mainly to the evaluation of the determinants of the form:

$$\left| \operatorname{tg}(a_r + b_s) \right|, \quad \left| \operatorname{cotg}(a_r + b_s) \right|, \quad \left| \begin{matrix} \operatorname{tg}(a_r + b_s) \\ \operatorname{tg}(a_r - b_s) \end{matrix} \right|, \quad \left| \begin{matrix} \operatorname{cotg}(a_r + b_s) \\ \operatorname{cotg}(a_r - b_s) \end{matrix} \right|, \quad \text{etc.}$$

REFERENCE

1. H. J. A. Duparc: 'On some determinants', *Proc. Akad. Wet. Amsterdam* 50 (1947), 157–165.

A11. If

$$f(z) = \frac{\sin(z - b_1) \cdots \sin(z - b_n)}{\sin(z - a_1) \cdots \sin(z - a_n)},$$

then

$$(1) \qquad \sum_{k=1}^{n} \operatorname*{Res}_{z = a_k} f(z) = \sin \sum_{k=1}^{n} (a_k - b_k).$$

However,

$$(2) \qquad \operatorname*{Res}_{z = a_k} f(z) = \frac{\sin(a_k - b_1) \cdots \sin(a_k - b_n)}{\sin(a_k - a_1) \cdots \sin(a_k - a_{k-1}) \sin(a_k - a_{k+1}) \cdots \sin(a_k - a_n)}$$

and equalities (1) and (2) can be used to derive some trigonometric identities.

The above result is due to Hermite [1] and it is a generalization of an identity given by J. Glaisher.

REFERENCE

Ch. Hermite: 'Sur une identité trigonométrique', *Nouvelles Annales de Mathématiques* (3) 4 (1885), 57. ≡ *Oeuvres*, tome IV, Paris 1917, pp. 206–208.

A12.   Consider the integral

$$I = \int_{-1}^{1} \frac{e^x}{x^2 + a^2} \, dx \qquad (0 < a < 1).$$

If a is very small, standard methods for approximate evaluation of I cannot be applied, but the result can be obtained by an application of the Cauchy residue theorem. Namely, if $\gamma = \{z \,|\, |z| = 1,\, \mathrm{Re}\, z > 0\}$, $l = [-1, 1]$, and $C = \gamma \cup l$, then

$$\oint_C \frac{e^z}{z^2 + a^2} \, dz = I + I_\gamma,$$

where

$$I_\gamma = \int_\gamma (e^z/(z^2 + a^2)) \, dz.$$

However,

$$\oint_C \frac{e^z}{z^2 + a^2} \, dz = 2\pi i \, \operatorname*{Res}_{z = ia} \frac{e^z}{z^2 + a^2} = \frac{\pi}{a} e^{ia},$$

and so

$$I = \frac{\pi}{a} e^{ia} - I_\gamma = \frac{\pi}{a} \cos a + i \frac{\pi}{a} \sin a - I_\gamma.$$

Since $I \in \mathbf{R}$, we have $I \approx (\pi/a) \cos a$, and the correction is contained in $\mathrm{Re}\, I_\gamma$. On the other hand, the integral I_γ can be evaluated by numerical methods since the pole $z = ia$ is not near the contour γ.

This example shows how the Cauchy method of residues can be applied to the integral $\int_a^b f(x) \, dx$ in the cases when a pole of f is very near the segment $[a, b]$.

The integral I is evaluated in detail in [1]. See also [2] and [3].

REFERENCES

1. D. Tošić: *Introduction to Numerical Analysis* (Serbian), Belgrade 1982, pp. 316–319.
2. Ph. J. Davis and Ph. Rabinowicz: *Methods of Numerical Integration*, New York, San Francisco, London 1975, pp. 139–142.
3. H. Laurent: *Traité d'analyse*, t. V, Paris 1890, pp. 208–210.

Name Index

The names are followed by the date of birth or the dates of birth and death, if known to the authors.

Since Cauchy is mentioned on almost all the pages, we have not given the numbers of those pages.

Subject Index